ISBN 978-0-265-76698-9
PIBN 10971831

FIELD PLOT TECHNIQUE

by

WARREN H. LEONARD, M. Sc. (Nebraska)
Professor of Agronomy
Colorado State College

and

ANDREW G. CLARK, M. A. (Colorado)
Professor of Mathematics
Colorado State College

1946 Printing

BURGESS PUBLISHING CO.
426 SOUTH SIXTH STREET
MINNEAPOLIS 15, MINN.

PREFACE

This manual has been the outgrowth of a set of lectures on Field Plot Technique given to seniors and graduate students at Colorado State College since 1930. It has been found practical in the classroom for a 2 to 4 credit combined lecture and laboratory course. The problems and questions have proved to be important aids to the student. While "Field Plot Technique" has been prepared primarily for class use, it is hoped that it will appeal to the technical worker in Agronomy as a reference to the more important statistical methods and tables. The large number of references quoted will give the reader a ready reference to the major papers on various phases of applied statistics.

The organization of the subject matter, and the manner in which the statistical methods are interwoven with the applications, differs somewhat from the conventional approach. The writers feel that the student of agronomic experimentation needs an elementary picture of the factors to be considered in a research program with special reference to the field experiment. For this reason, an attempt has been made to coordinate the historical and logical background of agronomic experimentation with statistical techniques and their application to the design of the practical types of field experiments. This also requires that the student be familiar with the mechanical procedures generally followed in routine experimental work.

The development of the various statistical techniques has been intuitive rather than rigorously mathematical. The aim has been to lead the student to understand the formulas he applies without necessarily being able to derive them mathematically. The symbolism employed in the text was chosen with regard to what appears to be the most common usage. Considerable effort has been spent in striving for consistency.

That it is impossible in an elementary text to present and interpret many of the complexities involved in some modern experiments is obvious. It is hoped that a sufficient foundation will be laid for the student so that he can intelligently study the more advanced treatises.

The writers are deeply indebted to Dr. F. R. Immer, Professor of Agronomy and Plant Genetics, University of Minnesota, for permission to make liberal use of his classroom material, especially in chapters 11, 17, and 18. They wish to express their appreciation to Dr. S. C. Salmon, Division of Cereal Crops and Diseases, U. S. Department of Agriculture, for criticisms and helpful suggestions. Dr. K. S. Quisenberry of the same division has assisted by his criticisms of chapter 21. The writers are particularly grateful to Professor R. A. Fisher and his publishers, Oliver and Boyd, for permission to reproduce the Table of χ^2 from "Statistical Methods for Research Workers." Professor G. W. Snedecor, Iowa State College, generously allowed us to include his table of "F and t". The writers also wish to express their thanks to Dr. C. I. Bliss for permission to use his table of angular transformations. The table of Neparian logarithms used in the manual is taken from "Four Figure Mathematical Tables" by the late J. T. Bottomley and published by Macmillan and Co., Ltd. (London). The writers are grateful to the publishers and to the representatives of the author for permission to use this table. To Dr. D. W. Robertson, one of their colleagues, they express their appreciation for various helpful suggestions.

Part I. Introduction to Experimentation

Part II. Statistical Analysis of Data

Part III. Field and Other Agronomic Experiments

Part IV. Appendix

FIELD PLOT TECHNIQUE

Part I

Introduction to Experimentation

CHAPTER I

STATUS OF AGRONOMIC RESEARCH

I. Rise of Agronomic Research

Although the art of agronomy has been practiced for centuries, the science of agronomy is only about 100 years old. The need for reliable information in this country has come about gradually as farmers have come to realize some of the many problems which confront the agricultural industry, problems in soil fertility, the control of diseases and pests, winter-hardiness in crops, among many others. In addition to the needs of the farmers themselves, the establishment of the Land-Grant Colleges under the Morrill Act in 1862 brought about an acute need for subject matter for the agricultural colleges. It soon became very apparent that the problems in agriculture were complex and that well-trained men were needed to solve them. In general, it may be said that agricultural research began with simple empirical tests, but has gradually developed until it has now attained a scientific basis. In the short space of 75 years, so much subject matter has been accumulated in the field of agriculture that no one man could hope to be familiar with all of it. This led to specialization within the field between 1900 and 1910 in America. The branches recognized in most agricultural colleges and experiment stations are: Agronomy, animal husbandry, horticulture, entomology, forestry, home economics, and veterinary medicine or pathology.

Agronomy as a science was developed from the old style variety trials, crop rotation tests, and soil culture experiments, when field culture was an empirical art. Research workers and others interested in the science of crops and soils formed the American Society of Agronomy in 1907. In regard to Agronomy, Carleton (1907) states: "As a science it investigates anything and everything concerned with the field crop, and this investigation is supposed to be made in a most thorough manner, just as would be done in any other science". Thus, agronomy is the laboratory and workship of many sciences: Agrostology, chemistry, botany, ecology, genetics, pathology, physics, physiology, and others concerned with the problems of crops and soils. Ball (1916) early observed that it has been necessary for the experimenter (in agronomy) to turn from the gross aspect to minute detail in order to solve some of its problems. Empirical knowledge has been rapidly supplemented by fundamental information as a result of organized research and the improvement in its technique.

II. Establishment of Experiment Stations

It is difficult to realize that the present large network of experiment stations in this country and in other parts of the world has been established in the past 100 years. In fact, the science of agriculture practically began with this movement.

(a) First Experiment Station

Jean Baptiste Boussingault established the first experiment station in 1834, being the first man to undertake field experiments on a practical scale. He farmed land at Bechelbronne, Alsace, where he carried on research of a high calibre. Boussingault set out to investigate the source of nitrogen in plants, and systematically weighed the crops and the manures applied for them. He analyzed both and prepared a balance-sheet. Furthermore, this investigator studied the effects on plants when legumes were in the rotation. He concluded that plants obtained most of their nitrogen from the soil.
(See Chapter 2.)

(b) Rothamsted Experimental Station

The Rothamsted Experimental Station was established by John Bennet Lawes on his farm in England in 1841. Hall (1905), in his account, states that "Rothamsted is now a household word wherever the science of agriculture is studied." Lawes found that phosphates were important fertilizers and discovered a method to make phosphate fertilizer by the application of sulfuric acid to phosphate rock. Formerly, bones were used as a sole source of phosphates. This significant discovery led to experimentation on a large scale. The systematic field experiments, begun in 1843 and continued to this day, have dealt particularly with soil fertilizers and crop rotation. These experiments long have been models for carefully planned experiments. Lawes was aided by Dr. J. H. Gilbert, who commenced work at Rothamsted in 1843. The two men worked together for 57 years. Recently, Dr. R. A. Fisher has brought about modifications in the field experiments to make them amenable to statistical treatment.

(c) American Experiment Stations

Some of the early history of American experiment stations is given by True (1937) and by Shepardson (1929). South Carolina went on record as favoring an experiment station in 1785, but the general movement for the establishment of experiment stations began about 1871 because of the attention attracted by the experiments of Lawes and Gilbert of England. In the meantime, the Morrill Act signed by Lincoln in 1862, provided for the so-called land-grant colleges for the study of agriculture and mechanic arts. California established the first experiment station in 1875, and began field experiments on deep and shallow plowing for cereals. A station was started in North Carolina in 1877, after which many others followed. The Hatch Act, passed by Congress in 1887, was the start of the present experiment stations. Twelve were in existence at that time. Increased funds were provided by the Adams Act in 1906, by the Purnell Act in 1925, and by the Bankhead-Jones Act in 1935. The United States Department of Agriculture has been of rather recent origin, the Secretary becoming a Cabinet member in 1889. At present, the federal government controls funds given to the states for experimental work. In general, the system has been satisfactory because it has proved to be participation and coordination rather than control.

III. Reasons for Public Support of Agricultural Research

There has been some criticism on the use of public funds for agricultural research, but their use has been justified on the grounds that the welfare of agriculture is basic to the nation. In addition, it would be almost impossible to place agricultural research on a self-supporting basis because the results of research are so difficult to control through patents or otherwise.

(a) Agricultural Welfare

There are 6,000,000 farmers and 360,000,000 to 365,000,000 acres of cultivated land in this country, some of which has been cultivated more than 300 years. The virgin fertility, in many cases, has been exhausted. The experiences and needs of these farmers are significant because the prosperity of the nation depends to a large extent upon agriculture. The production of food and fiber is fundamental to the public welfare, as research that leads to lower cost of production passes its benefits on to the consumer. Haskell (1923) calls attention to the fact that, in the case of crop losses due to diseases and other factors, the consumer ultimately pays a higher price for his food. He pays for depleted soil fertility in the same way. Thus, the state may actually gain more from the benefits of research than the farmer himself.

(b) Limitations of Farm Experiences

It has been impossible for several reasons to collect scientific information of much value from farm experiences. (1) Inadequate Farm Records: The results obtained by farmers are inaccurately and incompletely recorded from the experimental viewpoint. Their experiences are generally limited to acres and yields such as found in stories in the farm press. Farmers very often place undue emphasis on the unusual. (2) Failure to Consider all Factors: The essence of scientific progress is to determine "why". Among the many variables in agriculture, variation in season is exceedingly important and may over-shadow all other factors. The farmer is quite likely to base his judgment and conclusions on the results of one or two year's performance. Thorne (1909) states that many experiments which farmers attempt are valueless or misleading because of failure to observe some essential condition of experimentation. (3) Inadequate Training: As a rule, the farmer lacks the training or experience necessary for the evaluation of experimental results. Hall (1905) makes this statement: "Agricultural science involves some of the most complex and difficult problems the world is ever likely to have to solve, and if it is to continue to be of benefit to the farmer, investigations, so far as their actual conduct goes, must quickly pass into regions where only the professional scientific man can hope to follow them ...." (4) Inadequate Funds: Farmers lack the funds, help, and equipment necessary for experimental work. Experimentation is quite expensive since practical considerations are necessarily put aside. An experiment must be conducted with precision in order to obtain reliable results, rather than for financial return. For instance, Hall (1905) tells that some Rothamsted fields have grown wheat for 60 years, year after year, on the same land. As the modern farmer seldom grows wheat continuously, he looks upon this experiment as hopelessly impractical when it is pointed out to him on field days. Nevertheless, this very test furnished the bulk of the early proof that losses in yield would result from continuous wheat culture. The aim of the Rothamsted test, as it continues, is to find out how the wheat plant grows.

IV. Experiment Station Funds

Agricultural research in this country is publicly financed almost altogether. Federal and state agencies spent about 25 million dollars on agricultural research for the year 1927-28. This total sum represented approximately 0.20 percent of the gross income for agricultural products, a figure wholly within reason.

(a) The Hatch Act

The first federal subsidy for agricultural research was the Hatch Act, passed in 1887. It gave each state $15,000 per year, a wide latitude in the use of the funds being permitted. The Act made it possible to conduct original experiments or verify experiments along lines as follows: (1) physiology of plants and animals; (2) Diseases of plants and animals with remedies for the same; (3) the chemical composition of useful plants at different stages of growth; (4) rotation studies; (5) testing the adaptation of new crops and trees; (6) analyses of soils and water; (7) chemical composition of manures, natural and artificial, and their effect on crops; (8) test the adaptation and value of grasses and forage plants; (9) test the composition and digestibility of different foods for domestic animals; (10) research on butter and cheese production; and (11) examination and classification of soils. None of the funds can be used for the purchase or rental of lands or expenses for farm operations.

(b) The Adams Act

A similar amount of money was granted to the states by the Adams Act, passed in 1906. The funds must be used for original researches or experiments that bear directly on agriculture. Research of a fundamental nature is required under this fund. None of the money can be applied to substations, or to the purchase or rental of land.

(c) The Purnell Act

The Purnel Act passed in 1925 provided for additional funds which now amount to $60,000 per year for each state. These funds must be used on specific projects, but the requirements are less exact than for the use of Adams funds. The Act provides for investigations on the production, manufacture, preparation, use, distribution, and marketing of agricultural products.

(d) The Bankhead-Jones Act

Certain difficulties in the use of experimental funds for broad general projects led to the passage of the Bankhead-Jones Act in 1935 which will, in five years (1940), provide $5,000,000 for research. Its provisions have been described as follows: "To conduct scientific, technical, economic, and other research into laws and principles underlying basic problems of agriculture in its broadest aspects....". It also authorizes research for the improvement of quality of agricultural commodities and for the discovery of uses for farm products and by-products. The U. S. Department of Agriculture receives 40 per cent of this fund, while 60 per cent is allotted to the states on the basis of rural population. It is generally understood that the funds must be used for new lines of work.

V. The Personal Equation in Research

As for agricultural research in general, successful agronomic research depends upon the ability, permanency, and honesty of the workers. The personnel for investigational work must be well-trained in the basic sciences as well as leaders in agricultural thought. Their outlook must be broad.

(a) Education for Investigational Work

The amount of training necessary for research is great. The investigator must be skilled in the art of agronomy and trained in the closely related sciences. In fact, he should have an adequate educational background before research is even attempted. A good foundation in English, physics, and chemistry are basic for all research in agriculture. Biology adds the conception of organism, while mathematics is the common instrument. Thorough training in all branches of botanical science is desirable in agronomy. This includes taxonomy, anatomy, physiology, pathology, etc. Other sciences that are useful are: Geology, bacteriology, genetics, and statistics. Among the authorities who agree on this general type of background are Howard (1924), Wheeler (1911), Ball (1916), Carleton (1907), and Richey (1937). A practical viewpoint is necessary, but this is largely the result of boyhood training and common sense.

(b) Qualities in Successful Research Men

There is some question about the successful scientist necessarily being a genius. The term should be qualified to include perseverance, common sense, and infinite pains. Howard (1924) emphasized the qualities needed when he said: "Here the man is everything; the system is nothing." (1) Imagination: Some imagination is essential in the research worker but, of course, it must be scientific imagination. (2) Discrimination: An investigator must have the power of discrimination, that is, he must be able to recognize the essentials and non-essentials in research. He must select the features which are most worthwhile. It is possible to record too much data on a subject, and thus cloud the entire issue. (3) Accuracy: There is a great need for accuracy in experimental work. An investigator should record only those notes whose reliability is well established. One should never take measurements so fine that they imply false accuracy. The figures taken by an investigator should give him confidence in his work. (4) Honesty in observation: The investigator should always accept observations without regard to their agreement with his own preconceived ideas. One should record only the things he sees. (5) Fairness: The research man should give due credit to others, and keep within his own field unless a phase of his work calls for cooperation with others. (6) Enthusiasm: One should be enthusiastic about his work, being ready to put in long hours or extra time when necessary. Call (1922) says there must be a love for the work so great, in those engaged in research, that it will enable him to push forward in the face of obstacles which may seem insurmountable. (7) Courage: One should always have the courage of his convictions. He should not be afraid to try something new.

(c) Initiative in Experimental Projects

The project system has an enormous value in the coordination, continuity, and conclusion of agricultural experimental work because it requires the submission of an outline and its approval before any work is done. Success in experimental projects depends upon the leader, his scientific attitude, depth of motive, conception of the problem, and its requirements. Not all research is good. In fact, there is a chance for much waste. While partial failure is inevitable, it is possible for the investigator to gauge plausible success. Allen (1930), advises research workers to think scientifically, avoid adherence to routine, and keep abreast of the times. The investigators should avoid the belief that his own compartment is water tight and self-sufficient.

VI. Results of Agronomic Research

Many contributions have been made in crops and soils by the experiment stations, particularly during the past 25 years. Some of the more important advances in the past quarter of a century in field crops are summarized by Warburton (1933) while those in soils are given by Lipman (1933).

(a) Field Crops

Among the contributions in corn have been the discovery that the show-type ear is unrelated to its performance in the field, that ear-to-row breeding may not lead to improvement in corn yields, and that the combination of inbred lines in hybrids has resulted in higher corn yields. In wheat, the discovery of rust resistance and of physiologic races has enabled investigators to breed for resistant varieties. The same is true for bunt. The introduction and use of sorghums, as well as their improvement, has resulted in their production throughout the west. The cause of flax "sickness" has been discovered as due to wilt with the result that resistant varieties have been bred. Sweetclover, once a weed, has been found to be a valuable crop. Many improved varieties of crops have been developed for disease resistance, drouth resistance, high quality or high yield. Marquis wheat is one of the most widely known improved varieties.

Tillage has been shown to be beneficial because of weed control rather than moisture conservation from a dust mulch. Both Funchess (1929) and Richey (1937) have given similar lists of advances made in field crop science.

(b) Soils

A quarter-century ago, physical-chemical analyses of soil without other data were frequently erroneous as a basis for the estimation of the agricultural value of soils. In recent years some of the more valuable contributions have been as follows: (1) Use of mineral fertilizers to improve soil fertility; (2) ionic exchange in soil colloids that led to an explanation of alkali-soil formation; (3) soil classification and soil survey; (4) soil acidity in its relation to plant growth; (5) soil colloids and their properties; (6) soil bacteria and other organisms and their influence on soil fertility; and (7) soil erosion and its control. That soil productivity may be maintained for a long period of time by the use of sound rotation and manurial practices has been shown by the Morrow plots at Illinois. The results for 39 years have been summarized by De Turk, et al. (1927).

VII. Value of Early Agronomic Experiments

Some of the investigational work in agronomy before 1910 was of little value due to errors in the experiments, many of which were great enough to vitiate the conclusions. Contradictions were common. As Piper and Stevenson (1910) point out, results were sometimes suppressed because they failed to coincide with current theory. "In short, all scientific evils necessarily associated with experimental methods are too evident in the field work in agronomy." The same type of criticism applies to other agricultural branches at that time. There were many reasons for this situation. The "guess method" was widely used by the old school of experimenters for the accumulation of information. They usually lacked facts, lacked a broad outlook, were limited in their experiences and, in many cases, had wide differences in viewpoints. Some of the shortcomings have been due to pressure for information with the result that the conclusions were often based on too few data. Other weaknesses were due to the viewpoint in some quarters that empirical facts were preferable to fundamental information from a practical standpoint.

While many of the early experiments would be inacceptable today in the light of modern experimental standards, they nevertheless contributed to progress. Some of them were as well conducted as those of today. Early agricultural practices were determined quite as much by opinion as by experiment. It would have been a poor experiment indeed were it to be less reliable than unsupported opinion. These early experiments must be evaluated in relation to the knowledge of the time as well as their effects on agricultural science and practice. For example, the early Rothamsted investigations on the source of nitrogen in plants finally led to a solution of the problem even tho the field experiments conducted in connection with them would be considered today as inadequate.

Many of the weaknesses in early experiments have been met gradually through (1) wider application of modern statistical methods, (2) replication of plots or treatments, and (3) wider use of the inductive or scientific method in which general principles are sought rather than empirical facts.
(seek)

VIII. Present Trends in Agronomic Research

Some very definite trends are apparent in modern agronomic research, among them being the emphasis on design of experiments, long-time projects, and regional coordination of research.

(a) Design of Experiments

In recent years, a great deal of stress has been placed on the design of experiments. The field lay-out and the method of analysis of the data are coordinated so as to lead to more efficient experimental results. The emphasis on design has been made by the Rothamsted workers. Design focuses attention on the objects of an experiment that can be attained in no other way. This trend promises to reduce the number of situations where an experiment is conducted and data collected before a method of analysis is conceived.

(b) Long-time Projects

Another definite trend is toward the long-time project. According to Henry Wallace (1936), "The solution of problems related to crop production is a matter of years. The improvement of plants by breeding must extend through many generations. Varieties must be compared in a number of different kinds of seasons for correct evaluation. The same is true of tests of fertilizers, spraying practices, and cultural methods. To be productive, a program of plant research accordingly must be stable, with a concentration of effort until a given problem is solved or its solution found impractical for the time being."

(c) Regional Coordination of Research

The regional coordination of research work to reduce duplication of effort is being regarded more and more as essential. It has been stressed by Call (1934), Jarvis (1931), and others. Agronomic research began as isolated bits of investigation to solve local problems. Cooperation and coordination was developed later to reduce wasteful duplication. It also makes possible a comprehensive attack on intricate problems, as well as the elimination of artificial boundaries. Such effort encourages personal contacts and exchanges of ideas between different investigators. Various bureaus of the U. S. Department of Agriculture took the leadership in regional coordination. The most formal efforts on regional coordination are in the northeastern states on pasture investigations and soil organic matter studies. The limitation of initiative and individuality of investigators has sometimes been feared as a result of regional coordination, but for the most part appears to be unfounded.

References

1. Allen, F. W. Initiating and Executing Agronomic Research. Jour. Am. Soc. Agron., 22:341. 1930.
2. Ball, C. R. Some Problems in Agronomy. Jo. Am. Soc. Agron., 8:337-347. 1916.
3. Call, L. E. Increasing the Efficiency of Agronomy. Jour. Am. Soc. Agron., 14:329-339. 1922.
4. ___________ Regional Coordination of Agronomic Research from the Standpoint of the Station Director. Jour. Amer. Soc. Agron., 26:81-88. 1934.
5. Carleton, M. A. Development and Proper Status of Agronomy. Proc. Am. Soc. Agron., 1:17-24. 1907.
6. DeTurk, E. E., Bauer, F. C., and Smith, L. H. Lessons from the Morrow plots. Ill. Agr. Exp. Sta. Bul. 300. 1927.
7. Funchess, M. J. Some Outstanding Results of Agronomic Research. Jo. Am. Soc. Agron., 21:1117. 1929.
8. Hall, A. D. An Account of the Rothamsted Experiments (Preface and Introduction) 1905.
9. ___________ Fertilizers and Manures, pp. 1-38. 1928.
10. Haskell, S. B. Agricultural Research in its Service to American Industry. Jo. Am. Soc. Agron., 15:473-481. 1923.
11. Howard, A. Crop Production in India, pp. 185-195. 1924.
12. Jarvis, T. D. The Fundamentals of an Agricultural Research Program. Sci. Agr., 26:81-88. 1934.
13. Lipman, J. G. A Quarter Century of Progress in Soil Science. Jo. Am. Soc. Agron., 25:9-25. 1933.
14. Office of Experiment Stations. Legislation and Rulings Affecting Experiment Stations, Miscel. Publ. 202. 1934.
15. Pearson, Karl. The Grammar of Science, P. 30. 1911.
16. Piper, C. V., and Stevenson, W. H. Standardization of Field Experimental Methods in Agronomy. Proc. Am. Soc. Agron., 2:70-76. 1910.
17. Richey, F. D. Why Plant Research. Jour. Am. Soc. Agron., 29:969-977. 1937.
18. Thorne, C. E. Essentials of Successful Field Experimentation. Ohio Agr. Exp. Sta. Cir. 96. 1909.
19. True, A. C. A History of Agricultural Experimentation and Research in the United States. Miscel. Publ. 251. 1937.
20. Warburton, C. W. A Quarter Century of Progress in the Development of Plant Science. Jo. Am. Soc. Agron., 25:25-36. 1933.
21. Wheeler, H. J. The Status and Future of the American Agronomist. Proc. Am. Soc. Agron., 3:31-39. 1911.

Questions for Discussion

1. What conditions led to the subdivision of the agricultural field?
2. What are the functions of agronomy as a science? Why?
3. Who founded the first experiment station?. What results were obtained?
4. When was the Rothamsted Experimental Station established? Where? By whom? Why?
5. What has Rothamsted contributed to early agricultural science?
6. Where was the first American experiment station established? When? Upon what did it work?
7. Give some reasons to justify agricultural experimentation as a public duty.
8. Why is agriculture in America a national concern second to none?
9. Give several reasons why a farmer is generally unable to do experimental work.
10. Name the acts of Congress that contributed to the agricultural experiment stations, together with their dates of passage.
11. What special requirements is necessary for the expenditure of funds under the Adams Act? Bankhead-Jones Act? Hatch Act?
12. What kind of basic training is necessary for agronomic research?
13. What would you consider as some of the most important attributes of a successful investigator?
14. Discuss briefly the following characteristics in relation to research: (1) imagination, (2) classification, (3) discrimination, (4) accuracy, and (5) thoroughness.
15. Why has the project system been useful in research?
16. Name five contributions to crop knowledge made by experiment stations. Five contributions to soil science.
17. What are some reasons for the early contradictions in agronomic science?
18. What was the value of early agronomic experiments? What were some of their weaknesses?
19. Name and discuss three trends in agronomic research at the present time.

HISTORY OF BASIC PLANT SCIENCES

I. Early History of Basic Sciences

Agronomy as a science began with the establishment of the first expériment station by Jean Baptiste Boussingault in 1834, although many empirical facts were known before that time.

(a) Early Science

Science, in general, dates from Aristotle who was the founder of zoology and the forerunner of evolution. As one of the founders of the inductive method, he first conceived the idea of organized research. In fact, his principles might well be observed at the present time. After Aristotle, little progress was made for 2,000 years. Among his theories was the one that the universe was composed of four elements: air, earth, fire, and water. This was accepted for centuries because the habit was to assume some man as an authority rather than to investigate. At the beginning of the 17th century, Newton and Galileo began to base conclusions on facts. Francis Bacon wrote books which emphasized that theories should be based on facts rather than on authorities.

(b) Reasons for Slow Progress in Science

Progress in agricultural science has had to wait on discoveries in the basic sciences of physics and chemistry. There are many reasons for the slow development of science in past ages. (1) Slavery was the general rule, with the result that there was little stimulus to improve. (2) Experimentors lacked accurate instruments for measurement. (3) The mildness of the climate in the early civilized countries restricted industry. (4) Mathematical science was restricted. (5) The scientific method developed by Aristotle was seldom used. Instead, it was the habit to assume a general law. (6) Superstition and interference by the clergy discouraged experimentation.

II. Development of Agricultural Science

There was little activity in the sciences related to agriculture before 1800. Fundamental discoveries at the close of the 18th century, together with the appearance of several treatises on agriculture, started rapid development. Sir Humphrey Davy (1813) published a book entitled "Essentials of Agricultural Chemistry" in which he brought together many known facts. A von Thaer (1810) published a book on "Reasons for Agriculture" in which he emphasized the value of humus in the soil, from which he believed plants gained their carbon. In 1840, Justus von Liebig published his book on organic chemistry in relation to agriculture, in which he advocated that the soil need only be supplied with minerals. This latter work struck the scientific world as a thunderbolt. It has had a great deal of influence on modern agricultural research.

The establishment of the Rothamsted Experimental Station in 1838 also reflected the interest in agricultural science. The discoveries important to agriculture since 1850 have been: (1) The theory of evolution, (2) the discovery of anaerobic bacteria, (3) the source of nitrogen in plants thru the aid of bacteria, (4) Mendel's laws of heredity; (5) the chromosome theory of heredity as a physical basis for inheritance; and (6) the discovery of vitamins.

A -- Plant Nutrition

III. Early Plant Discoveries

Very little information was gathered on plant science from the time of the Greeks up to the Renaissance. (1) Theophrastus: Published a book on plants entitled "Enquiry into Plants." He classified plants into herbs, shrubs, and trees. Theophrastus also distinguished bulbs, tubers, and rhizomes from true roots. Plant adaptation was discussed. (2) Al Farbi: Discovered respiration in plants about 950 A.D. (3) Johann van Helmont: This worker, who lived in the 17th century, believed that water was transformed into plant material. He placed 200 lbs. of soil in a receptacle and grew a willow in it. Nothing was added but water. At the end of five years, he found the willow weighed 169 lbs. and 3 oz., while the original soil lost only 2 oz. from its original weight. He concluded that the growth came from the water alone, but failed to consider the air. (4) Jethro Tull: Believed that earth was the true food of plants and that they absorbed soil particles. Therefore, he believed it necessary to finely pulverize the soil through cultivation. Tull developed cultural implements and devised a system to plant crops in rows.

IV. Source of Nitrogen in Plants

The period from 1840 to 1885 was taken up largely with the Rothamsted-Liebig controversy on the source of nitrogen in plants.

(a) Earlier Work on Nitrogen

The element nitrogen was discovered in 1772. Joseph Priestly, followed by Jans Ingen-Hausz, settled the fundamental fact that green plants in sunlight decompose the carbon dioxide from the atmosphere, set oxygen free, and retain the carbon. This source of carbon accounts for the bulk of dry matter in plants. From his work in 1804, Theodore De Saussure concluded that plants were unable to assimilate free atmospheric nitrogen, but obtained it from the nitrogen compounds in the soil. The pot experiments carried out by J. B. Boussingault, who began his investigations in 1804, indicated that plants draw their nitrogen entirely from the soil or manure.

(b) Liebig-Rothamsted Controversy

Justus von Liebig in 1840 maintained that green plants, by the aid of sunlight, derive their total substance from carbonic acid, water, and ammonia present in the atmosphere, and from simple inorganic salts in the soil which are afterwards found in the ash when the plant is burned. Liebig believed combined nitrogen in the soil to be unnecessary in plant nutrition. This view was disputed by Lawes and Gilbert who began elaborate experiments at Rothamsted in 1857. They grew plants under glass shades, ammonia from the air being kept out. The earth, pots, manures, etc., employed in the experiment were burned to sterilize them. Carbon dioxide was introduced as required. Lawes and Gilbert made their trials both without manure and with ammonium sulfate. Their work was done so carefully that the possibility of nitrogen fixation by plants was excluded. While they concluded that plants require combined nitrogen from the soil, they were unable to account for the gain in nitrogen in some plants under field conditions. They found actual gains in nitrogen when leguminous plants were grown in the field, which was in agreement with the long experiences of practical farmers.

(c) Final Experiments on Nitrogen Relations

The final experiment on nitrogen assimilation by plants was performed by H. Helriegel and H. Wilfarth who found the symbiotic relationship between bacteria and legumes. When he grew plants in sand, Helriegel (1886) found that the Gramineae, Crucifereae, Chenopodiaceae, etc., grew almost proportionally to the combined nitrogen supplied. When absent, nitrogen starvation took place as soon as the nitrogen from the seed was exhausted. In legumes, he found that the plants were able to recover and begin luxurious growth. The roots always had nodules on them in such instances. However, legumes grown in sterile sand behaved the same as other plants, but recovery could be brought about when a watery soil extract was added to them. Renewed growth and assimilation of nitrogen was found to depend upon the production of nodules on the roots. Wilfarth (1887) found bacteria in the nodules and settled the point that bacteria are associated with nitrogen fixation. Later, these results were confirmed at Rothamsted and final proof was obtained on the role of nitrogen in plants. As Hall (1905) recounts, the "very vigor" of the Rothamsted laboratory prevented fixation of nitrogen by the exclusion of all possibility of inoculation. The legumes as a class were found to be an exception to the contention that plants could use only combined nitrogen from the soil. Both schools were partly right.

B -- Evolution and Genetics

V. Early Work in Genetics

Although many facts of inheritance were known previously, Genetics has been regarded as a science only since 1900. At that time, the work of Gregor Mendel, originally published in 1865, was brought to light. Early work is reviewed by Roberts (1919, 1936), Zirkle (1932, 1935), and by Cook (1937).

(a) Sex in Plants

The Bisexual nature of the date palm was recognized by the early Babylonians and Assyrians 5000 years ago. The ancients ascribed many monstrosities to hybridization. Many theories of heredity were in vogue, but no experimental data. Theophrastus and Pliny discussed sex in plants. Primitive men made improvements in crops, rice and maize being good examples.

About 1600 a new spirit of scientific skepticism began to be manifest. Many of the cumulative absurdities and theories were being put to experiment. The increased interest in biology culminated in the publication of the famous letter by Camerarius in 1694 on sex in plants. He gave convincing evidence that plants are sexual organisms. Sex in plants was demonstrated by actual experiments with spinach, hemp, and maize.

(b) Hybridization of Plants

This work was followed by the production of the first artificial plant hybrid by Thomas Fairchild in England, a short time before 1717. In the next 50 years there occurred a veritable wave of hybridizing. Crosses between more than a dozen genera were made by several investigators. This period culminated in the publication of the work of J. G. Koelreuter (1761-66) in which he reported the results of 136 experiments on artificial hybridization. In 1793, C. K. Sprengel observed cross pollination of plants by insects. However, Zirkle (1932) calls attention to the fact that insect pollination was observed by an American named Miller at a much earlier date. From 1760 to 1859 there followed many experiments on plant hybridization in attempts to determine the nature of inheritance. In 1822, John Goss (England) reported but failed to interpret dominance and recessiveness, and segregation in peas. A. Sageret (France) in 1826 classified contrasting characters in pairs, using muskmelons and cantaloupes. K. F. von Gaertner reported in 1835 on hybridizations made with 107 plant species. He noted plant vigor and the uniformity of the first generation after a cross. In 1863, C. V. Naudin (France) published a memoir on hybridization in which he almost discovered the laws of inheritance.

VI. The Theory of Evolution

The theory of organic evolution is one of the most profound theories expostulated in the past 300 or 400 years. It was brought to fruition in the publication of the "Origin of Species" by Charles Darwin in 1859. Hybrids are discussed extensively, but its contribution to genetics was mostly indirect. It marked the beginning of the modern experimental approach to biological problems.

(a) Evolution before Darwin

When Darwin published the "Origin of Species" spontaneous generation and special creation were the current theories. A great majority of naturalists believed that species were immutable productions specially created. Up to this time, empirical rather than scientific improvement had been made in plants. Darwin did not originate the evolution theory; he merely furnished evidence for its substantiation. Aristotle had expressed the central idea of evolution. Modern philosophy from Francis Bacon onward shows definiteness in its grasp and conception. Erasmus Darwin, grandfather of Charles Darwin, had a theory similar to that propounded in the "Origin of Species." J.B.P. de Lamarck, in his "Philosophie Zoologique" published in 1809, made the first attempt to produce a comprehensive theory of evolution. He added the idea of "use and disuse." Lamarck believed in the inheritance of acquired characters and attributed some influences to direct physical factors. In other words, all the principal factors of evolution had been worked out before the time of Darwin with the possible exception of "survival of the fittest" which he obtained from a book by Malthus on population.

(b) The Work of Charles Darwin

Darwin made an extended trip around the world in the Beagle, collecting voluminous facts and making extensive observations in support of his theory. He is given credit for the evolution theory because he was the first to gather facts. He attempted to show how and why new species arose. (1) Theory of Natural Selection: Present organic forms are believed to have evolved from more simple forms in past ages. The theory was founded on these facts: (a) Variations between individuals are universally present; (b) a struggle for existence takes place between individuals; (c) through natural selection these individuals with the most favorable variations survive; and (d) heredity tends to perpetuate the favorable variations from natural selection. (2) Reasons for Success: Darwin was successful because of his thoroughness, accuracy, hard work, honesty, ability to see, and because he was a stickler for details. He showed by example that disinterestedness, modesty, and absolute fairness are important attributes of character in intellectual work. Darwin (1859) himself states that his success was due to a love of science, unbound patience for long reflection on a subject, industry in the collection and observation of facts, as well as a fair share of invention and common sense.

VII. The Cell in Relation to Inheritance

Indepenaent progress was being made in other fields that were to have a profound influence on genetics after 1900. A. von Leeuwenhoek (Holland) discovered the microscope and saw mammalian germ cells in 1677. The cell theory was propounded in 1838-39 by M. J. Schleiden and T. Schwann (Germany). This was the first generalized statement that all organisms are made up of cells--one of the greatest generalizations of experimental biology. The union of sperm and egg cells, i.e., fertilization, was first seen in seaweed by G. Thuret (France) in 1849. A year later he showed that the egg would not develop without fertilization. The chromosomes were described in 1875 by E. Strassburger (Germany). During the same year Oscar Hertwig (Germany) proved that fertilization consists of the union of two parental nuclei contained in the sperm and ovum. W. Flemming (Germany) in 1879-82 describes the longitudinal splitting of the chromosomes, and later observed (1884-85) that the halves of split chromosomes went to opposite poles. Th. Boveri (Germany) in 1887-88 verified the earlier prediction of A. Weismann that reduction in the chromosomes takes place. In 1898, S. G. Navashin (Russia) discovered double fertilization in higher plants. Thus, the physical mechanism of inheritance was pretty well worked out by the time that the work of Mendel was discovered.

VIII. The Laws of Inheritance

The turn of the century proved to be an epochal year in the experimental study of heredity. The work of Gregor Mendel (Austria), an Augustinian monk, on inheritance in peas was rediscovered in 1900 by Hugo De Vries, C.F.J.E. Correns, and E. von Tschermak. The work had been published originally in 1866.

(a) Discovery of Principles of Heredity

Mendel made crosses of peas and observed carefully the resemblances and differences among different races. He began his work in 1857. The principles of heredity which he put forth were as follows: (1) single heredity units, (2) allelomorphism or contrasted pairs, (3) dominance and recessiveness, (4) segregation, and (5) combination. The last two are generally recognized as the distinct contributions of Mendel.

(b) Methods used by Mendel

There are several reasons for the success of Mendel. His work differed from that of his predecessors in several respects. (1) He made actual counts and kept records of each generation. (2) One pair of factors was studied at a time. (3) His material was carefully studied and selected. (4) He guarded against errors in accidental crosses. (5) He worked with large numbers. (6) The crosses were studied for seven generations. Roberts (1929) comments as follows on the work of Mendel: "Nothing in any wise approaching this masterpiece of investigation had ever appeared in the field of hybridization. For far-reaching and searching analysis, for clear thinking-out of the fundamental principles involved, and for deliberate, painstaking, and accurate following-up of elaborate details, no single piece of investigation in their field before his time will at all compare with it, especially when we consider the absolute absence of precedent and initiative for the work."

IX. Modern Developments in Genetics

The universality of Mendelian principles was verified in plants, animals, and man within three years. In 1902, Hugo De Vries advanced the mutation theory to explain sudden changes in plants that breed true, but which could not be accounted for by Mendelian inheritance. He found sudden changes in the evening primrose to breed true in certain cases. These mutations were believed to furnish the basis for evolution. This was soon followed by the pure-line concept, i.e., variations in the progeny of a single plant of a self-fertilized species are not due to inheritance. This was first put forth by W. L. Johannsen (Denmark) in 1903. H. Nilsson-Ehle (Sweden) advanced the multiple-factor hypothesis in 1908. The chromosome theory of heredity was announced by T. H. Morgan in 1910. His gene theory included the principle of linkage of genes resident on the same chromosome. This brilliant hypothesis has been upheld in many experiments. Much recent work has been concerned with polyploidy, the mechanism of crossing-over, and sterility.

The principles of genetics have enabled plant breeders to make definite contributions to improved varieties. Many new varieties are now grown on farms that have been made possible through application of the laws of inheritance. Many varieties are "made to order" to meet particular conditions.

C -- Other Basic Sciences

X. Development of Bacteriology

Great advances were made in the field of bacteriology between 1860 and 1880, it being definitely established that bacteria bring about putrefaction, decomposition, and other changes. The work of Louis Pasteur dominated the field during this period,

(a) Pasteur and his Work

Pasteur discovered anaerobic bacteria. Fermentation was commonly thought to be the result of a chemical change, but Pasteur proved it to be due to anaerobic bacteria. This wrecked the theory of spontaneous generation of life. Pasteur showed that the presence of bacteria could always be traced to the entrance of germs from the outside, or to growth already present. Other contributions of Pasteur included the discovery of the causes of many bacterial diseases, and the development of methods of immunization. Pasteur had several attributes that led to his success: (1) He established truth by experiment; (2) He was discerning with regard to the problem on which he worked; and (3) he worked on one problem at a time.

(b) Other Discoveries in Bacteriology

Many further developments in bacteriology depended upon the improvement of the microscope and the perfection of various technics. The oil immersion lense was developed about 1880. The agar-plate method for the study of growing colonies of bacteria was introduced by Robert Koch in 1881. The transformation of ammonia to nitrates was demonstrated by T. Schloesing and A. Muntz in 1877, but it remained for S. Winogradsky to isolate the organisms concerned. That nodules are formed on legumes as the result of inoculation with microorganisms was demonstrated by H. Helriegel and H. Wilfarth in 1886. M. W. Beijernick isolated non-symbiotic bacteria, i.e., the Azotobacter, in 1901. Among other contributions of bacteriology were sterilization technics, the classification of bacteria on a physiological basis (started by Ferdinand Cohn in 1872) the study of diseases due to filterable viruses, and studies in the nature of bacteriophagy.

XI. Plant Pathology

Like all natural sciences, plant pathology had its start with the dawn of civilization. The Hebrews mentioned plant diseases in the Bible, but only gave descriptions and mentioned damage. Little was known about plant diseases until the modern era which began about 1850.

One of the greatest early workers was Anton de Bary (German) who proved the parasitism of Fungi in 1853. A little later (1864) he proved heteroecism in rusts as illustrated by the relation of the aecidium on the barberry to the red and black rust stages on wheat. That bacteria may cause plant diseases was first proved by Thomas Burrill in 1879-81. He showed that a definite species, Bacillus amylovoris, was the causal agent of fire blight. The use of Bordeaux mixture as a fungicide was started in France in 1886. Since that time, many other fungicides have been used in plant disease control, the latest being the organic mercury compounds.

Biologic strains in rusts were discovered in 1894 by J. Eriksson (Sweden), while races within a variety of rust were demonstrated by E. C. Stakman and his coworkers in 1916. Another important discovery was made by J. H. Craigie in 1927 when he discovered sexuality in the rusts.

A rapid increase in the knowledge of so-called virus diseases of plants has taken place since the first proof of tobacco mosaic as an infectious disease in 1888. The role of insects in the transmission of the virus or active principle was soon recognized. Recently, W. M. Stanley (1937) has advanced strong evidence that the tobacco mosaic virus is due to a high molecular weight crystalline protein.

A great deal of attention is now being given to the production of disease-resistant and immune varieties of crop plants through the application of genetic methods.

References

1. Cook, R. A Chronology of Genetics. Yearbook of Agriculture, U.S.D.A. pp. 1457-1477. 1937.
2. Dampier-Whetham, W.C.D. A History of Science, pp. 33-40, 141-147, 150-169, and 285-287. 1931.
3. Darwin, Charles. Origin of Species. 1859.
4. Hall, A. D. The Book of the Rothamsted Experiments, preface and introduction, pp. 1-14. 1905.
5. Hall, A. D. Fertilizers and Manures, pp. 1-38. 1928.
6. Hayes, H. K., and Garber, R. J. Breeding Crop Plants, pp. 1-14. 1927.
7. Heald, F. D. Manual of Plant Diseases. pp. 17-20. 1933.
8. Roberts, H. F. The Founders of the Art of Plant Breeding, Jour. Her., 10:99-106, 147-152, 229-239, and 267-275. 1919.
9. Roberts, H. F. Plant Hybridization Before Mendel. 1929.
10. Russell, E. J. Soil Conditions and Plant Growth. pp. 1-31. 1932.
11. Stanley, W. M. Crystalline Tobacco Mosaic Virus Protein. Am. Jour. Bot., 24:59-68. 1937.
12. Theophrastus. Enquiry into Plants, Vol. I and II (Translated by Arthur Hort). 1926.
13. Vallery-Radot, R. The Life of Pasteur. 1928.
14. Weir, W. W. Soil Science, pp.3-26. 1936.
15. Whetzel, H. H. An Outline of the History of Phytopathology.
16. Zirkle, C. Some Forgotten Records of Hybridization and Sex in Plants. Jour. Her., 23:433-448. 1932.
17. Zirkle, C. The Beginnings of Plant Hybridization. 1935.

Questions for Discussion

1. Who is considered the "Father of Science"? Why?
2. What were the contributions of Galileo, Bacon and Newton to science?
3. Why did science develop slowly previous to the 16th century?
4. Name several discoveries important to agriculture since 1850.
5. Who was Theophrastus, and what did he do?
6. What were the views of these men on plant nutrition: Von Helmont, Jethro Tull, Thaer, Liebig?
7. What did deSaussure contribute to agricultural research? Sir Humphrey Davy? Boussingault?
8. What facts made the source of nitrogen in plants so important a problem during the 19th century?
9. Mention three theories that were proposed to account for the supposed extraction of nitrogen by plants from the air.
10. Describe the experiments at Rothamsted conducted to determine whether or not plants secure nitrogen from the air. What was the result of these experiments?
11. By whom, how, and when was the source of nitrogen of legumes discovered?
12. What important lessons are illustrated by the investigations relating to the source of nitrogen in plants?
13. What did these men contribute to early plant science: Camerarius, Kolreuter, Sprengel, and Naudin?
14. Who originated the theory of evolution? Why was it not accepted at that time?
15. What was the status of plant and animal improvement at the time of publication of the "Origin of Species"?
16. What did Darwin contribute to the theory of evolution? Why is he usually given credit for it?
17. What is the theory of natural selection?

18. Describe the work of Mendel and tell why he was successful.
19. What is the mutation theory? Chromosome theory of heredity? Multiple factor hypothesis?
20. What was the prevailing belief in spontaneous generation of life when Pasteur began investigating the subject?
21. What are the principal contributions of Pasteur?
22. Name several advances made in bacteriology since the time of Pasteur?
23. Name 5 important discoveries in plant pathology.

CHAPTER III

LOGIC IN EXPERIMENTATION

I. Scope of Science

Science is systematized knowledge. The function of science is the classification of observations and the recognition of their sequence and relative significance. Its scope is to ascertain truth in every branch of knowledge. Sound logic is just as fundamental to good science as accurate data. The thought process most important in science is induction, i.e., reasoning from the particular to the general. Generalizations may lead to laws and principles about natural phenomena.

II. Science among the Ancients

"Primitive peoples lived through thousands of years of myth and magic, while science was rising out of slow and unconscious observations of natural events, "Weir (1936) explains. Aristotle (384-322 B.C.), one of the first to stress science, taught that it can be developed only through reason. He set up a logical scheme, called the syllogism, which severely limited deductions made from generalizations. Jevons (1870) describes the syllogism as follows: "In a syllogism we so unite in thought two premises or propositions put forward, that we are enabled to draw from them or infer, by means of the middle term they contain, a third proposition called the conclusion." An example is as follows:

"All living plants absorb water; (major premise)
A tree is a living plant; (minor premise)
Therefore, a tree absorbs water." (conclusion)

The syllogism has been rejected for a long time because it leads to no new knowledge. It involves a deductive process, the conclusions being only as accurate as the premises upon which they are based. New generalizations can only be reached through induction, a process which affords a means to attack the premises themselves.

A -- Methods and Types of Research

III. Research

In the broad sense, the collection and analysis of data is research. However, there are different degrees of research value. Black, et. al. (1928) state that the mere accumulation of facts, computation of averages, or census-taking is not research. Fact-gathering alone is a mechanical procedure unless tied up with analysis. Moreover, projects designed to serve purely local or temporary needs, without some contribution to fundamental principles, ordinarily can have but little scientific value. General laws or principles are sought in research of the highest order. There are two methods of research, the empirical and the inductive. Black, et. al. (1928) state that "the essential difference between the two is that the first one accepts superficial relationships without inquiry as to antecedents, whereas the second one pursues antecedents a stage or two at least." An antecedent is a condition or circumstance that exists before an event or phenomenon.

IV. Inductive or Scientific Method

The process of induction is of special importance in experimental science in which general laws are established from particular phenomena. The inductive method is the scientific spirit of the day.

(a) Explanation of Induction

In induction one proceeds from less general, or even from individual facts, to more general propositions, truths, or laws of nature. In other words, it is the formulation of a principle from facts. Induction was the method of Francis Bacon, who held that general laws could be established with complete certainty by almost mechanical processes. Bacon advised that one begin by collecting facts, classifying them according to their agreement and difference. It is then possible to induce from their differences and similarities the possible reasons for the relationships exhibited and, from them, arrive at laws of greater and greater generality. Thus, the inductive method attempts to answer the question "why". A knowledge of causes enables the scientist to forecast with greater and greater assurance because, when he knows what is behind a set of relationships, he is in a much better position to know whether or not they will occur again. On the other hand, deduction is the inference from the general to the particular, i.e., some truth may allow individual facts to be sub-summed under it. Induction and deduction are used together in experimental work. For instance, a premature induction may be made to account for a phenomenon. A hypothesis is set up that may or may not be faulty. Next, an experiment is designed, a purely deductive process, to test this hypothesis. The investigator determines the particular instances he may create and observe by experiment to use as a basis of a re-generalization to establish the original hypothesis.

(b) Observation

The first requisite of induction is experience to furnish the facts. Such experience may be obtained by observation or experiment. Jevons (1870) makes this statement: "To observe is merely to notice events and changes which are produced in the ordinary course of nature, without being able, or at least attempting, to control or vary these changes." The botanist usually employs mere observation when he examines plants as they are met with in their natural condition. Progress of knowledge by mere observation has been slow, uncertain, and irregular in comparison with that attained in the controlled experiment. However, to observe well is an art that is extremely advantageous in the pursuit of the natural sciences. One should make accurate discrimination between what he really does observe and what he infers from the facts observed. The investigator should be "uninfluenced by any prejudice or theory in correctly recording the facts observed and allowing to them their proper weight", according to Jevons (1870).

(c) Experimentation

In the experimental method in its pure form, a special hypothetical plan becomes the basis of conclusions. The investigator varies at will the combinations of things and circumstances, and then observes the result. Fisher (1937) describes experimental observations as "only experience carefully planned, in advance, and designed to form a secure basis of new knowledge; that is, they are systematically related to the body of knowledge already acquired, and the results are deliberately observed, and put on record accurately." In actual practice, the effect of different factors is determined by holding all conditions constant or uniform except the one or ones whose "effects" are to be measured, a definite amount of change in this condition being balanced against a definite amount of change in the result. Black, et al, (1928) state that it is sometimes only the effect of the presence or absence of a condition that is noted. The method is qualitatively experimental instead of quantitatively in such cases. In many cases, it is impossible to hold all conditions but one constant or even uniform. So statistical analysis is combined with the experimental design to measure variation where it cannot be controlled. This is the practice in many agronomy experiments. For instance, when two or more wheat varieties are compared for yield, they are planted in the same field, at the same time, and at the same rate. Moreover, they are harvested at the same time, threshed by the same machine, and the seed weighed on the same balance. The conditions are thus uniform for the varieties rather than constant. The importance of the experiment is well summarized by Jevons: "It is obvious that experiment is the most potent and direct mode of obtaining facts where it can be applied. We might have to wait years or centuries to meet accidentally with facts which we can readily produce at any moment in a laboratory ...."

(d) Essentials of Good Scientific Method

The essentials in sound experimental method may be briefly summarized as follows:

1. The formulation of a trial hypothesis.
2. A careful and logical analysis of the problem generated by the hypothesis.
3. Use of the deductive method to design how to effect a solution of the problem. This involves a detailed outline of the experiment with costs, equipment, methods, etc. The factors should be expressed in quantitative terms when possible.
4. Control of the personal equation.
5. Rigorous and exact experimental procedure with the collection of data pertinent to the subject.
6. Sound and logical reasoning as to how the conclusions bear on the trial hypothesis and in the formulation of generalizations. A statement of the exact conclusions warranted from the cases examined should be made in accurate terms.
7. A complete and careful report of data and methods of analysis so that others can check them.

V. The Empirical Method

"When a law of nature is ascertained purely by induction from certain observations or experiments, and has no other guarantee for its truth, it is said to be an empirical law," according to Jevons (1870). Thus, knowledge is empirical when one merely knows the nature of phenomena without being able to explain the facts. It only answers the question "how". Formerly, the empirical method represented knowledge secured by trial, but today it means the haphazard "cut and try" method. A person who learns certain facts through repeated observations may know no reason for their being true, i.e., he cannot bring them into harmony with any other scientific facts. The method is valuable in spite of the criticisms against it. Empirical methods are most likely to be used when a science is new. Facts must be gathered before a notion of reasons can be formulated. The older crop rotation experiments were empirical. Recommendations are based on the results, i.e., certain rotation systems result in higher crop yields. Crop variety tests are generally empirical, since the chief concern is to determine what variety yields the highest.* Fully one-half the agronomic experiments in this country are haphazard in the nature of their relationship to the body of known knowledge in a given line. Too often they are not related to past experiments. (See Allen, 1930).

VI. General Types of Agronomic Experiments

Agronomic experiments can be divided into field and laboratory or greenhouse experiments. Questionaires and surveys are occasionally used to secure preliminary information.

*Note: In recent years, variety tests may involve more than empiricism. Crosses are often made to combine high yield with certain desirable quality factors or disease resistance. The yield trial determines whether or not the result has been accomplished.

(a) The Field Experiment

The field experiment involves the use of small plots, usually between 1/10 and 1/1000 - acre in size. The treatments are replicated, i.e., repeated on the experimental area in tests designed to remove the error due to soil heterogeneity. To make other conditions as uniform as possible, the varieties or treatments in the experiment are treated as nearly the same as possible except for the factor or factors under study. The field experiment has a wide application where yield is used as a criterion to measure treatment effect. Field experiments may be classified as follows: (1) Variety Tests: Such trials usually measure the yield of strains, varieties, and species. Various combinations of forage crops for hay or pasture are sometimes classified as variety tests. (2) Rate and Date Tests: These experiments are concerned with the yield response of a variety or crop when planted at different rates or on different dates. (3) Crop Rotation Tests: These trials include different series of rotations and crop sequences. (4) Cultural Studies: The time, manner, and frequency of field operations are considered in such tests. (5) Fertilizer Experiments: These experiments usually include tests to determine the needs of nitrogen, phosphorus, and potassium and their best combinations. Other considerations are ways to supplement farm manures, value of cover crops and green manures, and the amounts and methods of lime application. (6) Pasture Experiments: Field experiments with pastures are generally used to study methods to seed and fertilize new pastures, methods to renovate old pastures, and the influence of grazing on species survival. (See Noll, 1928).

(b) Laboratory and Greenhouse Experiments

Laboratory and greenhouse tests are often used to supplement field trials. These tests often involve potometer and lysimeter studies as well as those based on special techniques. Pot cultures are sometimes necessary for the study of the effect of one factor by the exclusion of the others, or by their exaggeration. However, the sole use of laboratory experiments may result in erroneous conclusions when applied to field conditions. (See Wheeler, 1907). The use of laboratories and greenhouses is on the increase because they have the advantage of controlled conditions. Some agronomic problems adapted to such conditions are: (1) artificial rust epidemics, (2) toxic effect of sorghums on crops that follow, (3) fertilizer cultures, (4) resistance of winter wheat to low temperatures, and (5) moisture, temperature, and light relationship studies. Equipment for the study of hardiness in crop plants has been described by Peltier (1931).

Potometers are pots filled with soil in which plants are grown for experimental purposes. To a greater or less extent the earlier investigators assumed the accuracy of such experiments when applied to field conditions. Lysimeters are modified soil tanks used to measure the magnitude of nutrient losses from the soil by leaching under various fertilizer and cropping conditions. Installation of lysimeter equipment is expensive but permanent. The principal feature is the measurement of drainage water. A description of lysimeter equipment is given by Lyon and Bizzell (1918) and by the American Society of Agronomy (1933).

(c) Questionaires and Surveys

Very little use is made of either the questionaire or survey in agronomic research. They are considered less desirable than the controlled experiment. The questionaire consists of a set of questions to be answered without the aid of an investigator (usually mailed). It is impossible to secure accurate answers on questions that are closely defined because the chances for misinterpretation are too great. Survey data are collected with the personal aid of an enumerator or investigator. Spillman (1917) assumes that careful analyses of the methods of a large number of farmers under essentially similar soil, climatic, and economic conditions, may be made to reveal the success of one person and the failure of others. He found that the discrepancy in the farmer's knowledge was small in large items, but increased as the importance to him decreased. Black, et al (1928) mentions some of the weaknesses of the survey: (1) It does not furnish enough detail for some types of problems; (2) It is not accurate enough for close analysis; and (3) It does not furnish a large enough sample for some purposes.

VII. Hypotheses, Theories, and Laws

The difference between the hypothesis, the theory, and the law, is in the degree of surety of the absolute.

(a) Explanation of these Terms

When an idea is suggested by observed phenomena it is spoken of as a hypothesis. It represents a desire to explain the phenomena such as, for example, the method by which plants take food from the soil. The hypothesis is important in the deductive method in that, to test this preliminary induction, it is replaced more or less completely by imagining the existence of agents which are thought adequate to produce the known effects in question. Thus, Jevons (1870) explains, the truth of a hypothesis altogether depends upon subsequent verification. A theory is a limited and inadequate verification of a hypothesis. Examples are the theory of the gene, and the theory of evolution. A theory becomes a law when it is proved to be a fact beyond a reasonable doubt. The Mendelian laws of heredity are good examples of laws.

(b) Formulation of a Hypothesis

There are certain advantages to the hypothesis: (1) It correlates facts; (2) it forecasts other facts; and (3) it allows for discrimination between valuable and useless information. Every experiment is the result of a tentative hypothesis thought out in advance of the actual test. The hypothesis is based on the recognition of coincident phenomena, or upon a familiarity with possible causes and effects. Hibben (1908) states: "Hypothesis and experiment to Charles Darwin were like a two-edged sword which he employed with rare skill and effect." The hypothesis is the precursor of the experiment which is merely an effort to solve the problem created by the hypothesis.

(c) Qualities of a Good Hypothesis

There are several qualities that a good hypothesis should possess. These are as follows: (1) It should be plausible. (2) It must be capable of proof, i.e., it should provide a susceptible means to attack the problem created thereby. (3) It must be adequate to explain the phenomena to which it is applied. (4) It should involve no contradiction. (5) A simple hypothesis is preferable to a complex one. There is little use to form a hypothesis on a complex basis unless it is possible to collect the data by which it may be proved. A multiple hypothesis is made up of several ideas. Occasionally it may be desirable to formulate several hypotheses. Salmon (1928) advises an investigator to at least give consideration to all observable hypotheses. They are useful even though wrong because they eliminate that particular idea from the problem. At any time, an investigator must be ready to abandon a hypothesis or theory when further data prove the previous views untenable.

(d) Null Hypothesis

In all experimentation the null hypothesis is characteristic. The term has been applied by R. A. Fisher (1937) in his "Design of Experiments." The basic assumption is that no difference exists between the treatments in the experiment, i.e., they are samples drawn from the same general population. For instance, in a variety test, the investigator makes the basic assumption that all varieties yield alike. He can never prove this assumption but he may disprove it in the course of experimentation. By the use of certain statistical arguments he may show a significant discrepancy from the hypothesis, i.e., the probability is that some of the varieties do differ in yield. Fisher (1937) states: "Every experiment may be said to exist only in order to give the facts a chance of disproving the null hypothesis."

(e) Crucial Tests

There may be two alternative conceptions or explanations which appear possible. A crucial test (experimentum crucis) is one by which two rival hypotheses can be tested so that if one is proved, the other is immediately disproved. This is the only means by which a hypothesis may be disproved. The first record of the application of the crucial test is attributed to Francis Bacon. A good example of a crucial test was the one applied by Richey and Sprague (1931) to test two theories for the cause of hybrid vigor in corn which is expressed when two inbred lines of reduced vigor are crossed to give the first generation hybrid. This additional vigor, Richey (1927) explains, has been attributed to the physiologic stimulation hypothesis in which heterogenous germplasm within the cells provides the stimulation. The other hypothesis is that of dominant growth factors in which it is believed that the maximum number of dominant growth factors are brought together in the first generation hybrid, and that linkages of favorable dominant growth factors with other less desirable factors prevented the recovery of individuals as vigorous as the F_1 in subsequent generations. Richey and Sprague (1931) applied a crucial test to the two hypotheses by the collection of data on the principle of convergent improvement, i.e., backcrossing the F_1 hybrid to each of the two inbred lines that went into the hybrid. It was hoped to transfer some of the favorable dominant growth factors from one of the lines and intensify them in the other. Thus, the two convergently improved lines would have less differences between them than was true of the original inbred lines. Lowered yields of the cross of the convergently improved lines, as compared to the cross of the original lines, would tend to support the physiological stimulation hypothesis. The same or higher yields from the cross of the convergently improved lines would lend support to the dominant growth factor hypothesis. The data collected gave support to the latter.

B -- Kinds of Evidence

VIII. Importance of Evidence

It is necessary to collect facts or data before generalizations can be made. There are different kinds of evidence, some kinds being more apt to lead to valid conclusions than others. However, plants are complex organic compounds with the result that it is more difficult to determine the elements of cause and effect than is ordinarily true in the more stable physical sciences. Environment has a tremendous influence on the plant. The more that experiments or observations are repeated with the same results, the more valid the evidence becomes in the minds of all normal human beings. For example, a large number of experiments show that weed control is the principal benefit derived from cultivation. The fact that a large number of investigators have found this to be true under different conditions adds to the assurance that the results are correct. Certain methods have been developed to deal with the evidence obtained by observation or experiment which may serve as guides to those in search of general laws of nature.

IX. Cause and Effect

Induction consists of inferring general conclusions from particular evidence. In some cases, generalizations relate to cause and effect. An antecedent is a condition which exists before the event or phenomenon, while a consequent follows after the antecedents are put together. Jevons (1870) makes this statement: "By the cause of an event we mean the circumstances which must have preceded in order that the event should happen. Nor is it generally possible to say that an event has one single cause and no more. There are usually many different things, conditions or circumstances necessary to the production of an effect, and all of them must be considered causes or necessary parts of the cause." It is certainly true that a multiplicity of causes is often involved in experiments in field crops and soils.

X. Qualitative Evidence

Qualitative evidence is that which can be measured only categorically. For example, seeds either germinate or fail to germinate. Classification by color is a common form of qualitative data.

(a) Method of Agreement

This method of induction is defined by Jevons (1870) as follows: "The sole invariable antecedent of a phenomenon is probably its cause." It is necessary to collect as many instances as possible and compare together their antecedents. The one or more antecedents which are always present when the effect follows is considered the cause. For example, when rust is present on wheat, low yields are obtained. Therefore, rust causes low yields. This method has a serious difficulty in that the same effect in different cases may be due to different causes.

(b) Method of Difference

In this method, the antecedent which is always present when the phenomenon follows, and absent when it is absent, is the cause of the phenomenon when other conditions are held constant. In other words, when the circumstances are all in common except one, i.e., the treatment, then the change that occurs is the effect of the treatment. This is probably the most widely used method in experimentation. The differences in crop yields under certain manurial treatments is an example of this method.

(c) Joint Method

In the words of Jevons (1870), the joint method of agreement and difference "consists in a double application of the method of agreement, first to a number of instances where an effect is produced, and secondly, to a number of quite different instances where the effect is not produced." For example, the experiments of Darwin on cross and self-fertilized plants may be cited. He placed a net around 100 heads to protect them from chance insect pollination. He also placed 100 heads of the same variety where they were exposed to bees. The protected flowers failed to yield a single seed, while the unprotected ones produced 2720 seeds. Thus, cross fertilization by means of insect pollination was proved to be a cause of seed set in this case.

XI. Quantitative Evidence

Every science and every question is first a matter of generalizations built upon qualitative evidence. The effort to more firmly substantiate such generalizations leads to the measuring of evidence quantitatively so that by degrees, the evidence becomes more and more precisely quantitative.

(a) Method of Concomitant Variations

This method can be applied where the phenomena can be measured. Every degree and quantity of the phenomenon adds new evidence in support of relationships that exist between antecedents and consequents of the phenomenon. The method which employs concomitant variations to determine the degree of such relationship is called correlation. For instance, an experiment with wheat results in a low yield under conditions of heavy stem rust infestation, with variations to the other extreme.

(b) Method of Residues

There may be several causes, each of which produces part of an effect, and where it may be desirable to know how much of the effect is due to each. This type of evidence consists in the analysis of a given phenomenon to determine the residue. For instance, manure contains something besides phosphorus, potash, and nitrogen as shown by the residues. In plants it has been determined that other than the so-called 10 essential elements are used because analyses of the plant ash show others to be present. The method of residues is constantly employed in chemical determinations.

XII. Relation to the Original Hypothesis

Some experiments fail in their objective in that there is insufficient evidence at hand to permit the investigator to draw positive conclusions. However, this evidence is valuable. It has been called "negative evidence," but in reality there is no such thing. Research would be much further along than it is today if all experiments had been reported in which the evidence was insufficient to prove the hypothesis that was originally set up by the investigator. Such evidence would have saved other workers from a repetition of the work.

XIII. Use of Analogy

Analogy is a form of inference in which it is reasoned that, if two (or more) things agree with one another in one or more respects, they will probably agree in still other respects. It is the simplest and most primitive form of evidence, its great weakness being the fact that the cases compared may not be parallel. Analogy may be tested by some inductive method. For example, the theory of evolution was suggested to Darwin from the "Essay on Population" by Malthus. It suggested to him that the struggle for existence is the inevitable result of the rapid increase in organic beings. The idea necessitated natural selection or "survival of the fittest." Another example might be cited in durum wheat. Durum wheat is adapted to Russia and so is Turkey wheat. Since Turkey wheat is adapted to the Great Plains in this country, durum wheat must be adapted to this region also. A common analogy made by agriculturists is that crops can be improved by systematic selection because live-stock breeders have succeeded in that way. Logic derived from analogy too often leads the inexperienced astray.

C -- Methods of Discovery

XIV. Work of other Investigators

An investigator seldom takes up work today that is entirely new. He secures valuable help from other research workers. The cooperative attitude among the workers on the Purnell corn projects is particularly commendable in this respect. They get together occasionally to talk over their problems freely and to offer suggestions. They have been unusually free with their preliminary data and unpublished results so far as fellow workers are concerned. This attitude has done much to advance research in corn improvement. The seed analysts have cooperated among themselves in a similar manner. Scientific meetings result in a more or less free exchange of ideas to the benefit of all. These get-togethers are a great aid and should be attended by research workers.

XV. Surprises and Accidental Discoveries

An important discovery is quite often made by accident. Several examples could be cited.

(a) Lemon Juice in Grasshopper Bait

Some 25 years ago, two workers in the U. S. Department of Agriculture were testing poison bran mash as a grasshopper bait in Kansas. These men had oranges in the lunch they took to the field with them. While eating their oranges, some of the juice accidentally came in contact with the bran mash. The men noticed that the grasshoppers preferred the mash that contained the orange juice. As a result of this discovery, Kansas came out with the lemon juice formula in 1911.

(b) Heterothalism in Stem Rust

Prior to 1927, it was believed that the pycnia on the upper surface of the barberry leaf had no function. Craigie (1927) got the idea that the mycelium, pycnia, and pycniospores of some of the pustules were plus sex strains and others minus sex strains. He happened upon the proof by chance. The first fly of the season appeared in the greenhouse on May 17. He watched it idly as it sipped nectar at one pustule and then at another. Professor Buller happened by and said at once: "The solution of the problem is an entomological one. Copy the fly. Take the plus pycniospores to the minus pycnia, and the minus pycniospores to the plus pycnia." Craigie followed this advice by mixing nectar from different pustules. The pycniospores germinated and brought on the development of aecia and aeciospores, the diploid phase. He repeated his test many times and found it to be true. Craigie proved his theory as follows: Flies were introduced in some cages containing barberry plants with pustules on the leaves, while flies were excluded from other barberry plants. Aecia were formed in five days where the flies were present, but none were formed where the flies were excluded.

XVI. Systematic Research

One of the principal methods of discovery is through systematic research where a problem is attacked from all conceivable angles. An example is the contribution of the Hawaiian Experiment Station on chlorosis. The pineapple industry was restricted to a small area because of a discoloration of the foliage that showed it to lack chlorophyll. The investigators on this problem first exhausted the possibilities of disease, after which they analyzed the soil and found it to contain considerable manganese. Next, the workers used this high-manganese soil on soil that would grow pineapples, and found that very little iron was taken into the plants. The manganese was thus found to inhibit iron absorption. The plants were then sprayed with iron salts and the chlorophyll deficiency corrected. Pineapple trees are now sprayed at the rate of 50 pounds of iron salts per acre, the yield of fruit being doubled as a result.

XVII. Other Methods of Discovery

Several other methods have resulted in significant discoveries. (1) Conflicting Results: Disagreement between different research workers in their results often leads to new discoveries. Pasteur became engaged in a controversy with Leibig on the spontaneous generation of life. As a result, Pasteur proved that all new life arose from forms that had already existed. Some of the most fertile fields for new ideas are the first new hypotheses, theories, and ideas. (2) Accurate Work: Accurate work is necessary to secure dependable facts on which to base conclusions. More information usually results from work done carefully than from that which has been unplanned and carried out in a haphazard manner. In addition, the work of investigators must be accurate to withstand the close scrutiny of other workers and of general opinion. Accurate work often has led to new discoveries. (3) Analogy: A fruitful source of new ideas that sometimes leads to new discoveries is analogy. It may suggest a hypothesis from the results secured in other experiments. (4) Ideas from Farmers: In agricultural research, the problems called to the attention of experiment station workers by farmers is an important source of discovery.

References

1. Allen, E. W. Initiating and Executing Agronomic Research. Jour. Am. Soc. Agron., 22:341. 1930.
2. Black, John D., et al. Research Method and Procedure in Agricultural Economics (mimeographed), pp. 1-20, 58-90, 113-126, and 298. 1928.
3. Craigie, J. H. Discovery of the Function of the Pycnia of the Rust Fungi. Nature, 120:765-767. 1927.
4. Fisher, R. A. The Design of Experiments. Oliver and Boyd. 2nd Ed. pp. 1-12, and 18-20. 1937.
5. Hibben, J. G. Logic: Deductive and Inductive. pp. 169-182, 222-277, and 291-329. 1908
6. Jevons, W. Stanley. Elementary Lessons in Logic. pp. 9-16, 116-117, 126-135, 201-210, and 218-276. 1870 (reprinted in 1928).
7. Lyon, T. L., and Bizzell, J. A. Lysimeter Experiments. Cornell Memoir 12. 1918.
8. Noll, C. F. The Type of Problem Adapted to Field Plot Experimentation. Jour. Am. Soc. Agron., 20:421-425. 1928.
9. Pearson, Karl. The Grammar of Science, pp. 1-151. 1911.
10. Peltier, G. L. Control Equipment for the Study of Hardiness in Crop Plants. Jour. Agr. Res., 43:177-182. 1931.
11. Richey, F. D. The Convergent Improvement of Selfed Lines of Corn. Am. Nat., 61:430-449. 1927.
12. ____________, and Sprague, G. F. Experiments on Hybrid Vigor and Convergent Improvement in Corn. Tech. Bul. 267, U.S.D.A. 1931.
13. Salmon, S. C. Some Limitations in the Application of Least Squares to Field Experiments. Jour. Am. Soc. Agron., 15:225-239. 1923.
14. ____________. Principles of Agronomic Experimentation. Kans. St. Agr. Col. (Unpublished Lectures). 1928.
15. Spillman, W. J. Validity of the Survey Method of Research. Dept. Bul. 529, U.S.D.A. 1917.
16. Standards for the Conduct and Interpretation of Field and Lysimeter Experiments. Jour. Am. Soc. Agron., 25:803-828. 1933.
17. Weir, W. W. Soil Science, pp. 11-16. 1936.
18. Wheeler, H. J. Some Desirable Precautions in Plot Experimentation. Jour. Am. Soc. Agron., 1:39-44. 1907.

Questions for Discussion

1. What is science?
2. What is the syllogism? Give an example.
3. Why has the syllogism been abandoned in experimental work?
4. What is research? Discuss different values of research.
5. How does the inductive method of science differ from the empirical?
6. Why is it considered desirable to determine basic or fundamental laws rather than merely to determine what happens?
7. Distinguish between induction and deduction.
8. What part does observation play in research work? What precautions are necessary in its use?
9. What is an experiment? Discuss its use.
10. What are the principal steps in the inductive method of science? Which ones are most often omitted?
11. Under what conditions is the empirical method justified?
12. Name some types of agronomic tests that are empirical in nature.
13. What serious limitation is true of the empirical method?
14. What are some reasons for criticism of the methods of research?

15. Classify field experiments and describe each class.
16. What place have laboratory and greenhouse tests in agronomic research?
17. How do questionaires and surveys differ?
18. Distinguish between potometer and lysimeter tests.
19. Distinguish between hypothesis, theory, and law.
20. Is it desirable to formulate hypotheses in experimental work? Why?
21. What qualities are necessary in a good hypothesis?
22. What is a working hypothesis?
23. What advantages are there, if any, in formulating multiple hypotheses?
24. What is the null hypothesis?
25. What is a crucial test? Explain one.
26. Why is research often more difficult in plant sciences than that in the physical sciences?
27. Distinguish between cause and effect.
28. Name, define, and illustrate five different kinds of evidence.
29. What is the most important inductive method in experimentation? Why?
30. What is analogy? Discuss its use and give an example.
31. What is the value of negative evidence?
32. Mention 4 ways in which discoveries are made.
33. How was the cause of chlorosis found in pineapples in Hawaii?
34. Mention 3 discoveries and tell how they originated.

CHAPTER IV

ERRORS IN EXPERIMENTAL WORK

I. Types of Experimental Error

Two kinds of error are common in experimental work, systematic errors and chance errors. The investigator needs to be familiar with both kinds. Such errors should be distinguished from mistakes and blunders. For example, a worker makes a mistake when he puts down a weight of 10 lbs. when the scale actually showed the weight to be 20 lbs.

(a) Systematic Errors

Systematic errors occur every time that an experiment is repeated in the same way. Most experimental plans involve some errors of this kind. For example, suppose that a large number of winter wheat varieties are arranged systematically in single-row plots. Some of the varieties kill out because of lack of hardiness. The varieties in the adjacent rows might yield abnormally high because of the additional space from which they could draw moisture. Such an error would be repeated every time the experiment is conducted in this manner. In this particular case, the competition effect could have been avoided by planting three-row plots for each variety and only the center row harvested for yield.

(b) Chance Errors

Errors which occur by pure chance with no definite assigned cause are known as chance errors. They are generally small fluctuations due to minor causes. Chance errors may accumulate to produce a sizeable deviation even though it be impossible to foresee and analyze all causes that contribute to them. The principal reason for statistical analysis in agronomic science is its very inexactness and the inability to control chance errors. In case the present theory of plot technique is acceptable, the variations in plot yields are due to chance errors and, in most cases, have been found by experience to be normally distributed. This means that there are a large number of small errors and a small number of large errors. Statistical methods are employed in field experiments to measure the effect of chance errors. In addition, some systematic errors can be removed by these methods as will be shown later.

II. Sources of Error in Experimental Work

Evidence gained by experiment is disputed, according to Fisher (1937) either on the grounds that the interpretation is faulty, or on the criticism that the experiment itself is poorly designed. Errors are always possible and seldom absent in experimentation.

(a) Faulty Design and Inferior Technique

Experimental designs are inadequate or faulty when they do not afford a proper opportunity for statistical analysis to analyze and measure experimental errors, both chance and systematic. Fisher states:, "If the design of an experiment is faulty, any method of interpretation which makes it out to be decisive must be faulty too." The investigator may fail to take certain variable factors into account. Aside from these, various personal errors may have been introduced, such as carelessness. Farrell (1913) lists a few sources of error in field experiments. Among the controllable ones are. Incorrect weights of crop products, faulty determinations of plot area, variations in quantities of products recovered and wasted, unobserved variations in field treatments, etc. Among the errors seldom controlled, he cites: Plant variation, soil irregularities, uneven distribution of soil moisture, and temperature variations. Frequently, the total effect from all causes is great enough to influence the conclusions of the experiment. It might be added that some of these errors can be measured and their influence on the conclusions removed.

(b) Improper Interpretation of Results

Two common types of misinterpretation of experimental results are drawing conclusions from too few data, and carrying the interpretation beyond the points actually tested. (1) Conclusions drawn from too few data: An experiment may be inadequately replicated in time and space to justify the conclusions drawn. Carleton (1909) warns that some experiments are defective because they are run for an insufficient length of time. Sometimes investigators are in too much of a hurry to obtain results. Another common mistake is to over-emphasize small differences. Statistical methods have done a great deal towards reducing invalid inferences due to too few data. (2) Interpretation carried beyond points tested: Sometimes the interpretation of the results of an experiment is carried beyond the points actually tested. Salmon (1923) believes that one of the chief sources of error in agronomic literature is the tendency to generalize from experiments limited in their scope. For instance, it should be quite obvious that laboratory tests may not always be applied to field conditions. Such generalization must be justified by a similarity of conditions. As an example, suppose phosphates were added to the soil in a fertilizer test in amounts of 100, 400, and 600 pounds per acre. One would be unable to draw conclusions on, say 1000 pounds, because it is beyond the amount tested in the experiment. It is obvious that a point may be reached where the addition may have a depressive effect. Sievers (1925) points out that recommendations based on variety tests conducted under different conditions as to soil, climate, and weather than those under which the farmer operates are unsatisfactory.

III. The Personal Factor

Individuals differ greatly in the way they attack problems and carry out the various details connected with them. For example, two men will seldom agree exactly when they make measurements on the same thing because they do not "see exactly alike". Such differences are apt to be more pronounced when personal judgment plays an important role. The mixing of materials illustrates a situation where individual workers may differ in the details of their procedures to an extent that the end-product is affected. Mechanical devices tend to do away with the personal factor.

When several individuals work on an experiment it is desirable for the same person to complete an entire operation, or at least for all the treatments in a single replicate. For example, in a variety test the same person should plant the plots, harvest them, and make the weights so far as possible. At least, the same crew should carry out the details uniformly for all plots or treatments, preferably for the entire test.

IV. Sources of Variation in Field Experiments

Certain limitations in plot work must be recognized. To quote Noll (1928): "The most serious are that the experiments must be made under constantly changing conditions as to moisture and temperature, and that the average results for a given soil in a given locality, no matter how carefully planned, are not necessarily applicable elsewhere." The most common variations in field experiments are those due to plants, those due to differences in seasons, and those due to the soil. The variations that cannot be balanced out can be measured in a well-designed experiment. Some of those due to defined causes, such as soil heterogeneity, can be removed or balanced in part but not entirely. The variations that are due to unrecognized causes are measured and assigned to experimental error.

(a) Errors Related to the Plant

Variation may be introduced due to differences in acclimatization unless this factor happens to be the one under study. Differences in stand may be a fruitful source of variation, particularly in crops like corn, sorghums, etc., where plant individuality is important. There are less corn plants on a unit area than wheat plants. A further source of variation due to plants is the difference in moisture content of the harvested crop. Correction to a uniform moisture basis is advocated under such conditions. Plant competition may introduce still further error in plot results.

(b) Variations in Seasons

Climate rather than soil may be the limiting factor in crop production. Some varieties are known to withstand extreme conditions like drouth or excessive moisture better than others. From uniformity trials with corn over a 3-year period, Smith (1909) concluded that more variation in yield could be expected in seasons unfavorable for the crop. For that reason, tests conducted for only one or two years may be very misleading. This situation may be remedied by the extension of a variety test over a number of seasons to determine the variety that thrives best in an average season. For a reliable average of seasonal conditions, a variety test should be conducted for at least three years and preferably more. Under dryland conditions, it takes at least 10 years to secure a reliable variety average. Variety comparisons should be strictly comparable, i.e., compared only for the same years under test. Usually this is accomplished by expressing yields in percent of the standard or check. Other factors that may cause the yields of varieties to vary from season to season are: (1) The plots may be damaged by windstorms one year and not in another. (2) Rodents may cause more damage in some years than in others. (3) Insects may be troublesome in certain seasons. (4) Rust in small grains may reduce yields more in some years than in others. (5) There may be an inaccuracy in scale weights from one season to another. (6) Carelessness in harvesting or threshing is another factor in some seasons. (7) Sometimes the planter fails to drill out to the end of a plot with a possible error in yield as a result. (8) Crooked rows may introduce errors in the yields of row crops.

(c) Errors due to Soil Variation

It is impossible to secure a perfectly uniform soil for field experiments. Differences in productive capacity commonly occur in different portions of the same field. In fact, soils vary in composition and productivity from foot to foot with the result that it is impossible to say that any soil is uniform, even on small areas. However, the investigator should secure as uniform a piece of ground as possible. Sedentary soils are usually more uniform than drift soils, and level land more likely to be uniformly productive than hilly land. Other factors that may introduce variation are: topography, under-drainage, sub-soil, and previous soil management practices.

V. Errors in Laboratory and Greenhouse Tests

There are many possibilities for error in tests of this kind. Probably the most serious one is to draw conclusions from laboratory tests for field conditions without a field test. Laboratory tests should supplement, rather than replace the field experiment.

(a) Errors in Greenhouse Tests

Some of the possibilities for error may be listed as follows: (1) The number of plants is small. Plant individuality assumes major importance, particularly when the investigator works with large plants. (2) There may be unequal distribution of water. It is difficult to get a uniform distribution of water through a heavy soil. (3) It is often important that the exact amount of water in a soil be known. This is particularly true in pots for freezing tests. (4) There may be a lack of uniformity in the soil itself. This may be alleviated by thoroughly mixing the soil in a homogenous mass. The mixed soil should be packed uniformly in all pots. (5) Some insects may be restricted only to greenhouse conditions. As a result, the behavior in the field may be entirely different so far as insects are concerned. (6) There may be a temperature or light differential under controlled conditions. A lack or over-balance of either or both may introduce a systematic error in the experiment. Le Clerg (1935), in a uniformity trial with 400 small pots in a greenhouse experiment, found the per cent of damping-off in sugar beets to be less in the border-row pots on a raised concrete bench than in those farther removed from the heat pipes. The effect was almost absent in a bench provided with wall boards to deflect direct heat. The unequal exposure to light or heat may be corrected in some instances by rotation of the pot table periodically.

(b) Comparison of Potometer and Field Trials

Data from pot experiments and field trials were found by Coffey and Tuttle (1915) to agree closely in fertilizer experiments. However, many fertilizer analogies from pot tests have led to errors in interpretation. Kezer and Robertson (1927) found no agreement between potometers and field plots in irrigation studies with wheat. Potometers with late irrigation treatments became so dry that the soil pulled away from the edge of the can. When water was added, most of it ran down the cracks and out of reacn of the root systems of the stunted plants.

VI. Statistical Methods in Relation to Variation

The statistical method is the mathematical means to measure and describe variation and to allocate its component parts to certain recognized sources. Variation can be measured quantitatively thru the medium of an experimental design that takes into account the recognizable sources of variation. The measurement of total variation makes it possible to obtain a measure of that due to all uncontrolled sources. The statistical method concludes its role when it gives the experimenter a means to compare the obtained quantitative measures of variation due to the recognized possible causal factors with the variation classified as error and also with each other. Thus, conclusions can be drawn in regard to the relative importance of the sources of variation.

VII. Classical Fallacies in Agronomy

A number of fallacies in agronomy have been listed by Salmon (1929). Many of these ideas were accepted as facts until rather recently. An analysis of these fallacies shows how each came to be accepted by agriculturists.

(a) Conservation of Moisture by the Dust Mulch

The effectiveness of the soil mulch in the conservation of soil moisture has been under discussion for many years. The early work, on which the dust mulch theory was based, was performed in the laboratory. Between 1885 and 1900, King (1907) showed that the dust mulch was quite effective in the reduction of water evaporated from the soil surface. In fact, the water loss was about one-half that from a bare soil. However, King worked in the laboratory with soil in tubes, the water table being only 22 inches from the soil surface. On the basis of this and similar experiments has rested the conviction that the soil mulch would reduce evaporation losses and materially aid in the conservation of moisture. This theory was believed and practiced until tests by the Office of Dry Land Agriculture (USDA) proved that it was without foundation. Call and Sewell (1917) showed that the soil mulch failed to increase the moisture in the soil. In fact, the mulched plots actually lost more water than bare undisturbed soil. The limit of capillary rise from a free water surface is only about 10 feet, according to the work of Shaw and Smith (1927). However, they found moisture losses to be quite rapid from unmulched soil where the water table was 4 to 6 feet from the surface. Other experiments in Illinois, Missouri, and Nebraska have shown that corn yielded almost as much where the weeds were scraped with a hoe as where the plots were cultivated (mulched). Shaw (1929) reworked King's experiment using soil tubes 4 feet high, and maintaining a constant water table at the base of each. The loss in the mulched tube was 38 per cent less than that from the tube in which the soil was left bare. This test merely confirmed the fact that the results from these soil tubes could not be applied to field conditions where the free water surface is usually more than 10 feet from the soil surface. Under dryland conditions where moisture conservation is extremely important, the water table is very often 200 to 400 feet from the surface.

(b) Deep Plowing for Moisture Conservation

The theory that very deep plowing will save moisture by an increase in the storage volume of the soil is an old one that dates back to about 1880. It was sometimes advocated that the soil be stirred from 14 to 18 inches deep. Deep tillage was widely advocated on the Great Plains along about 1910 by Hardy W. Campbell. Most of the implements used were soon allowed to rust out in fence corners. Experimentation very quickly showed that deep tillage (14 to 18 inches deep) was impractical or actually depressed the yields under dryland conditions. Brandon (1925) found that winter wheat grown on plots subsoiled every two years actually yielded 1.3 bushels per acre less as a 15-year average than wheat on land plowed at ordinary depths. Similar results were obtained in Wyoming by Nelson (1929).

(c) Continuous Selection of Small Grains

It was believed at one time that continuous selection was a means to invariably improve small grains. After 50 years of continuous selection, Vilmorin concluded that no improvement had resulted in wheat, a self-fertilized crop. The pure line theory worked out by Nillson-Ehle and by Johannsen showed that selection was effective only in heterozygous material. This old idea on the value of selection was probably due to a disregard of the difference between self and cross-fertilized plants.

(d) Selection of Seed Corn by Score Card Standards

Arbitrary score card standards were improvised in the early days as ideals for seed selection in corn. These standards laid stress on such points as shape of kernel, length of kernel, ears with well-filled butts and tips, percentage of grain on the cob, weight of ear, etc. Uniformity of ears was particularly stressed. The height of the belief in the "pretty ear" was reached about 1910 when the most "perfect" ear at the National Corn Show sold for several hundred dollars. When planted in the field in comparison with ordinary ears, it failed to surpass them either in yield or quality. This started a great amount of research on the relation of score card points to yield. It was generally proved that such arbitrary standards are of little value. In fact, close selection for type was generally shown to result in an approach to homozygosity with a reduction in yield and vigor as a consequence. Some of the investigators who aided in the upset of this theory were: Cunningham (1916); Love and Wentz (1917); Olson, Bull and Hayes(1918); Kiesselbach (1922); and Richey (1925)

(e) Calcium-Magnesium Ratio in Soils

A physiological balance seems to be necessary in nutrient solutions for a normal plant growth. In 1892, Loew proposed the calcium magnesium ratio hypothesis. He worked out the optimum ratio for a number of different plants in water cultures. He concluded that either calcium or magnesium used alone was toxic, but that the toxicity disappeared when these elements fell within certain limits. The ratios which Loew used varied from 1 CaO : 1 MgO to 7 CaO : 1 MgO. A large amount of investigation has been conducted on this ratio in which it has been shown that a rather definite ratio of CaO to MgO is required in nutrient solutions for optimum plant growth. The same applies to other nutrient elements as well. However, there appears to be little evidence to support the necessity for a definite ratio of CaO to MgO in soils. Recently, Moser (1933) reported that the ratio itself showed no relation to crop yields. The beneficial effect of lime added to the soil was attributed to the increase in replaceable calcium rather than to an alteration of the calcium-magnesium ratio. It is sufficient to state that Loew conducted his experiments with water cultures which probably react differently from soils.

(f) Addition of Burnt Limestone to the Soil

It is still believed by some farmers that the addition of burnt limestone to the soil results in a destruction of organic matter and an increase in the soil acidity. That burnt limestone increased the acidity was reported by the Pennsylvania Experiment Station. The theory, as taught, was based on small analytical differences in soil analyses.

(g) Acid Phosphate and Soil Acidity

The use of green manure and acid phosphate was at one time said to increase soil acidity. Grass and green material were known to decay and give an acid under laboratory conditions. Careful work under field conditions has shown that bacteria use up the organic acid formed. Acid phosphate was thought to increase soil acidity because of the name. It has been changed to superphosphate recently for psychological reasons.

References

1 Brandon, J. F. Crop Rotation and Cultural Methods at the Akron Field Station. Dept. Bul. 1304, USDA. 1925.

2. Call, L. E., and Sewell, M. C. The Soil Mulch. Jour. Amer. Soc. Agron. 9:49-61. 1917.

3. Carleton, M. A. Limitations in Field Experiments. Proc. Soc. for Agri. Sci., pp. 55-61. 1909.

4. Coffey, G. N., and Tuttle, H. F. Pot Tests with Fertilizers Compared with Field Trials. Jour. Am. Soc. Agron., 7:128-135. 1915.

5. Cunningham, C. C. The Relation of Ear Characters of Corn to Yield. Jour. Amer. Soc. Agron., 8:188-196. 1916.

6. Farrell, F. D. Interpreting the Variation of Plot Yields. Cir. 109, BPI, USDA, pp. 27-32. 1913.

7. Fisher, R. A. Design of Experiments, pp. 1-12. 1937.

8. Kezer, A., and Robertson, D. W. The Critical Period of Applying Irrigation Water to Wheat. Jour. Am. Soc. Agron., Vol. 19, No. 2. 1927.

9. Kiesselbach, T. A. Corn Investigations. Nebraska Agr. Exp. Sta. Res. Bul. 20. 1922.

10. King, F. H. Physics of Agriculture. 1907.

11. Le Clerg, E. L. Factors Affecting Experimental Error in Greenhouse Pot Tests with Sugar Beets. Phytopath., 11:1019-1025. 1935.

12. Lipman, Chas. B. A Critique of the Hypothesis of the Lime-Magnesia Ratio. Plant World, 19:83-105, and 119-135. 1916.

13. Love, H. H. and Wentz, J. B. Correlations Between Ear Characters and Yield in Corn, Jour. Amer. Soc. Agron., 9:315-322. 1917.

14. Moser, F. The Calcium-Magnesium Ratio in Soils and Its Relation to Crop Growth. Jour. Amer. Soc. Agron., 25:265-377. 1933.

15. Nelson, A. L. Methods of Winter Wheat Tillage. Wyo. Agr. Exp. Sta. Bul. 161, 1929.

16. Noll, C. F. The Type of Problem Adapted to Field Experimentation. Jour. Am. Soc. Agron., 20:421-425. 1928.

17. Olmstead, L. B. Some Applications of the Method of Least Squares to Agricultural Experiments. Jour. Amer. Soc. Agron., 6:190-204. 1914.

18. Olson, P. J., Bull, C. P., and Hayes, H. K. Ear Type Selection and Yield in Corn. Minn. Agr. Exp. Sta. Bul. 174. 1918.
19. Richey, F. D. Corn Judging and the Productiveness of Corn. Jour. Amer. Soc. Agron., Vol. 17, No. 6, 1925.
20. Salmon, S. C. Principles of Agronomic Experimentation (Unpublished lectures) Kansas State College. 1929.
21. ________. Some Limitations in the Application of the Method of Least Squares to Field Experiments. Jour. Amer. Soc. Agron. 15:225-239. 1923.
22. Shaw, C. F. When the Soil Mulch Conserves Moisture. Jour. Amer. Soc. Agron., 21:1165-1171. 1929.
23. ________, and Smith, A. Maximum Height of Capillary Rise Starting with Soil at Capillary Saturation. Hilgardia, 2.399-409. 1927.
24. Sievers, F. J. Outstanding Weaknesses in Investigational Work in Agronomy. Jour. Am. Soc. Agron., 17:88-89. 1925.
25. Smith, L. H. Plot Arrangement for Variety Experiments with Corn. Proc. Am. Soc. Agron., 1:84-39. 1909.

Questions for Discussion

1. Distinguish between chance and systematic errors.
2. What errors in field experiments can be controlled?
3. What kinds of errors in field experiments are not controlled? How are they minimized?
4. What errors can be made in the interpretation of experimental results?
5. How may the personal factor influence experimental results?
6. What are the general sources of variation encountered in field experiments?
7. What factors cause plot yields to differ from season to season?
8. What errors may occur in greenhouse tests?
9. How did the soil mulch theory originate and, in the light of present knowledge, how might the error have been prevented?
10. Is there any experimental or scientific basis for the belief that very deep plowing (10 inches or more) is profitable? Explain how this idea originated.
11. How did the belief that good seed corn is characterized by deep, rough kernels, and cylindrical ears originate?
12. What was the basis for the belief that a certain calcium-magnesium ratio was necessary for plant growth?
13. Explain the origin of the idea that burned lime decreases organic matter in the soil.
14. Is there any reason to believe that acid phosphate or green manure increases soil acidity? Why was it thought they did?
15. Make a general statement which will explain the sources of error that have occurred in agronomic science.

FIELD PLOT TECHNIQUE

Part II

Statistical Analysis of Data

CHAPTER V

FREQUENCY DISTRIBUTIONS AND THEIR APPLICATION

I. Measurements and Collection of Data

Quantitative data, collected as a result of measurements, are widely used in research work. To measure a quantity is to determine by any means, direct or indirect, its ratio to the unit employed in expressing the value of that quantity. (Weld, 1916). Every measure has some sort of linear scale, either straight or curved, on which the magnitudes are read. This is because the human eye can measure length far more accurately than it can most other magnitudes. However, the investigator should realize that there is no such thing as an exact measurement. Seldom will a re-weight or re-measurement give exactly the same quantity because of inaccuracies that arise from imperfect apparatus and judgment in estimation. An observer may tend to over-estimate, or his measurements may be prejudiced, or his judgment may fluctuate. Because it is next to impossible to arrive at a true value, measurements should be made as carefully as possible in order to obtain the closest approximation. The units of measure will depend upon the degree of precision required in the work. One should distinguish between errors and inaccuracies due to carelessness. These are more properly called mistakes. They consist of blunders like reading the wrong number on the scale, recording a figure in a notebook wrong, forgetting to deduct tare, etc. It is much easier to check the accuracy of weights when they are made more than once. Eternal vigilance and care are necessary to reduce mistakes to the minimum. The investigator should realize that it is impossible to evolve sound results from unsound or carelessly collected data merely thru the application of a formula.

II. Statistics in Experimental Work

After data are collected, it becomes desirable to describe them, interpret them, and induce from them. This is the realm of statistics.

(a) Statistics Defined

Statistics may be regarded as the mathematical analysis based on the theory of probability applied to observational data in an attempt to summarize and describe them so that conclusions can be drawn concerning the phenomena that supply the data. Fisher (1934) states that the original meaning of statistics suggests it was a study of populations of human beings living in political union. The methods developed, however, have little to do with political unity. In fact, they are applied to populations, animate or inanimate.

(b) Use of Statistics

Statistics are used in astronomy, biology, genetics, education, psychology, and many other fields. They are particularly applicable to data concerned with life or the products of life. Probably 75 to 80 per cent of the agronomic workers in agricultural experiment stations use statistical methods, although only about one-half of these apply statistics to other than yield data. However, statistical methods are being used more extensively as time goes on.

III. Some Typical Statistical Terms

The effort to characterize and describe the data mathematically leads to the calculation of various statistics. The simplest of these is the average or mean. It is natural for the first step to be an attempt to find a single measure which will best describe the sum total of the information expressed in a mass of data. The best single measure is the mean. However, it fails to tell the entire story. Among the other statistics are the median, mode, average deviation, standard deviation, coefficient, and correlation ratio. Among the derived values from these statistics are their standard errors and probable errors employed in the important problem of estimation and prediction.

(1) By a variable is meant any organ or character which is capable of variation or difference in size or kind. This difference may be measurable as in height, temperature, weight, etc., or indirectly as in the case of color, occupation, etc. Variation may be continuous or discrete*. For example, a temperature change from 60 to 61 degrees must pass continuously through every intermediate state between 60 and 61 degrees. On the other hand, variation may take place by integral steps without intermediate values, as in population which can never go up or down by less than one. (2) A variate (x) is an individual value of a variable, e.g., 3 feet, 200 grams, 15 pounds, etc. (3) The frequency (f) is the number of times a particular variate (x) occurs between two limiting values of a variable, i.e., the number of variates in any one class. (4) A population is the totality of individuals which are to be studied with regard to a character and may be finite or infinite. (5) A sample may be all or a part of a population. A random sample is a sample taken in such a way that all individuals which make up a population have an equal chance of being included in the sample.

IV. Rules for Computation

It is desirable to be consistent in the number of decimal places used in computations, and in the manner of dropping decimals. Suppose it is desired to retain two decimal places. For a number like 82.575, the value can be made 82.58 by raising the odd number to an even number. However, when the digit in the third decimal place is greater than 5, the number is added, but dropped when it is less than 5. For the square root of a quotient to be accurate to two decimal places, it is recommended that the quotient be carried to four decimal places. This is especially important where the square root is to be used in multiplications for other computations.

V. Arithmetic Average or Mean

Masses of unorganized data explain little or nothing. Individual measures are less significant than a typical value which stands for a number of measurements. An average or mean is such a value. It is the single constant most commonly employed to describe the sample.

(a) Simple Arithmetic Mean

The mean may be considered the center of gravity of a sample. It is equal to the sum of the individual measurements divided by their number.

$$\bar{x} = \frac{x_1 + x_2 + x_3 + \ldots\ldots\ldots\ldots x_n}{N} \quad \text{or} \quad \bar{x} = \frac{Sx}{N} \text{------(1)}$$

where Sx = the sum of all the variates, N = the total number of variates, $\bar{x}$ = the arithmetic mean, and x_1, x_2 ....x_n. the individual variates.

For example, the yields of Golden Glow corn on 3 plots were 84.8, 86.9, and 89.9 bushels per acre. The arithmetic mean would be:

$$\bar{x} = \frac{84.8 + 86.9 + 89.9}{3} = 87.2$$

*Note: This usage is somewhat different than that in Genetics where a discontinuous variation refers to a germinal change that breeds true, while a continuous variation applies to variations due to environment and non-heritable.

(b) Mean of Replicated Variates[1]

It must be remembered that the weight of each variate must be equal in the sample. When certain variates are repeated, the computation may be shortened by merely considering each distinct variate multiplied by the number of times it appears. Suppose 7 corn plants of variety "A" were measured for height in the first replication and were found to average 59 inches. In the second replication, 3 plants were measured, and averaged 67 inches. A total of 20 plants were measured for height in a third replication and found to average 54 inches. Suppose one desired to know the average height for the variety. A simple arithmetic mean of 59, 67, and 54, (i.e., 60 inches) would be incorrect because a different number of plants made up the original means in the different replications. The mean must be calculated so as to give due weight to each variate for the number of times that it occurs. For instance, the mean may be calculated as follows:

$$\bar{x} = \frac{(59 \times 7) + (54 \times 20) + (67 \times 3)}{30} = \frac{1694}{30} = 56.47 \text{ in.}$$

The same result may be obtained by the addition of the original 30 variates and dividing by 30.

VI. The Frequency Distribution

The mean for replicated variates may be calculated from a frequency table which is a simple device by which a considerable quantity of data may be organized in condensed and classified form. Some data presented by Goulden (1937) on the yields in grams of 400 barley plots will be used to illustrate the frequency table. The yields which follow represent an aggregate of data in which there are 400 variates. Each measurement is a variate, i.e., a particular measured value of the variable (x) yield.

Yields in Grams of 400 Square Yard Plots of Barley[2]

185	162	136	157	141	130	129	176	171	190	157	147	176	126	175	134	169	189	180	128
169	205	129	117	144	125	165	170	153	186	164	123	165	203	156	182	164	176	176	150
216	154	184	203	166	155	215	190	164	204	194	148	162	146	174	185	171	181	158	147
165	157	180	165	127	186	133	170	134	177	109	169	128	152	165	139	146	144	178	188
133	128	161	160	167	156	125	162	128	103	116	87	123	143	130	119	141	174	157	168
195	180	158	139	139	168	145	166	118	171	143	132	126	171	176	115	165	147	186	157
187	174	172	191	155	169	139	144	130	146	159	164	160	122	175	156	119	135	116	134
157	182	209	136	153	160	142	179	125	149	171	186	196	175	189	214	169	166	164	195
189	108	118	149	178	171	151	192	127	148	158	174	191	134	188	248	164	206	185	192
147	178	189	141	173	187	167	128	139	152	167	131	203	231	214	177	161	194	141	161
124	130	112	122	192	155	196	179	166	156	131	179	201	122	207	189	164	131	211	172
170	140	156	199	181	181	150	184	154	200	187	169	155	107	143	145	190	176	162	123
189	194	146	22	160	107	70	84	112	162	124	156	138	101	138	141	143	135	163	183
99	118	150	151	83	136	171	191	155	164	98	136	115	168	130	111	136	129	122	120
179	172	192	171	151	142	193	174	146	180	140	137	138	194	109	120	124	126	126	147
115	148	195	154	149	139	163	118	126	127	139	174	167	175	179	172	174	167	142	169
122	163	144	147	123	160	137	161	122	101	158	103	119	164	112	57	94	106	132	122
164	142	155	147	115	143	68	184	183	167	160	138	191	133	160	156	122	111	153	148
103	131	180	142	191	175	146	181	111	110	154	176	168	175	175	146	148	167	106	123
121	154	148	91	93	74	113	79	131	119	96	80	97	98	106	107	69	86	94	129

[1]This has been sometimes called a "weighted" mean.

[2]Data from Methods of Statistical Analysis by C. H. Goulden, p. 7, 1937.

(a) Grouping of Data into Classes

The above data are unwieldy in their present form, even though quite simple in nature. They may be condensed by grouping. First, find the highest and lowest values of the variates (barley yields in grams). The interval thus defined by these extreme values is known as the range. In this case it is 22 to 248. The next step in the formulation of a frequency table or distribution is to separate the range into classes. Although unnecessary, it is usually convenient for the classes to have equal range (interval) within themselves. The number of classes to be formed is the next question. Experience has shown that somewhere between 7 and 20 classes is a desirable number with which to work. The smaller the number of classes the greater is the error due to grouping. The approximate number of classes can be determined from a formula given by Yule (1929):

$$\text{Number of Classes} = 2.5 \sqrt[4]{\text{Number in Sample}} = 2.5 \sqrt[4]{400} = 11.18$$

Suppose 12 classes are decided upon. The quotient of the range divided by the number of classes is the approximate class interval, viz., 226/12 = 18. However, a class interval with an odd number is more convenient because the midpoint of the range does not require an additional decimal. Suppose 19 is selected as the class interval. The value of a class is taken at its mid-value. The barley data may be tabulated for a class interval of 19 as follows:

Class Range (gm)	Class Value (x) (gm)	Tabulation	Frequency (f) (No.)
22-40	31	1	1
41-59	50	1	1
60-78	69	1111	4
79-97	88	~~1111~~ ~~1111~~ 11	12
98-116	107	~~1111~~ ~~1111~~ ~~1111~~ ~~1111~~ ~~1111~~ ~~1111~~ 1	31
117-135	126		69
136-154	145	(This tabulation can be	80
155-173	164	continued in like manner	97
174-192	183	for the other variates.)	78
193-211	202		21
212-230	221		4
231-249	240		2
			N = 400

(b) Frequency Table

After the data are tabulated they are next arranged in a frequency table, i.e., the frequencies are entered to correspond to their class values.

x	f	fx
31	1	31
50	1	50
69	4	276
88	12	1,056
107	31	3,317
126	69	8,694
145	80	11,600
164	97	15,908
183	78	14,274
202	21	4,242
221	4	884
240	2	480
	N = 400	S(fx) = 60,812

The mean ($\bar{x}$) for this sample can be conveniently calculated from the frequency table. Each class value is multiplied by its frequency (f) to give fx. These values are summed to give S (fx) and divided by the total number in the sample. For the barley yields,

$$\bar{x} = \frac{S(fx)}{N} = \frac{60,812}{400} = 152.03$$

It should be evident that the classification of the data into a frequency distribution has distorted them from their original form.

(c) Graphical Representation of Frequency Table

A visible representation of a large number of measurements is afforded by either a histogram or a frequency polygon.

The histogram is most commonly used. The character to be measured is represented along the horizontal axis (abscissa), while the frequencies are represented vertically (Ordinate) to correspond to each class. For example, the barley yield data may be plotted as follows:

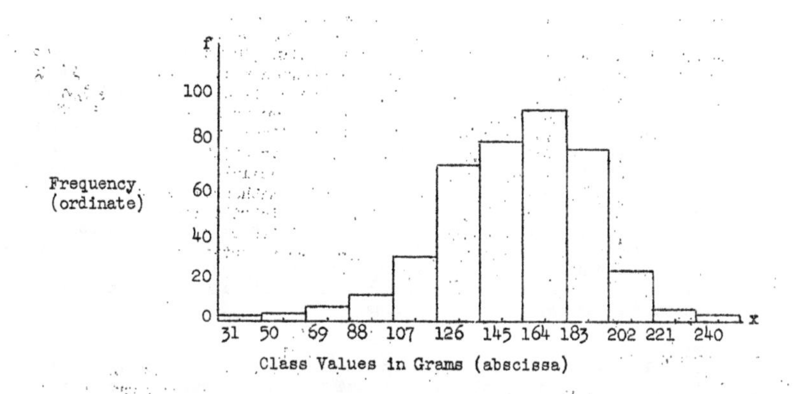

The frequency polygon is constructed by joining in sequence the midpoints of the tops of the bars of the histogram. Its shape tends towards the smooth curve of the population from which the sample was drawn. The frequency polygon for the barley yield data is as follows:

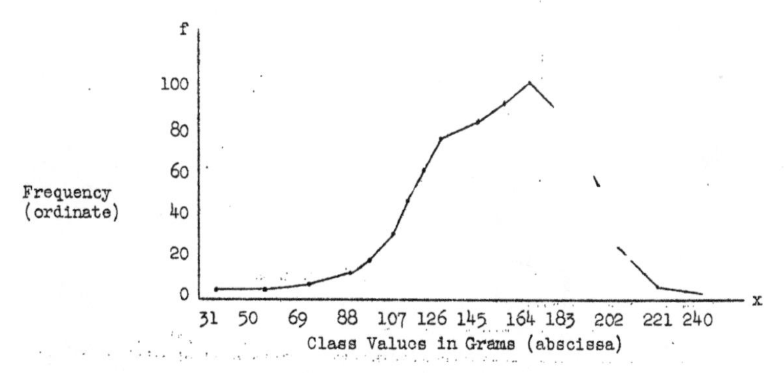

VII. Measures of Central Tendency

There are three measures of central tendency that must be defined at this point. (1) The arithmetic mean, already discussed, is the center of gravity of the population. (2) The median is the measure of the middle variate in an ordered arrangement of the variates according to magnitude. (3) The mode is the measure of the class of greatest frequency, or the point at which the most variates occur. In other words, it is the x-value at which the frequency polygon has the highest ordinate.

VIII. Types of Frequency Distributions

Before one goes further with the analysis to describe the nature of the aggregate of the data, it is necessary to roughly determine the type of frequency distribution. Some mathematical expression is essential corresponding to those types most often encountered in actual practice. (1) A great many frequency distributions found in practice are unimodal, i.e., have one peak. (2) There is a general tendency for them to be bell-shaped when the frequency polygon or diagram is smooth. It was early noticed that the curve derived from the theoretical distribution of the expansion of a binomial, $(a + b)^n$, possessed many of the same characteristics of frequency distributions met with in actual practice. However, the binomial distribution fails to represent continuous variation. An effort to find a mathematical equation for a curve which would well fit the points of a binomial distribution led to the discovery of what is known as the normal probability curve and its equation. Types of distributions most commonly approached in the graphical representation of data are the normal, binomial, and the Poisson distributions.

(a) Normal Distribution

The normal curve is a bell-shaped, symmetrical curve. It is characterized by the symmetrical arrangement of the items around the central value. The arithmetic mean, median, and mode coincide in the normal curve. As in the case of many frequency distributions, the small deviations from the central value (mean) occur more frequently and the larger deviations less frequently. Fisher (1934) gives the statures of 1375 women in a curve that closely approaches a normal curve.

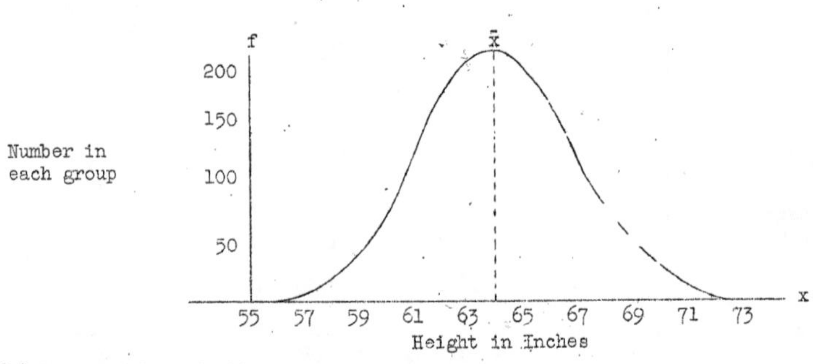

(b) Binomial Distribution

The binomial distribution is represented by the expansion of the binomial, $(p + q)^n$*. To understand the application of the binomial distribution to data, it is first necessary to make some study of probability. This subject will be treated later.

*Note: $(p + q)^n = p^n + n \cdot p^{n-1} q + \frac{n(n-1)}{1.2} p^{n-2}q^2 + \frac{n(n-1)(n-2)}{1.2.3} p^{n-3}q^3 + \ldots\ldots\ldots$

(c) Poisson Distribution

The Poisson distribution is biometrically unsymmetrical, i.e., it is extremely skew. This type of distribution results from an attempt to represent the expansion of $(p + q)^n$ when "p" is extremely small. This type seems particularly applicable to purity and germination counts in seed testing, as well as many other applications.

(d) Other Types of Distributions

Sometimes two or more factors influence the shape of a frequency distribution so that it has two peaks. This would be a bimodal curve. When the data which provide two unimodal frequency distributions with two substantially different means are combined into one frequency distribution, the distribution that results may be bimodal due to the fact that nonhomogenous data over-lap.* This happens occasionally in genetic data.

IX. Some Constants used to Describe Distributions

There are several constants or statistics used to describe distributions. Those of position or central tendency (mean, mode, median) have been discussed already. The constants commonly used to measure dispersion of the variates are the standard deviation, quartile deviation, and the average deviation.

(a) Standard Deviation

The standard deviation of the sample (s') is most frequently used in statistical work to measure dispersion. It is sometimes called the standard error of a single observation. The squared standard deviation $(s')^2$ is the sum of the squares of the deviations from the mean divided by the number. This is sometimes called variance, or the second moment about the mean.

$$(s')^2 \text{ (variance)} = u_2 = \frac{S(x - \bar{x})^2}{N} \qquad \text{------------(2)}$$

where u_2 is the second moment.

The standard deviation (s') is the square root of the variance. The formula for the standard deviation may be expressed as follows:

$$s' = \sqrt{\frac{d_1^2 + d_2^2 + d_3^2 + \ldots d_n^2}{N}} = \sqrt{\frac{Sd^2}{N}} \text{ or } \sqrt{\frac{Sfd^2}{N}} \qquad \text{------------(3)}$$

where d is the deviation from the mean, e.g., $d_1 = x_1 - \bar{x}$.

The above formula gives the standard deviation of the sample about its mean. When it is desired to use this result as an estimate of the standard deviation of the population (s) about its mean ($\bar{m}$), N-1 should be used in the denominator instead of N. This makes little difference in the result when the sample is large, but N-1 should be used when the sample is small, i.e., when N is less than 50 as an arbitrary rule.

As an example, the calculation of the standard deviation of a sample (s') can be illustrated with the barley yields as grouped in VI (b) above. The deviations for each class are taken from the actual means, i.e., 152.

*Note: Pearson's generalized frequency curves or the Gram-Charlier method of curve-fitting should be used for a finer method of analysis for such distributions.

x	f	fx	d	fd	fd^2
31	1	31	-121	-121	14641
50	1	50	-102	-102	10404
69	4	276	-83	-332	27556
88	12	1056	-64	-768	49152
107	31	3317	-45	-1395	62775
126	69	8694	-26	-1794	46644
145	80	11600	-7	-560	3920
164	97	15908	12	1164	13968
183	78	14274	31	2418	74958
202	21	4242	50	1050	52500
221	4	884	69	276	19044
240	2	480	88	176	15488

$N - 400 \quad S(fx) = 60812 \qquad Sfd^2 = 391050$

$$\bar{x} = \frac{Sfx}{N} = \frac{60812}{400} = 152.0$$

$$s' = \sqrt{\frac{Sfd^2}{N}} = \sqrt{\frac{391050}{400}} = 31.27$$

Thus, 31.27 is the standard deviation (s') of this sample. However, the best estimate for the standard deviation for the population (σ) from which this sample was drawn, would be:[1]

$$s = \sqrt{\frac{Sfd^2}{N-1}} = \sqrt{\frac{391,050}{399}} = 31.31$$

Another formula for the calculation of the standard deviation of the sample (s') has been recommended by J. Arthur Harris for machine calculation:

$$s' = \sqrt{\frac{Sx^2}{N} - \left(\frac{Sx}{N}\right)^2} \quad \text{------------------------------------(4)}$$

This formula is essentially the same as the one given above except that the variates themselves are used rather than their deviations from the mean.[2]

The calculation of the standard deviation of the sample (s') by this formula is illustrated with the barley yield data as follows:

Note:[1] The estimate (s) of the population standard deviation (σ) may be computed from the sample standard deviation (s'):

$$s = s'\sqrt{\frac{N}{N-1}} = \sqrt{\frac{N}{N-1}} \cdot \sqrt{\frac{Sx^2}{N} - \left(\frac{Sx}{N}\right)^2}$$

Note:[2] $d = x - \bar{x} \qquad d^2 = (x - \bar{x})^2 = x^2 - 2x\cdot\bar{x} + \bar{x}^2$

$Sd^2 = Sx^2 - 2\bar{x}Sx + N\bar{x}^2$ (remembering that $S\bar{x} = N\bar{x}$)

$\frac{Sd^2}{N} = \frac{Sx^2}{N} - \frac{2\bar{x}Sx}{N} + \frac{N\bar{x}^2}{N} = \frac{Sx^2}{N} - 2\bar{x}^2 + \bar{x}^2$ (remembering that $\frac{Sx}{N} = \bar{x}$.)

Therefore, $\sqrt{\frac{Sd^2}{N}} = \sqrt{\frac{Sx^2}{N} - \bar{x}^2} = \sqrt{\frac{Sx^2}{N} - \left(\frac{Sx}{N}\right)^2}$

(1) x	(2) f	(3) fx	(4) fx^2
31	1	31	961
50	1	50	2500
69	4	276	19044
88	12	1056	92928
107	31	3317	354919
126	69	8694	1095444
145	80	11600	1682000
164	97	15908	2608912
183	78	14274	2612142
202	21	4242	856884
221	4	884	195364
240	2	480	115200
	N = 400	S (fx) = 60812	9636298 = S (fx^2)

$$\bar{x} = \frac{Sfx}{N} = \frac{60,812}{400}$$

$$= 152.03$$

Note: Multiply column No. 1 by column No. 3 to obtain Sfx^2.

$$s' = \sqrt{\frac{Sfx^2}{N} - \left(\frac{Sfx}{N}\right)^2} = \sqrt{\frac{9636298}{400} - \left(\frac{60812}{400}\right)^2} = 31.27$$

$$= \sqrt{24,090.7450 - 23,113.1209} = 31.27$$

(b) Coefficient of Variability (C.V.) is the standard deviation of the sample (s') expressed in percentage of the mean. This gives a relative measure of dispersion so that variation may be compared in features expressed in different units of measurement. It would be often impossible to compare the variabilities of two experiments unless it was expressed in a common unit. The formula is as follows:

C. V. (Coefficient of Variability) $= \frac{100\ s'}{\bar{x}}$ ----------------------------------(5)

For the barley data, it is as follows:

$$C.\ V. = \frac{(31.27)(100)}{152.03} = 20.57.$$

X. Sheppard's Correction for Grouped Data

An error is introduced by grouping variates into classes due to the fact that the midpoint of the class is likely to deviate from the mean of the distribution by more than the mean of the variates grouped in the class in question. This is particularly true for the extreme classes. The majority of the variates in a class are grouped on the side nearest the mean of the distribution. This error can be compensated for mathematically by the use of Sheppard's correction. This correction is equal to 1/12 of the class interval (C), and is subtracted from the value of the squared standard deviation $(s')^2$ as ordinarily obtained, i.e., $(s')^2 - C^2/12$. However, Sheppard's Correction is applicable only to large samples where the variables are continuous.

To calculate the standard deviation without Sheppard's Correction, is to assume that the variates in each class are grouped with the highest frequency at the mean of the class as shown in the diagram. To do this evidently leads to an error in that s' will be computed larger than it actually is. Sheppard's Correction compensates for this type of error which results from grouping data in a frequency distribution.

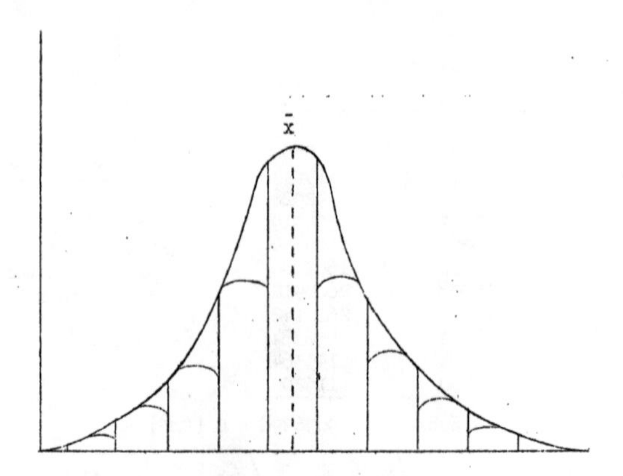

XI. Short-Cut Methods for Computation of Statistics

So far the statistics for simple frequency distributions have been calculated. Several short-cut methods are used which greatly reduce the labor of computation. These methods give the same results. Usually the computations are made from an arbitrary origin or guess mean ($\bar{w}$), with the guess mean corrected to give the true mean ($\bar{x}$) of the sample. The guess mean can be taken at any position.[1] Usually it is taken at the middle of the range or at the lowest class.

The method of computation by use of an arbitrary origin, or guess mean, can be shown with the barley yield data.

x	f	d	fd	fd^2
31	1	0	0	0
50	1	1	1	1
69	4	2	8	16
88	12	3	36	108
107	31	4	124	496
126	69	5	345	1725
145	80	6	480	2880
164	97	7	679	4753
183	78	8	624	4992
202	21	9	189	1701
221	4	10	40	400
240	2	11	22	242
	N = 400		Sfd = 2548	Sfd^2 = 17314

[1]Note: It can be readily proved that a guess mean can be used provided a correction is applied to obtain the true mean. Let $\bar{x}$ = the true mean, $\bar{w}$ = the guess mean, C = a constant (class interval), d = the deviation from the guess mean, and N = the number.

$$x = Cd + \bar{w}$$

$$\bar{x} = \frac{Sfx}{N} = \frac{Sf\,(Cd + \bar{w})}{N} = \frac{C}{N}\,S(fd) + \frac{\bar{w}}{N}\,S(f)$$

$$= C\bar{d} + \bar{w}\ ,\ \text{since } S\,(f) = N$$

Symbols: $\bar{w}$ = guess mean; d or Sfd = correction to the guess at the mean, and C = class interval.

$$\bar{d} = \frac{Sfd}{N} = \frac{2548}{400} = 6.37$$

$$\bar{x} = \bar{w} \text{ (guess mean)} + C\bar{d}\ .$$

$$= 31 + (19)\ (6.37) = 31 + 121.03 = 152.03$$

$$s' = C \sqrt{\frac{Sfd^2}{N} - \bar{d}^2} \quad \text{or} \quad C \sqrt{\frac{Sfd^2}{N} - \left(\frac{Sfd}{N}\right)^2}$$

$$= 19 \sqrt{\frac{17314}{400} - \left(\frac{2548}{400}\right)^2} = 19 \sqrt{43.2850 - 40.5769}$$

$$= (19)\ (1.6456) = 31.2664.$$

Sheppard's Correction:

$$s' \text{ (corrected)} = \sqrt{(s')^2 - \frac{C^2}{12}} = \sqrt{977.5878 - \frac{361}{12}}$$

$$= \sqrt{977.5878 - 30.0833} = \sqrt{947.5045} = 30.7816$$

$$C.\ V. = \frac{100\ s'}{\bar{x}} = \frac{(30.7816)\ (100)}{152.03} = \frac{3078.16}{152.03} = 20.2471$$

The arbitrary origin in this case was taken at the first class. The calculation involves larger numbers than when taken near the center of the range, but all numbers are positive.

XII. General Applicability of Statistical Methods

Knowledge of the frequency distribution leads to an elementary insight into the statistical process. The methods of statistics must be applied with caution to experimental data.

(a) Mathematical Basis for Application

The methods of statistics comprise the application of the solutions affected by the calculus of probability to precisely stated mathematical problems in the attempt to answer questions connected with actual experiments. For the methods of statistics to validly apply to the practical problems connected with experimental work it is necessary that a high degree of correspondence exist between the realities observed in phenomena and the abstract but very definite concepts upon which the mathematical solution of the problem is based. The one possible way to be certain of a correspondence is to carry out repeated random experiments. Statistical methods may be employed to answer questions and test hypotheses that concern phenomena observed in experimental work when this correspondence is satisfactory. The principal cause of the misapplication of the statistical method is the fact that it is often merely assumed that a correspondence exists between measurements and observations concerned with phenomena that result from experiments and the abstract concepts of the probability theory employed to produce the statistical method used in the interpretation of the experimental results. (See Neyman, 1937).

(b) Value of the Statistical Method

There are many advantages attributed to the use of the statistical method. (1) It provides a sound basis for the formulation of experimental designs. Goulden (1937) makes this statement: "The experiment that has been correctly designed gives maximum efficiency, an unprejudiced estimate of the errors of the experiment, and yields results not only on the primary factors with which the experiment is concerned, but also on the important inter-relations of these factors." (2) It tends to eliminate the personal equation, i.e., it does away with differences in personal interpretation. (3) The statistical method is useful in the reduction and condensation of data. Fisher (1934) states that no human mind is able to grasp in its entirety the meaning of any considerable quantity of numerical data. It allows one to express relevant information by means of comparatively few numerical values. (4) It affords a means to measure and evaluate chance errors. This is probably the outstanding contribution of statistics. (5) The statistical method affords one of the best measures of concomitant variations, i.e., correlation. (6) It gives a quantitative measure of variation, including chance variation. Statistics are widely used in genetics for this purpose.

(c) Reliability of the Statistical Constant

The reliability that can be placed on statistical constants depends, in many cases, on the type of data being analyzed. However, several factors contribute to reliability. (1) Reliability depends on the accuracy of the measurements. (2) Quantitative data are likely to be more accurately measured than qualitative data. (3) Samples collected at random are usually more reliable than those selected by other means, although samples by design in planned arrangements are very good. (4) A large sample is more likely to be representative than a small one. Arbitrarily, populations of less than 100 individuals or variates ordinarily are considered small samples to which special precautions should be applied. (Fisher, 1934). Conclusions drawn from many of the older field experiments are questionable because there were too many different kinds of treatments and too little replication or repetition. Statistical methods have done much in recent years to increase the reliability of field experiments. The difficulty of small samples has been alleviated in many instances by the calculation of a generalized standard error based on all the plots of the experiment. Harris (1930) claims that many agronomic experiments can be organized to "make possible the application of the powerful methods of biometric description and analysis."

(d) Some Misconceptions of the Statistical Method

There is little question about the value of the statistical method as such, but much question as to its application. The statistical method cannot correct poor technic or be applied indiscriminately. The standard error of a statistical constant fails to measure the accuracy of an experiment unless all errors (personal equation) have been eliminated except those due to chance. The statistical method may eliminate some systematic errors, but to no great extent. An effective way to eliminate systematic errors, or at least to discover them, is to repeat the experiment in a different manner. Statistics may lend support to a hypothesis but does not necessarily prove it. Several years ago, arguments on the use of statistical methods in agricultural research were quite common. The mathematical foundations of the statistical formulae are now regarded as well established, but argument on the proper application of certain statistical measures will continue much as it does in experimental technic generally. Blind application of statistical procedures, as with any other technic, is harmful. Common sense and good judgment are vital in all phases of experimental work. Salmon (1929) points out that statistical treatment in itself is seldom satisfactory because: (1) The observed result may not be due to the assigned cause. (2) The laws of chance are often an unsatisfactory basis for action or for specific advice. (3) Many experiments do not furnish results which readily lend themselves to statistical treatment because of bias, lack of randomness, or paucity of the observations. (4) Most experiments furnish evidence supplementary to the main issue which is of the greatest value for the arrival at a reasonable interpretation of the results. This type of statement is answered by Goulden (1937) who "doubts very seriously the contention that all really worthwhile effects are obviously significant. At any rate this is at best a dangerous concept as evidenced from scores of examples in published papers where conclusions have been drawn that can be proved by the data to have very little foundation.............Thus, the experimentalist who states that his results are so obvious that they do not require tests of significance is merely stating that in his experience with such experiments, differences as great as those obtained are very unlikely to have arisen by chance variation. We have no quarrel with this reasoning in that it is exactly the type of reasoning employed in tests of significance. Our contention is merely that a determination of probability based on a measure of variability furnished by the experiment itself is sound experimental logic and vastly superior to any method based on pure guesswork."

References

1. Fisher, R. A. Statistical Methods for Research Workers. Oliver and Boyd. (5th edition) pp. 1-7, and p. 49, 1934.
2. Goulden, C. H. Methods of Statistical Analysis. Burgess Publishing Co., pp. 1-8. 1937.
3. Harris, J. Arthur. Mathematics in the Service of Agronomy. Jo. Am. Soc. Agron., 20:443-454. 1928.
4. ______ Criticism of the Limitations of the Statistical Method. Jour. Am. Soc. Agron., 22:263-269. 1930.
5. Love, H. H. The Importance of the probable Error Concept in the Interpretation of Experimental Results. Jo. Am. Soc. Agron., 15:217. 1923.
6. Neyman, J. Lectures and Conferences on Mathematical Statistics. Graduate School, U.S.D.A. 1937.
7. Salmon, S. C. Why We Believe. Jo. Am. Soc. Agron., 21:854-859. 1929.
8. ______. The Statistical Method: A Reply. Jo. Am. Soc. Agron., 22:270-271. 1930.
9. Tippett, L. H. C. The Methods of Statistics. Williams and Norgate. 2nd Ed. pp. 19-42. 1937.
10. Treloar, A. E. An Outline of Biometric Analysis. Burgess Publishing Co. pp. 4-20. 1935.
11. Weld, L. D. Theory of Errors and Least Squares. Macmillan pp. 1-30. 1916.
12. Yule, G. U. An Introduction to the Theory of Statistics. 9th Ed. pp. 211-213. 1929.

Questions for Discussion

1. Explain why there is no such thing as exact measurement in quantitative data.
2. Distinguish between errors and blunders in measurements.
3. Define statistics. Why are some typical statistical constants?
4. In what branches of science have modern statistical methods been most extensively used? Why?
5. Define these terms: variable, variate, frequency, population, and sample.
6. What is the mean? How does it differ from the so-called weighted mean?
7. What is a frequency distribution or frequency table?
8. What is a class interval? How would you determine it for an array of data?
9. What is the difference between a histogram and a frequency polygon?
10. Give 3 measures of central tendency and distinguish between them.

11. What is a normal curve? Skew curve? Bimodal curve?
12. Distinguish between the binomial, normal, and Poisson distributions.
13. Define the standard deviation of the sample. Population.
14. What is the best estimate of the standard deviation of the population as obtained from the sample?
15. Prove that

$$\sqrt{\frac{Sd^2}{N}} = \sqrt{\frac{Sx^2}{N} - \left(\frac{Sx}{N}\right)^2}$$

16. What is the coefficient of variability? When is it correctly used?
17. What is meant by an arbitrary origin? Where can it be taken? Why?
18. Explain Sheppard's Correction and the reason for its use.
19. What are some of the specific things that statistical methods are expected to do when properly applied to data?
20. What are some of the difficulties likely to be encountered in applying statistical methods to field experiments?
21. What factors contribute to the reliability of statistical constants?
22. Is the evidence afforded by statistical analysis of data negative or positive? Explain.
23. Why have statistical methods been only partially used in agronomy? Name 3 men who have advocated such methods in this field.
24. What are the principal arguments of that school of opinion which favors (or insists) on the application of modern statistical methods to field experiments?
25. What are the principal arguments of those who do not favor the use of such methods?
26. What is generally indicated when "common sense" and interpretation based on statistical methods do not agree?

Problems

1. In determining the moisture content of corn by the Brown-Duvall moisture tester, the common practice is to base the moisture percentage on the total or wet weight (corn plus moisture) of the corn. The moisture content of hay, however, is often expressed as a percentage of the dry weight of the hay.

   (a) A variety of corn produced 17.2 lbs. of shelled corn that contained 14.0 per cent moisture on a 12-hill plot. The hills were 3 x 3 feet apart. Calculate the yield of shelled corn in bushels per acre on a 15.5 per cent moisture basis.

   (b) A twentieth acre plot of hay produced 120 pounds of field cured hay. Samples taken when the hay was weighed showed that it contained 20 per cent moisture. Express the true yield in tons per acre on a 15 per cent moisture basis.

2. Head counts were made on a number of fields in a township as follows:

FIELD	NO. HEADS COUNTED	NO. HEADS SMUTTED
1	50	5
2	1000	1
3	100	1
4	500	15
5	400	20
6	800	4
7	1000	10
8	600	12
9	200	6
10	10000	50

   What percent smut may be expected in the wheat delivered to the elevator from this township?

3. These data were taken from several fields to determine the probable losses from smut for a community:

FIELD (No.)	PCT. SMUTTED (X) (Heads)	SIZE OF FIELD (f) (Acres)
1	1.0	100
2	15.0	20
3	0.5	250
4	20.0	10
5	0.0	300
6	0.5	500
7	3.0	50
8	2.5	125
9	0.1	225
10	5.0	150

What percent smut may be expected?

4. Some Iowa data were collected to determine the relation of certain ear characters in corn. The yields from the very short ears, when used for seed, were as follows:

Year	No. ears Used	Yield-Bu. per acre
1916	25	47.18
1917	24	42.70
1918	3	26.33

Determine the average yield for the very short ears for the 3-year period.

5. The table that follows gives the heights of plants of buckwheat in a study of variation at Cornell University. Plot the frequency curve on cross-section paper.

Height in Centimeters	25	35	45	55	65	75	85	95	105	115	125	135	145	155
Number of Plants	2	2	3	5	10	12	60	99	144	85	65	18	2	1
											Total	508		

Does this seem to approximate a normal curve? What can be said as to the position of the mean in a normal curve? The mode? The median? Is a normal curve symmetrical? Is a symmetrical curve necessarily normal?

6. The table that follows gives the average yield of wheat per plant in certain studies at Cornell University. Plot the frequency curve as in the previous example.

YIELD PER PLANT (grams)	NUMBER PLANTS
0.5	37
1.5	59
2.5	88
3.5	41
4.5	45
5.5	29
6.5	26
7.5	5
8.5	8
9.5	6
10.5	8
11.5	3
12.5	3
13.5	1
14.5	1
15.5	2
16.5	2
17.5	2
Total	366

In what respect does it differ from that of the previous example. What name is given to frequency curves of this kind? Do the mean, median, and mode coincide in this curve?

7. The number of stalks were measured on two different kinds of Colsess barley plants grown in 1930 at the Colorado Experiment Station. One kind was a normal green (AcAc) and the other heterozygous for a lethal factor (Acac). Plot the frequency curves. Does the lethal seem to be detrimental to growth?

No. Stalks per Plant	Heterozygous Plants (Frequency)	Green Plants (Frequency)
1	5	7
2	14	9
3	51	28
4	62	33
5	63	31
6	41	19
7	21	12
8	12	4
9	5	1
10	1	0
11	0	1
12	0	0
13	1	0
Totals	276	145

(Note: Calculate the frequencies of the green plants on a basis of N = 276 in order to make the two sets of data readily comparable.)

8. Some data were collected by Emerson (1913) for the study of size inheritance in corn. Classify the data for hybrid 60 x 54. Prepare a frequency table for these data and calculate the mean of the sample using a guess mean. Continue and find the standard deviation (s') and the coefficient of variability. The measurements are given as lengths of ears in centimeters:

Hybrid 60 x 54

15	13	10	12	13	10	13	15	11	10
10	13	15	12	13	14	14	14	11	10
13	12	11	12	11	12	10	13	14	12
11	11	14	10	9	10	11	13	13	14
12	11	10	14	11	13	12	13	13	10
11	12	12	11	13	12	10	13	12	10
11	13	14	13	12	15	14	12	13	

9. Calculate the standard deviations (s') for height of plants in problem 5, using (a) deviations from a true mean and (b) deviations from a guess mean.

10. Some 1930 data on black hulless barley plants were compiled by the Colorado Experiment Station to determine the variation in number of kernels per plant. The data are grouped in classes. (a) List the class boundaries, and calculate the mean, standard deviation, and coefficient of variability. (b) Apply Sheppard's correction to the standard deviation. (c) Is the number of group classes sufficient according to Yule's formula? Calculate.

x (class center)	15	45	75	105	135	165	195	225	255	285	315	
f (frequency)	2	12	11	26	38	26	18	13	13	3	1	= 163

Note that the origin is taken at the class center below 15.

CHAPTER VI

TESTS OF SIGNIFICANCE

I. Statistics as a Basis for Generalization

So far, the discussion has dealt with a sample and its statistical description. The investigator may desire to apply the information collected from the samples to describe the general population. Before he can do that, he must take into consideration the chance or random errors introduced in the actual taking of the sample. Chance errors result from the operation of a great many factors, none of which is dominant, and all of which are relatively similar, equal, and independent. When only chance errors operate, the data are said to be random and follow the law of great numbers.

Two kinds of error exist, chance and systematic. Errors due to chance may not be entirely eliminated but can be submitted to mathematical treatment. Systematic errors can be largely eliminated when an experiment is properly planned.

II. Theory of Probability

In the analysis of chance errors, it is necessary to introduce some of the fundamental concepts of mathematical probability.

(a) Single Probabilities

The probability of the occurrence of an event can be defined from two viewpoints.

(1) Mathematical Probability: The mathematical or a priori probability of an event is the ratio of the number of ways the event may occur to the total number of ways it may either occur or fail to occur, assuming all such ways are equally likely. Thus, the probability of drawing any individual card from an ordinary deck is 1/52, while that of drawing any card of a given suit is 13/52 or 1/4. Probabilities are sometimes stated in terms of odds, e.g., suppose the probability of the occurrence of an event is 1/25. The odds are 1:24 in favor of its occurrence, or 24:1 against its occurrence. To be more explicit, the occurrence of the event is expected just once in 25 trials.

(2) Statistical Probability: Suppose an experiment is repeated a great number of times. When it terminates in a particular manner a certain number of times, the ratio of this latter number to the total number of trials defines an estimate of the probability of the particular termination. Suppose N^1 and N represent the number of successes and the number of trials (both successes and failures), respectively, then

$$\lim_{N \to \infty} \frac{N^1}{N} \text{ will be defined as the probability}$$

of a success. Thus this probability can be approached but never attained in practical work with infinite populations. The permanency of the value N^1/N for N large is the law of great numbers. This permanency results from randomness in the experimental trials and is the necessary property that statistical data must possess to admit valid treatment by mathematics. As an illustration of statistical probability, in a frequency distribution, any particular class frequency divided by the total number of observations in the distribution gives an estimate of the probability that any individual observation made at random will fall in that particular class.

It is evident from either definition that the probability of the occurrence of an event may vary between zero (0), i.e., certainty that the event will not happen, and one (1), i.e., certainty that the event will happen.

(b) Several Probabilities

When several probabilities are to be dealt with simultaneously, it becomes necessary to consider two fundamental theorems.

(1) Theorem I: When a number of mutually exclusive events have certain probabilities of occurrence, the probability of occurrence of some one or other of these events, is the sum of their individual probabilities. For example, the probability that an observation in the barley yield data (Chapter 5, pages 39 and 40) will fall in class x = 88 is 12/400, while the probability that one will fall in class x = 107 is 31/400. The probability that an observation will fall in either class 88 or 107 is 12/400 + 31/400 = 43/400, i.e., P = 0.11.

(2) Theorem II: When a number of independent events have certain probabilities of occurrence, the probability of all occurring together is the product of their individual probabilities. In the above example, the probability that the first and second observations will fall in classes x = 88 and x = 107, respectively is 12/400 times 31/400 = 372/160,000, i.e., P = 0.0023.

A -- Large Sample Theory

III. Probability and the Normal Curve

Statistical data that possess the property of randomness often are distributed in a manner closely expressed by a normal distribution. Many of the sample statistics of large samples can be mathematically proved to have distributions extremely close to normal. Therefore, the application of probability to the normal curve is important in practical work. The area below the curve is taken as one unit. Hence, the area between any two ordinates may be considered as the probability that an individual observation will fall within the range defined by the two ordinates. Now, by theorem I, the probability that an individual observation will fall within any range, is the sum of the probabilities that it will fall in all sub-divisions of that range.

Thus, for characters which are distributed normally, it is possible to estimate the probabilities of their occurrence in any given range. This is done by finding the areas beneath the normal probability curve that correspond to the given range. Mathematical tables of such areas, called probability integral tables, have been constructed. (See Table I in appendix).

Some of the most important probabilities and ranges are given below with the aid of a figure.

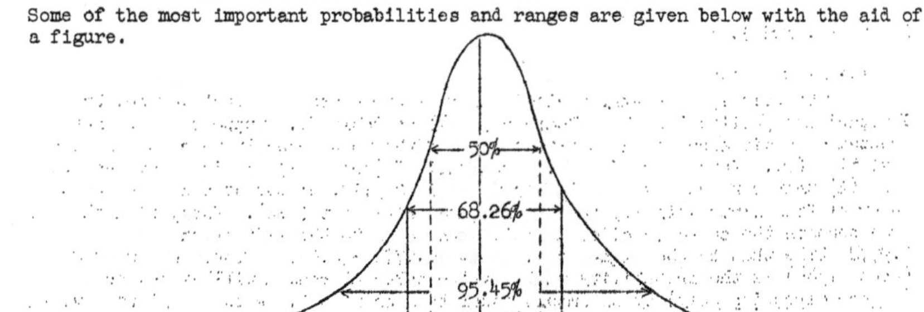

(* = -0.6745 and + 0.6745, respectively)

In the case of a normally distributed variable, it is clear that the probability that an individual observation will fall within a range of σ on either side of the mean ($\bar{x}$) is approximately 0.68; within a range of 2 σ it is approximately 0.95; while within a range of 3 σ it is 0.997. Thus, the probability that the observation will differ from $\bar{x}$ by at least 2 σ is 1.00 - 0.95 or about 0.05. In other words, the chances that an individual will fall outside a range of 2 σ are approximately 5:95 or 1:19. This means that such a situation may be expected about once in 20 times due to chance alone. In like manner, the probability that the observation will differ from $\bar{x}$ by at least 3 σ is 1.000 - 0.997 or 0.003. Such a result, then, may be expected to happen only once in 333 times. Therefore, when an observation differs from the mean by "too much" there arises the important question as to whether or not this abnormal result might not be due to some special cause acting in the case of this individual. When some special affecting condition is known to exist, common sense leads one to the conclusion that the extreme abnormality of the observation is more likely due to the affecting condition than to be expected on the basis of probability.

IV. Levels of Significance

What constitutes an abnormality which is "too much" is a matter of arbitrary decision. Common usage in this country considers an abnormality of twice the standard deviation (standard error in this sense) as being sufficient to warrant the statement that the abnormality of difference from the mean is a real or significant difference.[1] This does not mean that an individual observation taken at random and showing a significant difference does not belong to the general population. However, in such a case one would inquire as to whether the individual case in question was of a special nature, either inherently or by reason of treatment. Should such a condition be substantiated, it is quite proper to attribute the abnormality to special cause or condition and not to chance.

Some workers in the field of statistics use a difference of 3 σ as a criterion for a significant difference. This allows the worker to place more confidence in a conclusion derived from a "significant" observation, but this advantage is over-shadowed by a possible tremendous loss of information due to the imposition of a too stringent criterion.

V. Different Kinds of Probability Tables

There are two kinds of probability tables, viz., one-way and two-way tables. The use of a particular one depends upon the nature of the statistical hypothesis to be tested. The results obtained in one can be readily explained in terms of the other (See Livermore, 1934).

(a) One-Way Tables

The principal one-way table for normal curve areas is that devised by Sheppard and published by Karl Pearson (1914) as Table II. Suppose an ordinate is erected at a distance on the positive side of the mean, exactly twice the standard deviation (σ). Thus t or $d/\sigma = 2$. From Table I (appendix), it is found that the area (A) that corresponds to t (or d/σ) = 2 is 0.9772, or the area defined by the interval from minus infinity to the assigned value of t (d/σ). Thus, with the total area beneath the curve considered as 1.0, the area to the left of the ordinate is 0.9772 while that to the right is 1.0000 - 0.9772 = 0.0228. Thus, P = 0.0228 (about 1/44) is the probability that a value taken at random will exceed the mean (in one direction only) by an amount equal to 2 or more times the standard deviation (σ).

[1]This approximates 3 times the probable error.

Sometimes probabilities are expressed as odds:

Area inside the ordinates divided by area outside the ordinates is equal to the odds against the occurrence of a deviation as great or greater than the designed one due to chance alone. In the above example, 0.9772/0.0228 = 43:1 (approximately)

In this case the odds are 43:1 that a value will not exceed the mean to the extent of two or more times the standard deviation due to chance alone. Table I (appendix) is a one-way table.

(b) Two-way Tables

Suppose one inquires as to the probability of selecting a variate at random so that it shall fall outside the limits of plus or minus twice the standard deviation. Two ordinates are erected, one at t or $d/\sigma = -2$ and one at t = + 2. The problem is to find the area in both tails of the curve. This will be (1.0000 - 0.9772) times 2 = 0.0456, or double that in the one-way table. This means that the probability that a single variate selected at random will deviate by an amount equal to or greater than $\pm 2\sigma$ is 0.0456, or approximately 1/22.

The values on a two-way basis can be expressed as odds as follows: 0.9544/0.0456 = 21:1. Thus, the odds are 21:1 against the occurrence of a deviation as great or greater than the designated one (plus or minus twice the standard deviation) due to chance alone. A typical two-way table for large samples is Table IV given by Davenport (1936).

In summary, it should be clear that the one-way interpretation or the use of a one-way table gives the probability or odds that an obtained value shows a certain discrepancy from the mean in a stated direction whereas the two-way interpretation does not state the direction which the discrepancy must take in a statement of probability or odds.

(c) Transformation of Values

The probability values obtained in one type of table can be readily transformed into terms of the other to meet the experimental argument at hand. Probability values in a one-way table can be doubled to give the results obtained from a two-way table, and vice versa.

The transformation of odds is as follows:

$$\text{Odds in two-way table} = \frac{\text{odds in one-way table} - 1}{2}$$

$$\text{Odds in one-way table} = \left[(\text{odds in two-way table})\ (2) + 1.\right]$$

VI. Standard Errors of Statistical Constants

Each statistical constant or estimate has its own standard error. The standard error of a statistic derived from a sample is the standard deviation of the distribution of that statistic thought of as resulting from many samples. The distributions of many statistics are nearly normal, particularly when the basic sample is large.

(a) Standard Error of a Single Observation

The "best"[1] estimate of the standard deviation of a single observation (σ) is the standard error (s) derived from the sample. Some data on the total weight of

[1]Note: The best unbiased estimate is simply called "best". See more advanced treatments of mathematical statistics.

grain in grams for non-competitive Colsess barley plants as follows:

Class Center	1	3	5	7	9	11	13	15	17	19	21	23	25	27	29	31	33
Frequency	3	11	21	35	43	55	51	52	47	35	21	10	9	11	5	1	1

$N = 411 \quad \bar{x} = 13.8 \qquad Sfd^2 = 14776.64$

$$s = \text{standard error of a single observation} = \sqrt{\frac{Sfd^2}{N-1}} \quad \text{------(1)}$$

In the above example, it would be calculated as follows:

$$s = \sqrt{\frac{14,776.64}{411 - 1}} = 6.003 \text{ grams}$$

This value, $s = 6.003$ grams, is the standard error of a single variate in this sample. For instance, the value of the mean, $13.8 \pm 2\ (6.003)$ indicates that the odds are 21:1 that a single individual taken at random will not deviate from the mean by more than 2 σ in either direction, where the normal distribution of the population affording the sample data is assumed.[1]

(b) Standard Error of the Mean

Suppose a second sample were taken. One could hardly expect to get exactly the same result for the mean ($\bar{x}$) as in the sample in question. Thus, the mean ($\bar{x}$) obtained from a single sample is merely an estimate of the true mean ($\bar{m}$) of the whole population. The latter is unknown and necessarily must remain so. In case it were possible and practical to take and analyze a greater number of samples, finding the mean ($\bar{x}$) for each, one would expect the mean of all the sample means to be very close, indeed, to the mean of the population ($\bar{m}$). Since this is not feasible, one can only ask how good an estimate of the population mean ($\bar{m}$) is the mean ($\bar{x}$) computed from a single sample. The answer to this question can only be given in terms of probabilities. It can be shown mathematically that the mean computed from a large number of large samples are distributed nearly normally with standard deviation, $\sigma_{\bar{x}}$, which is theoretically equal to the ratio of the standard deviation of the population to $\sqrt{N}$, the number of observations that make up the sample.[2] However, the standard deviation of the population (σ) is unknown and in its stead its estimate (s) derived from the sample is used. Therefore the standard deviation of the hypothetical distribution of means of a large number of samples will be estimated as follows:

$$\sigma_{\bar{x}} = \text{standard error of the mean} = \frac{s}{\sqrt{N}} \quad \text{------(2)}$$

The greater the number of observations in the sample, the smaller will be the standard errors of the various statistical constants. Hence, the statistical constants derived from a large sample are more likely to represent the true constants of the general population than those derived from a small sample. When the sample is small, the argument is the same except that the distribution for $\bar{x}$ deviates from normality and needs special interpretation.

---

[1] It should be noted that $s = \sqrt{\frac{Sfd^2}{N-1}}$, which best estimates the standard deviation of the population, closely approximates $s' = \sqrt{\frac{Sfd^2}{N}}$ which is the standard deviation of the sample, and is the estimate of σ given by the maximum likelihood principle.

[2] This is sometimes expressed as "S.E. of the mean."

In the above example the standard error of the mean ($\sigma_{\bar{x}}$) is:

$$\sigma_{\bar{x}} = \frac{s}{\sqrt{N}} = \frac{6.003}{\sqrt{411}} = 0.296 \text{ grams.}$$

The mean is 13.8 $\pm$ 0.296 grams. Therefore, the odds are 21:1 that $\bar{x}$ (13.8 grams) does not differ from the unknown true mean ($\bar{m}$) of the general population by more than 2 $\sigma_{\bar{x}}$, or 2(0.296) = 0.592 grams.

(c) Standard Error of the Standard Deviation

Next, it is desired to discover how reliably the standard deviation of a single sample (s') estimates the unknown standard deviation of the population (σ). Mathematically, it has been found that the best estimate of a hypothetical distribution of standard deviations derived from a great number of samples is as follows:

$$\sigma_{\sigma} = \text{standard error of standard deviation} = \frac{s}{\sqrt{2N}} \text{ (approximately)} \quad \ldots\ldots\ldots(3)$$

From the example used above,

$$\sigma_{\sigma} = \frac{s}{\sqrt{2N}} = \frac{6.003}{\sqrt{2(411)}} = 0.209 \text{ grams}$$

Therefore, the odds are 20:1 that s = 6.003 grams does not differ from the unknown true standard deviation of the general population (σ) by more than 2 σ_{σ} or 2 (0.209) = 0.418 grams.

(d) Standard Error of the Coefficient of Variability

By use of the same argument, the standard error of the coefficient of variability is:

$$\sigma_{c.v.} = \frac{C.\,V.}{\sqrt{2N}} \left[1 + 2\left(\frac{C.V.}{100}\right)^2\right]^{\frac{1}{2}} \text{ when C. V. is greater than 10} \quad \text{-----------}(4)$$

$$\sigma_{c.v.} = \frac{C.\,V.}{\sqrt{2N}} \text{ when C.V. is less than 10} \quad \text{-----------}(5)$$

In the example used,

$$\sigma_{c.v.} = \frac{43.5}{\sqrt{2(411)}} \left[1 + 2\left(\frac{43.5}{100}\right)^2\right]^{\frac{1}{2}} = 1.781$$

A table has been worked out by Brown (1934) to shorten the computation necessary to secure the standard error of the coefficient of variability when C.V. is greater than 10.

(e) Standard Error of an Average of Averages

The standard error of an average of averages is given by the formula:

$$\sigma_a = 1/N \sqrt{\sigma^2_{\bar{x}_1} + \sigma^2_{\bar{x}_2} + \ldots\ldots\ldots\ldots \sigma^2_{\bar{x}_n}} \quad \ldots\ldots\ldots\ldots(6)$$

where N equals the number of separate means and $\sigma\,\bar{x}_1$, $\sigma\,\bar{x}_2$, etc., represent their separate standard errors.

(f) Standard Error of a Difference

Suppose two samples are measured with respect to a common character. From the data, let two similar statistical constants be compared, e.g., the two means or the two standard deviations. The question arises as to whether the two constants differ significantly. Its answer depends upon the standard error of the difference which is as follows:

$$\sigma_d = \sqrt{\sigma^2_1 + \sigma^2_2} = \sqrt{\frac{s^2_1}{N_1} + \frac{s^2_2}{N_2}}^{1} \qquad \text{.......................(7)}$$

where s_1 and s_2 are the standard errors of the two like statistical constants derived from the two samples. Where a significant difference results in the case of two samples drawn from the same population, it would indicate probable improper sampling technique leading to lack of randomness.[2] The principal use of this method lies in its test as to whether or not a factor known to exist in the case of one sample, and not in the other, is really a causal factor to which an abnormal difference can be attributed, e.g., the difference between two yields in a yield trial.

For example, suppose it is desired to determine the standard errors of the difference of the mean yields of Kanred and Turkey wheats, and also for Manchuria and Minnesota 445 barleys.

(1)

Wheat Variety	Yield (Bu.)
Kanred	25 ± 0.7
Turkey	24 ± 0.6
Difference	1 ± 0.92

$\sigma_d = \sqrt{(0.7)^2 + (0.6)^2} = 0.92$

(2)

Barley Variety	Yield (Bu.)
Manchuria	38.9 ± 0.9
Minnesota 445	48.5 ± 1.2
Difference	9.6 ± 1.5

$\sigma_d = \sqrt{(0.9)^2 + (1.2)^2} = 1.5$

VII. Significant Differences

After a difference is obtained for two statistical constants, as in the above example, it is desirable to test this difference for statistical significance. An investigator may arbitrarily choose whatever level of significance he desires, but should state the level chosen. He must use care in attributing differences to causal factors when the differences approach the level of significance that he has chosen. To determine the significance of two statistical constants, their difference divided by the standard error of the difference (σ_d) is commonly employed. For example, in the case of Kanred and Turkey wheats cited above,

$$t = \frac{d}{\sigma_d} = \frac{1}{0.92} = 1.09$$

When the level of significance is taken as $d/\sigma_d = 2$, this difference is not significant. On the basis of probability, the odds are a little more than 2:1 that this difference is a real or significant difference. Hence, it may be ascribed to chance, or to put it in another way, one would not claim superiority for Kanred because the probability is too large that such a statement is incorrect.

In a comparison of Manchuria and Minnesota 445 barley,

$$t = \frac{d}{\sigma_d} = \frac{9.6}{1.5} = 6.4$$

---

[1]When $\sigma_1 = \sigma_2$, $\sigma_d = \sigma\sqrt{2}$.

[2]Relation (7) holds strictly only where the two variable statistics are normally distributed and derived from uncorrelated data.

Since $t = d/\sigma_d$ is far greater than 2, the difference in yield between the two varieties is said to be real and not due to chance. In this case the claim is made that Minnesota 445 is superior to Manchuria in yield ability. The probability of the incorrectness of this statement is insignificantly small. It would be a miracle if this claim were really incorrect.

VIII. Probable Errors.

The quantity 0.6745σ, which gives the range that contains half the observations, is termed the probable error. For example, an average yield of 15.0 ± 1.5 bushels would mean that the chances are 50:50 that the true value of the average for an infinite population lies between 13.5 and 16.5 bushels. It also indicates that the chances are even that it may lie outside this range.

(a) Use of Probable Errors

Historically, the probable error was used before the standard error. It is still widely used in this country in the statistical treatment of biological data, but the tendency is to use the standard error and think in terms of it. It is generally felt that "the probable error is an unmitigated nuisance," and has nothing to recommend except its previous usage.

(b) Formulae for Probable Errors

The probable error is approximately two-thirds of the standard error. It can be obtained when each standard error value is multiplied by 0.6745. These formulae may be briefly summarized:

(1) $$\text{P.E. single determination} = \pm 0.6745 \sqrt{\frac{Sfd^2}{N-1}} \qquad (8)$$

(2) $$\text{P.E.}_{\bar{x}} = \frac{0.6745\, s'}{\sqrt{N-1}} \quad \text{or} \quad \pm \frac{0.6745\, s}{\sqrt{N}} \qquad (9)$$

(3) $$\text{P.E.}_{\sigma} = \frac{0.6745\, s'}{\sqrt{2N}} \qquad (10)$$

(4) $$\text{P.E.}_{\text{c.v.}} = \pm \frac{0.6745\ \text{C.V.}}{\sqrt{2N}} \left[1 + 2\left(\frac{\text{C.V.}}{100}\right)^2\right]^{\frac{1}{2}} \qquad (11)$$

where C.V. is greater than 10.

(5) $$\text{P.E.}_{\text{c.v.}} = \pm \frac{0.6745\ \text{C.V.}}{\sqrt{2N}} \qquad (12)$$

where C.V. is less than 10.

(c) Levels of Significance for Probable Errors

The level for significance for the probable error is commonly taken as $D./P.E._d = 3$. This is equivalent to odds of about 22:1. Some workers use 3.2 times the probable error, for which the odds are approximately 30:1. A table of odds for probable errors is given by Hayes and Garber (1927) in "Breeding Crop Plants."

(d) Relation of Standard Errors to Probable Errors

Based on the normal curve the quartile lines, Q_2 and Q_3, give the probable error of a single variate, or $Q = 0.6745\ \sigma$.

The intervals:

$\bar{x} \pm 1\,\sigma$		68.3 % of variates.	$\bar{x} \pm 1$ P.E.		50.0 % of variates
$\bar{x} \pm 2\,\sigma$	include	95.5 " " "	$\bar{x} \pm 2$ P.E.	include	82.3 " " "
$\bar{x} \pm 3\,\sigma$		99.7 " " "	$\bar{x} \pm 3$ P.E.		95.7 " " "

B -- Special Case of Small Samples

IX. Use of Small Samples in Biological Research

The methods heretofore explained relate to the determination of significance based on the normal distribution for large samples, but it is not always possible to obtain large samples. This is often the case in agricultural or biological experiments. When the investigator can be certain that the populations which afford small samples approximate the normal distribution in form, he may feel that the interpretation of the statistical analysis was valid. Therefore, the material that follows is given on the basis of small populations whose distributions approach that of the normal curve. Statistical treatment of small samples, from populations far from normal in distribution, may probably be inadequate. Too often it may lead to incorrect conclusions. Statistical analysis of a single sample with less than 20 cases is hazardous. In samples of 20 to 100 cases, the near-normality of the under-lying population should be known. This places a severe limitation on the use of small samples, but fortunately in agricultural and biological experiments, most of the populations with which the experimenter deals, are near normal. The importance of the small sample, together with its statistical treatment, has been discussed by Fisher (1934).

X. Degrees of Freedom

The reliability of a statistic (estimate of a population parameter) will obviously depend upon the number of variates in the sample. This dependence is also affected by the number of restrictions placed on the aggregate observations in the determination of an estimate of a population parameter. The total number of observations diminished by the number of restrictions which they in aggregate must submit to has been termed "degrees of freedom" by Fisher (1934).

It has been stated that the best estimate of the variance of a population as derived from the sample is as follows:

$$s^2 = \frac{S(x - \bar{x})^2}{N-1} \quad (13)$$

In this case, the number of individual observations (N) is diminished by one to give the degrees of freedom. The number of statistical constants of the sample which are directly used in the computation are subtracted. The mean or total fixes one value in the above formula, so that only N-1 observations are free to vary. This is of little importance when a large sample is analyzed, but very important in small samples.

XI. Probability Determinations with Small Samples

The distribution of $\bar{x}/\frac{s}{\sqrt{N}}$ is not sufficiently close to normal for small samples. The nature of the distribution of $\bar{x}/s'$ was found by "Student" in 1908. He prepared a series of tables based on the distribution of s' (where $s' = \sqrt{S(x - \bar{x})^2/N}$) which he designated as "Z". He showed that the "Z" distribution, now more commonly called

Student's distribution, was the same as the Pearson Type III curve. More recently, he has prepared tables for the distribution of "t" which is designated as $\bar{x}/\sigma_{\bar{x}}$ or $\bar{x}\sqrt{N}/s$ by Fisher (1934). For a given value of "t" that corresponds to a given number of degrees of freedom one can read the probability in an analogous manner to the way the tables of areas of the normal curve are used.

The "t" table devised by Fisher (1934) is a two-way table. A probability of 0.05 is Fisher's 5 per cent point for which the odds are 19:1; a probability of 0.01 is the one per cent point for which the odds are 99:1. In addition several one-way tables are in use. These are as follows: (1) Student's "t" table, (2) Livermore's modification of Student's "t", (3) Student's "Z", and (4) Love's modification of "Z". For example, suppose t = 4.604 for 4 degrees of freedom. The probability as found in a one-way table is equal to 0.995. The calculated odds would be 199:1. They are calculated as follows: 1-P = 1.000 - 0.995 = 0.005. 5/1000 = 1/200. P = 1/200 is equivalent to odds of 199:1.

XII. Significance of Means

When d is the difference between the mean of the sample and any value ($\bar{m}'$) assumed to be the mean ($\bar{m}$) of the population, it has been stated that the difference, $d = \bar{x} - \bar{m}'$, is significant when $d/\sigma_{\bar{x}}$ exceeds 2. When this occurs, the hypothesis ($\bar{m} = \bar{m}'$) is rejected. This procedure holds when $d/\sigma_{\bar{x}}$ is nearly normally distributed as in large samples. As this distribution is not close to normal for small samples, the "t" table should be used in such cases. When the 5 per cent point is used as the level of significance, a value of $t \geq d/\sigma_{\bar{x}}$ that corresponds to P = 0.05 is considered as significant. In this test for the significance of the mean one determines the probability of drawing a sample with a mean equal to $\bar{x}$ from a population whose true mean ($\bar{m}$) is assumed to be some particular value ($\bar{m}'$).

XIII. Means of Two Independent Samples

One of the most important problems in statistics is to test the significance of a difference between two means, i.e., $\bar{x}_1 - \bar{x}_2 = d$. Previously, it has been stated that the standard error of the difference of the means of two samples is $\sigma_d = \sqrt{\sigma^2_{\bar{x}_1} + \sigma^2_{\bar{x}_2}}$. Should there be any reason to suspect that the standard deviations of the two underlying populations are different, one should form t/σ_d with σ_d as given here.

(a) Samples with Different Numbers of Observations

When it can be assumed that the standard deviations of the populations are the same, or that the samples have been drawn from the same population, then the best estimate (s) of the population standard deviation (σ) is:

$$s = \sqrt{\frac{S(x_1 - \bar{x}_1)^2 + S(x_2 - \bar{x}_2)^2}{(N_1 - 1) + (N_2 - 1)}} \quad (14)$$

Here N_1 and N_2 are the numbers of observations in the two samples while the denominator evidently denotes the degrees of freedom. This method to determine s as an estimate of σ is particularly important in the case of small samples.

The "t" value, equivalent to d/s_d, is calculated as follows:

$$t^{1/} = \frac{\bar{x}_1 - \bar{x}_2}{s} \sqrt{\frac{N_1 N_2}{N_1 + N_2}} \quad \text{- (15)}$$

(b) Samples with Same Number of Observations

The above formulae are simplified when the number of observations are the same in each sample, i.e. $N_1 = N_2$.

The standard error (single observation) is as follows:

$$s = \sqrt{\frac{S(x_1 - \bar{x}_1)^2 + S(x_2 - \bar{x}_2)^2}{2(N-1)}} \quad \text{- - - - - - - - - - - - - - - - - - - (16)}$$

The value of "t" is as follows:

$$t = \frac{\bar{x}_1 - \bar{x}_2}{s} \sqrt{\frac{N}{2}} \quad \text{- -(17)}$$

Some data presented by Immer (1936) may be used to illustrate the computation. Single plots of Velvet and Glabron barley were grown side by side in single plots on 12 different farms. The yields in bushels per acre are given below:

Farm No.	Glabron (x_1)	Velvet (x_2)	Sum
1	49	42	91
2	47	47	94
3	39	38	77
4	37	32	69
5	46	41	87
6	52	41	93
7	51	45	96
8	57	56	113
9	45	42	87
10	45	39	84
11	48	47	95
12	64	39	103
	$S(x_1) = 580$	$S(x_2) = 509$	1089
	$\bar{x}_1 = 48.3333$	$\bar{x}_2 = 42.4167$	$\bar{x} = 45.3750$
	$S(x_1^2) = 28,620$	$S(x_2^2) = 21,979$	100,269
	$\frac{(Sx_1)^2}{N_1} = 28,033.31$	$\frac{(Sx_2)^2}{N_2} = 21,590.10$	

1/ This can be readily proved as follows

$$t = \frac{\bar{x}_1 - \bar{x}_2}{\sqrt{\frac{s_1^2}{N_1} + \frac{s_2^2}{N_2}}} = \frac{\bar{x}_1 - \bar{x}_2}{\sqrt{\frac{N_2 s_1^2 + N_1 s_2^2}{N_1 N_2}}} = \frac{x_1 - x_2}{s\sqrt{\frac{N_2 + N_1}{N_1 N_2}}} = \frac{\bar{x}_1 - \bar{x}_2}{s} \sqrt{\frac{N_1 N_2}{N_1 + N_2}}$$

where "s" is an estimate derived by pooling the two samples, based on the hypothesis that the two populations have a common standard deviation (σ).

The computations are as follows:

$$s^2 = \frac{\left[S\,(x_1{}^2) - (Sx_1)^2/N_1\right] + \left[S\,(x^2{}_2) - (Sx_2)^2/N_2\right]}{2\,(N-1)}$$

$$= \frac{(28,620.00 - 28,033.31) + (21,979.00 - 21,590.10)}{22}$$

$$= \frac{586.69 + 388.90}{22} = \frac{975.59}{22} = 44.3450$$

$$s = \sqrt{44.3450} = 6.6592$$

$$t = \frac{\bar{x}_1 - \bar{x}_2}{s}\sqrt{\frac{N}{2}} = \frac{48.33 - 42.42}{6.6592}\sqrt{\frac{12}{2}} = 2.1769$$

The "t" table is entered for t = 2.1769 for 2 (n-1) = 22 degrees of freedom. P lies between 0.05 and 0.02. It may be concluded that the odds are in excess of 19:1, that the difference between the mean yield of these two varieties is not due to chance.

XIV. Means of Paired Samples

In this case, the variables are paired, i.e., each value of x_1 is associated in some logical way with a corresponding value of x_2. As a result, there will be the same number of variates in the two samples. When there are N pairs there will be N-1 degrees of freedom available for the comparison. This is widely known as Student's Pairing Method.

(a) Student's Pairing Method

This method is devised to compare two results on a probability basis. It is used primarily for small samples it not being necessary to assume a normal population. Partial mathematical proof of the method was first published by Student (W.S.Gossett) in 1908. Differences between paired values are dealt with directly, with the result that the correlation between paired values is taken into account. The method was brought to the attention of American agronomists in 1923 by Love, et al. (1923, 1924).

The variance (s^2) and "t" values are calculated as follows:

$$s^2 = \text{variance} = \frac{S(d^2) - (Sd)^2/N}{N - 1} \quad \text{- (18)}$$

$$t = \bar{d} \Big/ \sqrt{\frac{s^2}{N}} \quad \text{- (19)}$$

Here "t" is used to test an obtained value, $\bar{d}$, in accordance with the hypothesis that the mean of the population of differences is zero. A significant result would mean the rejection of the hypothesis and would warrant a statement that the mean of one of the basic populations exceeded that of the other.

(b) Method of Computation

The method of computation can be illustrated from the Glabron vs Velvet barley yields mentioned above. The computation follows:

Farm No.	Glabron (x_1)	Velvet (x_2)	d(Velvet from Glabron)
1	49	42	7
2	47	47	0
3	39	38	1
4	37	32	5
5	46	41	5
6	52	41	11
7	51	45	6
8	57	56	1
9	45	42	3
10	45	39	6
11	48	47	1
12	64	39	25
Sum	$S(x_1) = 580$	$S(x_2) = 509$	$S(d) = 71$
Mean	$\bar{x}_1 = 48.3333$	$\bar{x}_2 = 42.4167$	$\bar{d} = 5.9167$

$S(d^2) = 929 \qquad (Sd)^2/N = 420.0857$

$$s^2 = \frac{S(d^2) - (Sd)^2/N}{N - 1} = \frac{929.0000 - 420.0857}{11} \doteq 46.2649$$

$$t = \bar{d} \Big/ \sqrt{\frac{s^2}{N}} = 5.9167 \Big/ \sqrt{\frac{46.2649}{12}} = 5.9167 \Big/ \sqrt{3.8554} = 3.0133$$

The value of "t" is then looked up in the t-table (Fisher, 1934) for 11 degrees of freedom (N-1 paired values) where it is found that the observed value lies between P = 0.02 and P = 0.01.

The "Z" table devised by Student is sometimes used. He designed "Z" as the ratio of the mean difference to the standard deviation of the mean difference, i.e., $Z = \frac{\bar{x}}{s'}$ where $s' = \sqrt{\frac{S(x-\bar{x})^2}{N}}$

Student (1926) calls attention to the fact that the "Z" table should be entered with N-1 degrees of freedom. As mentioned previously, his "Z" table is a one-way table. The Z-value can be transformed to "t" as follows:

$$t = Z \Big/ \sqrt{N-1}$$

(c) Application of the Pairing Method

The application of this method is highly desirable for making comparisons between pairs of varieties or treatments when the scope of the experiment is limited to a few pairs of observations. It is useful for simple tests such as treated vs. untreated where only two or three things are being compared. In plot work, the method can only be used to remove soil heterogeneity where the plots are physically paired, i.e., adjacent.

References

1. Brown, Hubert M. Tables for Calculating the Standard Error and the Probable Error of the Coefficient of Variability. Jour. Am. Soc. Agron., 26:65-69. 1934.
2. Davenport, C. B., and Ekas, M. P. Statistical Methods in Biology, Medicine, and Psychology. John Wiley and Sons. pp. 35-40, and pp. 166-172. 1936.
3. Fisher, R. A. Statistical Methods for Research Workers (5th edition). Oliver and Boyd, pp. 112-125. 1934.

4. Goulden, C. H. Methods of Statistical Analysis. Burgess Publ. Co., pp. 9-11, and 20-26. 1937.
5. Hayes, H. K., and Garber, R. J. Breeding Crop Plants. McGraw-Hill, p. 42 & 86-92. 1927.
6. Immer, F. R. Manual of Applied Statistics. University of Minnesota. 1936.
7. Livermore, J. R. The Interrelations of Various Probability Tables and a Modification of Student's Probability Table for the Argument "t". Jour. Am. Soc. Agron., 26:665-673. 1934.
8. Love, H. H. The Importance of the Probable Error Concept in the Interpretation of Experimental Results. Jour. Am. Soc. Agron., 15:217-225. 1923.
9. Love, H. H., and Brunson, A. M. Student's Method. Jour. Am. Soc. Agron., 16:60. 1924.
10. Love, H. H. A Modification of Student's Table for Use in Interpreting Experimental Results. Jour. Am. Soc. Agron., 16:68-73. 1924.
11. Pearson, Karl. Tables for Statisticians and Biometricians. Part I. Cambridge U. Press. pp. 2-8 (Table II). 1914.
12. Student. The Probable Error of the Mean. Biometrika 6:1-25. 1908.
13. ______. New Tables for Testing the Significance of Observations. Metron, 5:18-21. 1925.
14. ______. Mathematics and Agronomy. Jour. Am. Soc. Agron., 18:703-720. 1926.
15. Tippett, L. H. C. The Methods of Statistics. Williams and Norgate. (2nd edition). pp. 110-121. 1937.

Questions for Discussion

1. What is the basis for using statistics for generalization?
2. Distinguish between _a priori_ and _statistical_ probability.
3. Give two basic theorems where several probabilities are involved.
4. What is the geometrical significance of the standard error? Its significance in practical problems?
5. Why is a difference said to be statistically significant when it is two or more times the standard error?
6. Is it correct to say that standard error is a measure of experimental error? Explain.
7. What is the difference between a one-way and two-way table in the calculation of probability? Interpret probabilities calculated from each kind of a table.
8. How do odds differ in one-way and two-way tables?
9. How can odds be transferred from a one-way to a two-way basis? Explain the difference in interpretation.
10. Explain the difference between the standard error of a single observation and the standard error of the mean. Give the formula for each.
11. What is the formula for the standard error of an average of an average? Standard error of a difference?
12. What is the relation of the standard error to the probable error? Why do most statisticians prefer to use standard error?
13. Why are special methods used for small samples?
14. What is meant by "degrees of freedom"?
15. Who was "Student"? What were some of his contributions to statistics?
16. What is the meaning of Fisher's "t"?
17. What was Student's "Z"? How can it be transformed to "t"?
18. How is the standard error calculated for the means of two independent small samples drawn from populations with equal standard deviations?
19. What is Student's pairing method? How does it differ from other methods of calculating standard errors?
20. Under what conditions can Student's pairing method be used? What are some of its limitations?

PROBLEMS

1. (a) If the mean of a population is 21.65, and σ = 3.21, determine the probability that a variate taken at random will be greater than 28.55 or less than 14.75.

   (b) Determine d/σ for P = 0.01, 0.05, and 0.50.

2. Suppose the odds in a 1-way table are 87:1. Transform them to a 2-way basis.

3. In a wheat variety test, yields in bushels per acre were as follows:

   Kanred : 54.6, 53.7, 68.0, 55.2, 58.5, 62.1, 56.7
   64.6, 52.2, 57.5. $\bar{x}$ = 58.3

   Cheyenne: 66.3, 60.9, 64.8, 67.6, 65.8, 62.2, 68.4
   60.6, 67.2, 55.3. $\bar{x}$ = 64.3

   Calculate: (a) The standard error for a single plot (s), and the standard error of the mean ($\sigma_{\bar{x}}$) for each variety. (b) The standard error of the difference between the two varieties (σ_d); and (c) Determine whether or not the difference between the varieties is statistically significant. Assume the population standard deviations are different.

4. The yields of two varieties in bushels per acre are as follows for several replications:

   Variety A: 38,40,40,42,39,35,32,28,42, and 44.
   Variety B: 37,37,40,40,32,30, and 31.

   Compute a pooled estimate of the standard error (s) of the two varieties, compute t, and determine whether or not the varieties differ significantly in yield by reference to Table II in the appendix.

5. Two varieties of small grain, Big Four and Great Northern, were grown each year in adjacent plots from 1912 to 1920. The yields are given below.

Yields in Bushels per Acre

Year	Great Northern	Big Four
1912	71.0	54.7
1913	73.9	60.6
1914	48.9	45.1
1915	78.9	71.0
1916	43.5	40.9
1917	47.9	45.4
1918	63.0	53.4
1919	48.4	41.2
1920	48.1	44.8

Which varieties yield higher? Is this difference significant as shown by: (a) means of two independent samples? (b) Paired Samples?

6. The grain yields in grams per plot for spring wheat irrigated at the tillering and jointing stages were as follows for 1921 to 1923 (incl.):

Year	Plot	Tillering	Jointing
1921	A	155	281
	B	232	202
	C	248	271
	D	257	265
1922	A	459	366
	B	332	408
	C	341	396
	D	312	366
1923	A	513	602
	B	561	635
	C	563	593
	D	346	539
3-Year Average		360	410

Determine whether or not irrigation at tillering results in a significantly higher yield than irrigation at the jointing stage. Consider the values paired.

CHAPTER VII

THE BINOMIAL DISTRIBUTION AND ITS APPLICATIONS

I. The Binomial Distribution

Suppose that "p" is the probability that an event will occur in one trial, and "q" the probability of failure of that event to occur. Then, it can be shown by means of the two theorems on probability that the successive terms of the binomial expansion will give the respective probabilities that, in "n" trials, this event will occur exactly N, N - 1, N - 2, ... or 0 times. The binomial expansion is as follows:

$$(p + q)^N = p^N + N \cdot p^{N-1}q + \frac{N(N-1)}{1.2} p^{N-2}q^2 + \frac{N(N-1)(N-2)}{1.2.3} p^{N-3}q^3 + \ldots\ldots\ldots q^N$$

where evidently $p + q = 1$

Then, the probability of exactly x-occurrences in N trials is:

$$P_x = \frac{N!}{x!\,N - x!} p^x \cdot q^{N-x}$$

where $N! = 1.2.3.\ldots\ldots\ldots.N$.

This expansion is called the Bernoulli series or distribution. When p = q, the binomial distribution is symmetrical. This distribution is similar to the normal distribution for large values of N but it is unsuited for continuous variables because the distribution itself is discontinuous.

(a) Relation to Probability

Suppose a die is thrown 20 times. In this case, the 21 terms of the expansion $(1/6 + 5/6)^{20}$ will give the various probabilities of a particular face, say six, appearing 20, 19, 18, 17 ...... or 0 times.

Now suppose that the problem is more complicated. Let 4 dice be thrown 20 times and the sixes counted that appear on each throw. In any one throw the probabilities of getting 4,3,2,1 or 0 sixes are given by the terms of $(1/6 + 5/6)^4$. To secure the most probable results of the experiment, multiply each of these probabilities by 20. The probability for sixes,

$P_{(4)} = 1/1296$ times 20 = 0.016

$P_{(3)} = 20/1296$ times 20 = 0.308

$P_{(2)} = 150/1296$ times 20 = 2.310

$P_{(1)} = 500/1296$ times 20 = 7.710

$P_{(0)} = 625/1296$ times 20 = 9.640

Then, the most probable outcome of the experiment is: No sixes, 10 times; 1 six, 8 times; 2 sixes, 2 times; 3 sixes, 0 times; and 4 sixes, 0 times.

(b) Constants of the Binomial Distribution

The formulas for the more important constants of the binomial distribution are as follows:

Mean number of occurrences, $\bar{x} = Np$ - - - - - - - - - - - - - (1)

Variance, $\sigma^2 = Npq$ - (2)

Standard error, $\sigma = \sqrt{Npq}$ - - - - - - - - - - - - - - - - - - (3)

Probable error, P.E. $= \pm 0.6745 \sqrt{Npq}$ - - - - - - - - - - - - (4)

Mean proportion of occurrence, $p = \dfrac{N_1p_1 + N_2p_2}{N_1 + N_2}$ - - - - - - - (5)

Variance of proportion of occurrences, $\sigma^2 = \dfrac{pq}{N}$ - - - - - - - (6)

Standard error of proportion of occurrences, $\sigma = \sqrt{\dfrac{pq}{N}}$ - - - (7)

II. Applications of the Binomial Distribution

The Binomial distribution may have a variety of uses in comparisons of observed data with an a priori hypothesis or the comparisons of two samples.

(a) Comparison of Observations against an a Priori Hypothesis.

Suppose in a sample of N-trials of an experiment the number of occurrences of a given phenomenon is x. Let it be desired to test this result in accordance with an accepted standard outcome of such experiments, the expected proportion of occurrences being p. The expected number of occurrences ($\bar{x} = Np$), and the discrepancy will be the numerical value of $x - Np = d$. Then the probability that corresponds to $t = d/\sigma = d/\sqrt{Npq}$ may be found with the aid of the t-table when N is small, or with the table of normal curve areas when N is large. It would be equivalent to test the proportion (x/N) against the expected proportion (p) by the formation of $t = d/\sigma = \dfrac{(x/N) - p}{\sqrt{\frac{pq}{N}}}$

A very common application is the comparison of observed data for monohybrid Mendelian ratios with the theoretical. (See III below).

(b) Comparisons of Samples from Different Populations

It may be desirable to compare the proportion of occurrences in two samples from admittedly different populations. Then the samples provide the following information:

Sample I		Sample II
N_1	Number of cases or trials	N_2
x_1	Number of occurrences of a given phenomenon	x_2
$p_1 = \dfrac{x_1}{N_1}$	Proportion of occurrences	$p_2 = \dfrac{x_2}{N_2}$
$\sigma_1^2 = \dfrac{p_1q_1}{N_1}$	Variance of the proportion	$\sigma_2^2 = \dfrac{p_2q_2}{N_2}$

The differences in proportions $= d = p_1 - p_2$.

The standard error of the difference $= \sigma_d = \sqrt{\sigma_1^2 + \sigma_2^2}$

$$= \sqrt{\frac{p_1q_1}{N_1} + \frac{p_2q_2}{N_2}}.$$

Thus, $t = \frac{d}{\sigma_d} = \frac{p_1 - p_2}{\sqrt{\frac{p_1 q_1}{N_1} - \frac{p_2 q_2}{N_2}}}$

(c) Comparison of Samples from Same Populations

The difference between this problem and that in (b) above consists in the hypothesis that but a single population is being considered. As a result, the data afforded by both samples are combined to give estimates of p and σ for the population. Thus, the estimated proportion of occurrences will be: $p = \frac{N_1 p_1 + N_2 p_2}{N_1 + N_2}$, and the estimated standard error of the population will be $s = \sqrt{\frac{pq}{N_1 + N_2}}$, where $q = 1-p$.

Then $t = \frac{p_1 - p_2}{s}$ may be interpreted as in previous cases.

III. Standard Errors of Mendelian Ratios

In the analysis of genetic data, it is necessary to test the significance of the observed with the calculated counts obtained when certain theoretical conditions are postulated. With monohybrid ratios, the general practice is to use the binomial distribution, which is sometimes referred to as the probable error of a proposition. The ratios which may be calculated in this work by the binomial distribution are: 1:1, 3:1, 9:7, 13:3, 15:1, 63:1, and 27:37.

(a) Formula for Mendelian Ratios

The standard error of a Mendelian ratio is:

$$\sigma = \sqrt{p\ (1-p)N} \quad \text{or} \sqrt{Npq} \text{ - - - - - - - - - - - - - - - - - - - (8)}$$

where N = the number of individuals, p = the proportion of one group as a decimal fraction, and 1 - p = the proportion of the other group as a decimal fraction, (1 - p = q). Some writers use the formula, $S.E. = \sqrt{p.q.N}$, where "p" and "q" represent the proportions in decimal fractions.

(b) Use of Method

The binomial method can be used in genetic data only when two phenotypic classes are grouped, other methods being used for three or more classes. In the F_2 generation of a barley cross, 200 green and 72 white seedlings were counted. It is desired to test these data for a 3:1 ratio.

	Green	White	Total (N)
Observed numbers	200	72	272
Calculated 3:1 ratio	204	68	272
Deviation		4	

To obtain the calculated number for a 3:1 ratio, divide the total number observed by the combined possible number of classes which is 4 in this case, e.g., 272/4 = 68. This gives the calculated value for the white (or 1) class. For the green (or 3) class, multiply 68 by 3. This gives 204

$\sigma = \sqrt{p\ (1-p)N} = \sqrt{0.75 \times 0.25 \times 272} = 7.1460$

Next, the deviation divided by the standard error is computed:

$d/\sigma = 4/7.1460 = 0.56$

The observed ratio fits the calculated 3:1 ratio very well, indicating that a simple Mendelian factor pair is responsible for the production of green and white seedlings. It is to be noted that d/σ is less than 2, which indicates that the fluctuation of the observed ratio from the calculated may be considered as due to chance. In any event, there is no reason to reject the theoretical ratio hypothesis.

Another example may be given for green and white barley seedlings.

	Green	White	Total	d	σ	d/σ
Observed	183	161	344			
Calculated 3:1 ratio	258	86	344	75	8.0356	9.33
Calculated 9:7 ratio	193.5	150.5	344	10.5	9.2068	1.14

It is apparent that the data do not fit a 3:1 ratio as shown by the high value of d/σ. However, they fit a 9:7 ratio very well, indicating that there are two factor pairs involved in the production of green vs. white seedlings in this cross.

(c) Short-Cut Tables for Computations

Tables published by Cornell University give the Probable Errors for Values of N from 11 to 1000. Another set of tables occurs in "Mendelian Inheritance in Wheat and Barley Crosses," by Kezer and Boyack (1918). The probable error values obtained from such tables can be converted to standard errors by the division of the probable error value by the factor, 0.6745.

IV. The Poisson Distribution as A Special Case

As a rather special case of the binomial distribution, there is an approximation of what is known as a Poisson distribution. This occurs when p, the probability of the occurrence of an event, is very small and N, the number of trials, is very large so that Np becomes appreciable. In a Poisson distribution the probability, P_x, of exactly x occurrences in N (N = very large) trials, is given by:

$$P_x = \frac{e^{-Np}(Np)^x}{x\,!}$$

Where e is a constant (2.718) and p is the probability of occurrence in a single trial, and $x! = 1.2.3.........x$.

Although there are tables published of these probabilities, their use is unnecessary in the more common types of application.

(a) Constants of the Poisson Distribution

For the Poisson distribution, the mean and variance are equal.

Mean = $\bar{x}$ = Np

Variance = $\sigma^2 = Np$, so that $\sigma = \sqrt{Np}$

(b) Use of Poisson Distribution

The Poisson distribution gives a basis for the solution of many problems that involve the maintenance of certain standards. Suppose that registered seed regulations state that red clover seed must not contain over a given percentage of noxious weed seeds in order to gain certification. Suppose that from a lot of seed, a sample is taken of such size that a count of 10 noxious weed seeds corresponds to the allowable percentage. In this case, the mean $\bar{x} = Np = 10$. The standard error, $\sigma = \sqrt{Np} = \sqrt{10} = 3.1$. However, 18 weeds may have occurred in the sample analyzed. The whole lot is rejected for registration because the deviation from the mean, 18 - 10 = 8, exceeds twice the standard error, i.e., $2\sigma = 6.2$. Suppose that 16 noxious weed seeds are counted in a sample from another lot. Now a decision becomes questionable. Suppose that a second sample is taken and 14 weed seeds counted. Now consider the two samples as one. The mean, $\bar{x} = Np = 20$, and the standard error, $\sigma = \sqrt{Np} = \sqrt{20} = 4.5$. Thus, 14 + 16 = 30 which differs from the mean by 10. However, this lot would be rejected because the deviation from the mean, 10, exceeds $2\sigma = 2(4.5) = 9.0$.

References

1. Anonymous. Tables of Probable Error of Mendelian Ratios. Department of Plant Breeding, Cornell University (mimeographed).
2. Fisher, R. A. Statistical Methods for Research Workers (5th edition), pp. 55-72. 1934.
3. Kezer, Alvin, and Boyack, B. Mendelian Inheritance in Wheat and Barley Crosses. Colorado Exp. Sta. Bul. 249. 1918.
4. Miles, S. R. A Very Rapid and Easy Method of Testing the Reliability of an average and a Discussion of the Normal and Binomial Methods.
5. Robertson, D. W. The Effect of a Lethal in the Heterozygous Condition on Barley Development. Colorado Exp. Sta. Tech. Bul. 1. 1932.
6. Sinnott, E. W., and Dunn, L. C. Principles of Genetics, McGraw-Hill, pp. 371-373. 1932.
7. Tippett, L. H. C. The Methods of Statistics. Williams and Norgate, pp. 30-33. 1931.

Questions for Discussion

1. Give the binomial expansion of $(p + q)^N$.
2. What type of distribution is the binomial distribution? What are its limitations?
3. What is the genetic application of the binomial distribution? Its limitations?
4. How does the Poisson distribution differ from the binomial distribution?
5. Under what conditions might the Poisson distribution prove useful?

Problems

1. Colsess, a white-glumed barley was crossed with Nigrinudum, a black-glumed barley. The segregation in the F_2 was 785 black-glumed plants and 255 white-glumed plants. What ratio best fits these data? Calculate d/σ.

2. The F_2 segregation of a Colsess (hooded) by Minnesota 90-8 (awned) cross gave 229 hooded plants and 83 awned plants. Determine the ratio that best fits these data, and test its fit.

3. In a cross between Colsess II and Colsess III, 183 green seedlings and 101 white seedlings were observed in the F_2. Determine the ratio that best fits these data and test the fit.

CHAPTER VIII

THE χ^2 TESTS FOR GOODNESS OF FIT AND FOR INDEPENDENCE

I. The χ^2 Test

So far, statistics like the sample mean ($\bar{x}$) and the standard deviation (s') have been used to express differences between distributions, either an observed against a hypothetical distribution, or one observed distribution against another. However, in such cases the general form of the distribution (normal, binomial, Poisson) has been assumed and comparisons have been limited to values of parameters of the distribution. The use of moments such as these might be adequate for an accurate comparison of distributions were a sufficient number of higher moments employed. However, this method has the principal disadvantage of being tedious as well as involving questions as to the validity of the sampling errors of higher moments.

Many times it is desired to compare or test observed data with those expected on the basis of some hypothesis. This has been referred to as a test for "goodness of fit." Again, individuals may be measured or classified categorically with respect to two separate characters or conditions. It may be desired to test these characters for association. Both of these general problems can be attacked by use of a statistic known as χ^2 (Chi-squared) calculated from the data afforded by the sample.

II. The χ^2 Distribution

The χ^2 test, to measure "goodness of fit" of observed results to those expected, was advanced by Karl Pearson in 1900.

(a) Formula for χ^2

The theoretical distribution must be adjusted to give the same total frequency as the observed. Then, when O is the number observed in any one group or category of the experimental distribution, and C the theoretically calculated number for the same group, based on the hypothesis that the data follow some certain distribution, the formula for χ^2 is as follows:

$$\chi^2 = S\left[\frac{(O - C)^2}{C}\right] \qquad (1)$$

where "S" is the summation extended over all the groups or classes. It is obvious that the more closely the observed number agrees with the calculated the smaller χ^2 will be. Further, all differences in frequency (O-C) are squared, whether positive or negative. Thus, χ^2 is always a positive quantity, its size being clearly dependent on the number of groups into which the distribution is separated and degree of agreement between the several values of "O" and the corresponding values of "C". Therefore, in the ordinary application of χ^2, the number of degrees of freedom will be the number of groups diminished by the number of restrictions imposed on the theoretical distribution that supplies the values of C. When the only restriction imposed is that the total frequencies of the observed and theoretical distributions shall be equal, the degrees of freedom are one less than the number of groups. In other words, where the frequencies are determined for all groups but one, the frequency of that one is automatically determined by subtraction from the total.

(b) Sampling Distribution

The sampling distribution of χ^2 has been worked out so that it is possible to find the probability (P) of obtaining from a hypothetical population with a given distribution, a sample that shows a distributional variation from that of the population which would result in a χ^2 value as large or larger than that exhibited by the sample in hand. For every value of χ^2, in conjunction with any optional number of degrees of freedom P = 1.00 for $\chi^2 = 0$ and, as χ^2 increases, P diminishes. Since the mathematical relationships between χ^2 and P are complex, it is necessary to have tables that relate P, χ^2, and the number of degrees of freedom, for practical use.

(c) Grouping Data

It is unwise to group too finely or to apply this test where the data are so insufficient that, for certain of the groups, the expected frequency is small. This condition very easily might cause that part of χ^2 contributed by such groups to unduly affect the total χ^2. This is obvious from the mathematical form, $(O - C)^2/C$, where C is small. Fisher (1934) recommends that each group should contain at least five individuals for the test to apply. Sometimes the tail groups with very low frequencies should be combined.

III. Probability Tables for χ^2

As has been stated the probabilities for χ^2 values are obtained from tables. In order to use them, it is first necessary to know "n", the number of degrees of freedom in which the observed series may differ from the hypothetical. It is equal to the number of classes, the frequencies in which may be filled arbitrarily. When only the totals, have been made equal, n = n' - 1, where n' is the total number of classes or groups. In contingency tables, where tests for independence are being made, the number of degrees of freedom is the product of rows and columns minus one in each case (r - 1) (c - 1) because the hypothetical and observed classifications are forced to conform both for row and column totals. To quote Tippett (1931): "Suppose, in an extreme case, there are n' groups and we fitted a curve involving n' constants which were calculated from the data; then the two distributions would agree exactly and χ^2 would be zero because sampling errors would have had no play." The importance of degrees of freedom in looking up the probabilities that correspond to χ^2 has been emphasized by Fisher (1922, 1923, 1934).

(a) Elderton "Table of Goodness of Fit"

A table was prepared by Elderton with the values of "P" (probability) that a deviation as great as or greater than the observed may be expected on the basis of random sampling. These values correspond to each integral value of χ^2 from 1 to 30. This table is available in "Tables for Statisticians and Biometricians" by Karl Pearson (1914). The user must be careful with this table because n' is equal to the number of degrees of freedom (n) plus one. The probability of intermediate χ^2 values can be obtained approximately by interpolation.*

(b) Fisher "Table of χ^2"

More recently, Fisher (1934) has published a table of χ^2 which uses degrees of freedom (n) directly. It gives values of χ^2 that correspond to special values of "P". Fisher (1934) states: "In preparing this table we have borne in mind that, in practice, we do not want to know the exact value of 'P' for any observed χ^2, but in the first place, whether or not the observed value is open to suspicion. If 'P' is between 0.1 and 0.9 there is certainly no reason to suspect the hypothesis tested.

---

*Note: For example, the probability for $\chi^2 = 4.12$ determined from 4 classes can be interpolated as follows:

When $\chi^2 = 4$,		P = 0.261464
	4.12	
$\chi^2 = 5$,		P = 0.171797
Difference	0.12	0.089667
Product	0.12 x 0.089667 =	0.010760
"P" value	0.261464 - 0.010760 =	0.250704

If it is below 0.02 it is strongly indicated that the hypothesis fails to account for the whole of the facts. We shall not often go astray if we draw a conventional line at 0.05 and consider that higher values of χ^2 indicate a real discrepancy." The table given by Fisher has values of "n" up to 30. Beyond this point it will be found sufficient to assume that $\sqrt{2\chi^2} - \sqrt{2n-1}$ is distributed normally with unit standard deviation about zero. For example:

$$\chi^2 = 35.62,\ n = 32,\ \sqrt{2\chi^2} = 8.44,\ \sqrt{2n-1} = 7.94,\quad \text{Difference} = 0.50.$$

Thus, where $\sqrt{2\chi^2} - \sqrt{2n-1}$ is materially greater than 2, the value of χ^2 is not in accordance with expectation.

(c) Normal Probability Integral Table

In the special case for one degree of freedom (n - 1), the probability can be obtained from the table of the normal probability integral because χ is normally distributed for one degree of freedom. (See Table II, "Tables for Statisticians and Biometricians") For example, suppose it is desired to find the probability that corresponds to $\chi^2 = 3.200$

$$\chi = \sqrt{\chi^2} = \sqrt{3.200} = 1.7889$$

In the table opposite t = 1.7889, the value of the probability that corresponds to it is found to be 0.9632. The value of the probability for the one tail will then be 1.0000 - 0.9632 = 0.0368. On the basis of a 2-tailed table it would be 0.0368 x 2 = 0.0736.

A -- Goodness of Fit

IV. Uses of χ^2 for Goodness of Fit

The χ^2 test for "goodness of fit" can be applied to data grouped into classes where it is desired to compare them with a theoretical or hypothetical ratio. The great advantage of this test for goodness of fit is that no limitations or conditions are imposed upon the form of the distribution under investigation. Historically, the χ^2 test was first used to test the goodness of fit of an observed frequency distribution to a normal distribution of the same total frequency, the same mean, and the same standard deviation. It is still used effectively for this purpose when the number in the sample is large. One sacrifices a fit in the tails of the distribution by use of the χ^2 test, but often the investigator is only interested in the central range which the data cover. The χ^2 test is particularly useful in genetics to test F_2 and later segregations where two or more phenotypic classes are involved. J. Arthur Harris (1912) first called attention to the value of the χ^2 test for genetic data.

V. Computation of χ^2 for Goodness of Fit

In Mendelian ratios from F_2 progenies and later generations, the common practice is to summate the numbers in each phenotypic class and to formulate a hypothesis on the basis of the ratio obtained in order to establish the number of genetic factors involved. The χ^2 test is used to determine whether the deviations of the observed numbers from the calculated numbers are not due to chance.

(a) General Method of Computation

In a cross that involves two independently inherited Mendelian factor pairs, a 9:3:3:1 ratio is expected in the F_2 generation. A segregation in the F_2 generation of a barley cross that involved long vs. short-haired rachilla (Ss) and covered vs. naked seeds (Nn), gave results as follows: (Data from Robertson)

Long-Haired Rachilla		Short-Haired Rachilla		Total
Covered Seeds	Naked Seeds	Covered Seeds	Naked Seeds	
2061	645	675	256	3637
(SN)	(Sn)	(sN)	(sn)	

The calculated ratio for a 9:3:3:1 is calculated so that the total of the theoretical values equal the total in the sample, i.e., 3637. The value 3637 is divided by 16 (9 + 3 + 3 + 1) to give the expected number in the class with short-haired rachillas with naked seeds, i.e., 3637/16 = 227.3125. The values on the basis of expectancy for the 3-classes can be computed by multiplying 227.3125 by 3 = 681.9375, etc. The results can be put down as follows:

Classes	Ratio	Observed No. (O)	Calculated No. (C)	O - C	$(O - C)^2$	$(O - C)^2/C$
SN	9	2061	2045.81	15.19	230.7361	0.1128
Sn	3	645	681.94	36.94	1364.5636	2.0010
sN	3	675	681.94	6.94	48.1636	0.0706
sn	1	256	227.31	28.69	823.1161	3.6211
Totals		3637	3637.00			$\chi^2 = 5.8055$
n = 3			P = 0.1233			

Hence, the deviations from the calculated ratio cannot be regarded as significant.

(b) Method for Two Classes

The χ^2 value may be calculated directly where "A" is the number in one class, "a" the number in the other, and "N" is the total number in the sample (A + a). These formulae are given by Immer (1936) and represent a transformation from the standard method for the computation of χ^2 for goodness of fit.

Ratio A : a	χ^2 Value	
1 : 1	$\frac{(A - a)^2}{N}$	(2)
3	$\frac{(A - 3a)^2}{3N}$	(3)
9 : 7	$\frac{(7A - 9a)^2}{63N}$	(4)
m : n	$\frac{(nA - ma)^2}{mnN}$	(5)

The computation may be illustrated with data which appear to fit a 3 : 1 ratio. A = 2903, a = 936, and N = 3839.

$$\chi^2 = \frac{(A - 3a)^2}{3N} = \frac{(2903 - 2808)^2}{3 \times 3839} = 0.7840. \quad P = \text{Value close to 1.}$$

(c) The χ^2 Test Applied to Several Genetic Families

In genetic data, Kirk and Immer (1928) show that the total class frequencies obtained by summation are composite results which may easily mask a serious lack of consistency in numerical ratios of the separate families with respect to agreement with expectation. To summate the numbers in each class of all progenies is to rely on mean values and thereby disregard deviations from the ratio expected to occur in each family. This applies particularly where the numbers are small. The smaller the number in each progeny, the greater the opportunity to err when the summations are taken as an indication of the genetic constitution. In such cases, a goodness of fit test like χ^2 is required which involves in its calculation deviations from expectancy for each class of each progeny. It should be mentioned that χ^2 values cannot be averaged. However, they are additive provided the number of degrees of freedom are properly taken into account.

VI. Fit of Observed Data to the Normal Curve

The χ^2 criterion is useful to determine whether or not observed data give an acceptable fit to the normal curve or any other assumed form of distribution. It is useful where the sample is large and where the requirements for χ^2 are fulfilled. First, the range of measures is divided into an arbitrary number of classes so as to meet the number of measures in the separate classes which a valid use of the χ^2 criterion demands. Data on number of culms counted on 411 wheat plants at the Colorado Experiment Station are used to illustrate the computation. The data are as follows:

X (Class center)	1	3	5	7	9	11	13	15	17	19	21	23	25	27
f (Frequency)	2	24	52	85	114	69	36	18	6	1	2	1	0	1 = 411

$\bar{x}$ = 8.9172 s' = 3.4715. s' (corrected for grouping) = 3.4231.

(1) The data are regrouped in order to have a larger number of cases in the tail classes.

Classes	Class Range	f
less than 4		26
5	4.0 to 6.0	52
7	6.0 to 8.0	85
→ 9	8.0 to 10.0	114
11	10.0 to 12.0	69
13	12.0 to 14.0	36
15	14.0 to 16.0	18
more than 16		11
	N =	411

The class range is reduced to σ-units, viz., 2/3.4231 = 0.5843

The correction to the mean above 8.0 = 0.9173/3.4231 = 0.2680 σ-units.

(2) The next step is to calculate the end points of units for the class intervals in σ-units.

Class range	Area Range	Unit Frequency	Calculated Frequency
Less than 4.0	- ∞ to - 1.43	0.08	32.9
4.0 to 6.0	- 1.43 to - 0.85	0.12	49.3
6.0 to 8.0	- 0.85 to - 0.27	0.19	78.1
→ 8.0 to 10.0	- 0.27 to + 0.32	0.24	98.6
10.0 to 12.0	+ 0.32 to + 0.90	0.19	78.1
12.0 to 14.0	+ 0.90 to + 1.48	0.11	45.2
14.0 to 16.0	+ 1.48 to + 2.06	0.05	20.6
more than 16.0	+ 2.06 to + ∞	0.02	8.2
Total			411.0

The σ- value for the class that contains the mean is: 0.5843 - 0.2680 = +0.3163 σ. This value is the ordinate for 10.0 while -0.27 is the σ-ordinate for 8.0, these values being within the range, 8.0 to 10.0. The other area ranges are calculated by the addition of 0.58 to determine the next higher or lower range. For example, it is 0.32 + 0.58 = + 0.90 for the range 12.0.

(3) It is now necessary to compute the unit per cent frequency for each class by reference to a table of the probability integral, (Table I, appendix). For example, the unit frequency for the range, -0.27 + 0.32 is computed as follows:

For t = -0.27, P = 0.61 - 0.50 = 0.11

t = +0.32, P = 0.63 - 0.50 = 0.13

The frequency per cent for the distance, -0.27 to +0.32, is equal to 0.11 + 0.13= 0.24. This means that 24 per cent of the frequencies would be within this range on the basis of the normal curve. The other values can be calculated in a similar manner, except that the two values are subtracted. The last class, from 2.06 to include the remainder of the curve, is computed as follows:

For t = 2.06, P = 0.98 1.000 - 0.98 = 0.02

(4) The next step is to multiply the per cent frequencies by the number in the sample (N) to obtain the calculated frequencies, e.g., (0.08)(411) = 32.9, etc.

(5) The observed and calculated frequencies are now compared by use of the χ^2 criterion.

Class range	Observed Frequency	Calculated Frequency	O-C	$(O-C)^2$	$(O-C)^2/C$
less than 4.0	26	32.9	-6.9	47.61	1.4471
4.0 to 6.0	52	49.3	2.7	7.29	0.1479
6.0 to 8.0	85	78.1	6.9	47.61	0.6096
8.0 to 10.0	114	98.6	15.4	237.16	2.4053
10.0 to 12.0	69	78.1	-9.1	82.81	1.0603
12.0 to 14.0	36	45.2	-9.2	84.64	1.8726
14.0 to 16.0	18	20.6	-2.6	6.76	0.3282
more than 16.0	11	8.2	2.8	7.84	0.9561
Totals	411	411.0		χ^2 =	8.8271
				P =	0.1172

There are 8 classes, but only 5 degrees of freedom available because 3 constants have been used in fitting the data to the normal curve. It is obvious in this case that the probability (P) is greater than 0.05. Thus, the underlying distribution of the data may have been normal. This method applies to fitting observed data to any hypothetical distribution.

VII. Partition of χ^2 into its Components

When a discrepancy in a theoretical genetic ratio on the basis of independent inheritance occurs, it may be produced either by linkage or a departure from the 3 : 1 ratios. Fisher (1934) has suggested a method whereby χ^2 can be partitioned into its components to determine the source of the discrepancy. In a barley cross, the F_2 data were as follows for non-tipped and tipped lateral spikelets (Tt), and for hoods and awns (Kk):

	TK (a)	Tk (b)	tK (c)	tk (d)	Total
Observed No.	1496	515	550	216	2777
Calculated No.	1562.06	520.69	520.69	173.56	2777.00

χ^2 = 14.8835 P = very small

To determine whether or not the discrepancy is due to linkage, the χ^2 value is partitioned into its components as follows:

x = non-tipped vs. tipped = (a + b) - 3(c + d)
= (1496 + 515) - 3(550 + 216) = -287

y = hoods vs. awns = (a + c) - 3(b + d)
= (1496 + 550) - 3(515 + 216) = -147

z = interaction or linkage = a -3b - 3c + 9d
= 1496 - 3(515) - 3(550) + 9(216) = +245

The χ^2 values can be computed for each component as follows:

1. non-tipped vs. tipped:

$$\chi^2 = \frac{x^2}{3n} = \frac{(287)^2}{3(2777)} = 9.8870$$

2. hoods vs. awns:

$$\chi^2 = \frac{y^2}{3n} = \frac{(147)^2}{3(2777)} = 2.5938$$

3. interaction (or linkage):

$$\chi^2 = \frac{z^2}{9n} = \frac{(245)^2}{9(2777)} = 2.4017$$

The data can be brought together in a summary form as below:

Factor Pairs	d.f.	χ^2	P
Non-tipped vs. tipped (Tt)	1	9.8870	0.0016
Hooded vs. awned (Kk)	1	2.5938	0.1074
Interaction	1	2.4017	0.1212
Totals	3	14.8825	very small

Thus, the 3 : 1 ratio for non-tipped vs. tipped is found to account for a large part of the high χ^2 value. There is no indication of linkage.

B -- Test for Independence

VIII. Independence and Association

When observations have been classified in two ways, it may be desirable to determine whether or not the two variables are associated. The χ^2 test for independence has been used for this purpose. Two variables are said to be associated when the numbers in the cells of the contingency table are not randomly distributed. Contingency tables may be manifold, there being (r - 1) (c-1) degrees of freedom where there are "r" rows and "c" columns. In tests for independence, the subtotals of the classes into which the variates are distributed are used to determine the theoretical frequencies with the result that the subtotals must be considered as constants in the determination of degrees of freedom. For example, the degrees of freedom in a 2 by 2 contingency table are one. The value of χ^2 is referred to a χ^2 table to determine the value of "P" that corresponds to it for the number of degrees of freedom in the contingency table. A "P" value greater than 0.05 indicates lack of proof of association between two variables, i.e., they may be independent. The χ^2 criterion has proved useful as a test for the independence of two genetic factor pairs.

IX. Calculation of Independence or Association

The test for independence can be made when the data are compiled either in simple 4-fold (2 by 2) or manifold contingency tables.

(a) The Manifold (m by n) Contingency Table

The computation can be illustrated by some F_2 data (Hayes) in an oat cross, Bond x D.C., where it was desired to learn whether or not there was any association between the reaction to stem rust and to crown rust. The data are:

		Stem Rust Reaction			
		Resistant	Susceptible	Totals	Ratio
Crown	Resistant	50(57.2494)	22(14.7550)	72	0.2951
Rust	Susceptible	119(112.1126)	22(28.8950)	141	0.5779
Reaction	Intermediate	25(24.6380)	6(6.3500)	31	0.1270
	Totals	194	50	244	1.0000

In case that the amount of stem rust infection has no influence on the amount of crown rust infection, the 244 observations would be expected to be distributed at random in the 6 cells of the contingency table, with the restriction that they must add up to give the totals in the table (See Tippett, 1931, p. 69). The probability that an observation will fall in row No. 1 is 72/244, and that it will fall in column No. 1 is 194/244. Then, the probability that an observation will fall in the first cell is (72/244) (194/244). The expected number of individuals in that square on the basis of independence is the probability multiplied by the total number, i.e., (72/244) (194/244) (244) = 57.2494.

The various steps in the computation are as follows:

(1) The ratio of rows, for row No. 1, is 72/244 = 0.2951.
(2) The theoretical frequencies can be obtained by the multiplication of each of the ratios for rows by each of the subtotals for columns, e.g., 0.2951 times 194 = 57.2494 for cell No. 1. The other values are computed in a similar manner. In this case, it is necessary to compute the value for only one other cell, i.e., 0.5779 times 194 = 112.1126. The other values can be obtained by subtraction from the marginal totals.
(3) The observed and theoretical values are then compared by use of the χ^2 criterion.

Observed No.	Calculated No.	O-C	$(O-C)^2$	$(O-C)^2/C$
50	57.2494	7.2494	52.5538	0.9180
119	112.1126	6.8874	47.4363	0.4231
25	24.6380	0.3620	0.1310	0.0053
22	14.7550	7.2450	52.4900	3.5574
22	28.8950	6.8950	47.5410	1.6453
6	6.3500	0.3500	0.1225	0.0193
244	244.0000		χ^2 =	6.5684
n = (n - 1) (m - 1) = 2			P =	0.0387

Thus, the indications are that there is an association between the reactions to stem rust and to crown rust.

(b) <u>The 2 by 2 or 4-Fold Table</u>
The 4-fold table is often used to test the independence of two genetic factor pairs. The independence of the two 3 : 1 ratios can be tested as follows:

	K	k	Total
V	a = 142	b = 43	a + b = 185
v	c = 49	d = 15	c + d = 64
Totals	a + c = 191	b + d = 58	N = 249

The value of χ^2 can be determined by the method outline in (a) above, or it can be computed by a short-cut formula given by Fisher (1934).

$$\chi^2 = \frac{N\ (ad - bc)^2}{(a + c)\ (b + d)\ (a + b)\ (c + d)} \quad \text{- - - - - - - - - - - - - - - (6)}$$

$$= \frac{249\ [(142)(15)\ -\ (43)(49)]^2}{(191)(58)(185)(64)}$$

$$= \frac{(249)(529)}{(191)(58)(185)(64)} = \frac{131,721}{131,163,520} = 0.0010$$

when X^2 = 0.0010, P = value close to 1.

(c) Inadequacy of X^2: Correction for Continuity

When the several categories are represented by relatively small frequencies, the value of X^2 often gives inaccurate results because the corresponding probability of occurrence is too small. This is particularly the case in a 2- by -2 classification. Yates (1934) has developed a correction that should be applied in such cases. This correction simply amounts to the reduction of each numerical value of each (O-C) determination by 1/2. Thus, in the example above, the correction applied to X^2 is as follows:

$$\chi^2\ \text{(corrected)} = \frac{N(ad - bc - N/2)^2}{(a + c)(b + d)(a + b)(c + d)} \quad \text{- - - - - - - - - - - - - - (7)}$$

$$= \frac{249\ [(142)(15)\ -\ (43)(49)\ -\ 249/2]^2}{(191)(58)(185)(64)}$$

$$= \frac{(249)\ (-106.5)^2}{131,163,520} = \frac{2,824,220.25}{131,163,520}$$

$$= 0.0215 \qquad P = \text{value close to 1.}$$

In this case even tho the frequencies may be fairly large, it is quite proper to introduce the correction. However, the larger the number of categories, the less important is the correction.

X. The Null Hypothesis and X^2

It is important to understand something about the philosophical and logical bases for the making of inferences from the X^2 as well as from other criteria for significance. The basic premise involved in every test for significance is a negative premise and has been termed the null hypothesis by Fisher (1937). It is simply a tacit assumption of agreement, such as agreement between standard deviations of distributions, and agreement between distributions as a whole. In association and correlation studies the null hypothesis is construed to mean independence or lack of association between characters or conditions under investigation. This tacit negative premise can never be proved. For example, it is impossible to prove statistically that two samples came from the same population, or that the population which afforded the samples under comparison possess the same means or other statistics. It is impossible to prove statistically that two characters or conditions are independent or devoid of association. To draw such conclusions would simply be to reiterate what was originally only assumed to be true.

Therefore, definite conclusions can be drawn only when the criteria for significance have been met, such conclusions being positive in nature. The investigator is able to prove differences to exist, association to be present, etc. In short, he is abl to prove the falsity of the null hypothesis but never its truth.

References

1. Fisher, R. A. On the Interpretation of χ^2 from Contingency Tables, Calculation of P. Jour. Roy. Stat. Soc., Vol. LXXXV, Part I, 1922
2. ____________ Statistical Tests of Agreement between Observation ar Economica, 3:139-147. 1923.
3. ____________ Statistical Methods for Research Workers (5th Editior and 274-275. 1934.
4. ____________ The Design of Experiments, pp. 18-20. 1937.
5. Goulden, C. H. Methods of Statistical Analysis, pp. 88-113. 1939.
6. Harris, J. A. A Simple Test of Goodness of Fit of Mendelian Ratios, 46:741. 1912.
7. Immer, F. R. Applied Statistics Manual (mimeographed) 1936.
8. Kirk, L. E., and Immer, F. R. Application of Goodness of Fit Tests Class Frequencies. Sci. Agr., 8:745-750. 1928.
9. Pearson, Karl, Tables for Statisticians and Biometricians, pp. 2-3 a 1914.
10. Tippett, L. H. C. The Methods of Statistics, pp. 63-88. 1931.
11. Yates, F. Jour. Roy. Stat. Soc., Suppl. I, No. 2. 1934.
12. Youden, W. J. Statistical Analysis of Seed Germination thru the Use Square Test. Contrib. Boyce Thompson Inst., 4:219-232. 1932.
13. Yule, G. Udny. An Introduction to the Theory of Statistics (9th edi pp. 370-378. 1929.
14. Yule, G. Udny. Probability Values for One Degree of Freedom. Jour. Soc., 85:95-104. 1922.

Questions for Discussion

1. What are the uses of the χ^2 criterion?
2. What conditions must be fulfilled in the use of the χ^2 test?
3. What is the range of χ^2 values? "P" values?
4. Give a rule for the number of degrees of freedom in a "goodness of What is it for a contingency table?
5. What precautions are necessary in the grouping of data for a "goodn test? Why?
6. How do the Elderton and Fisher tables for χ^2 differ? What precauti necessary in the use of each?
7. Interpret "P" = 0.50 on the basis of goodness of fit.
8. In what special case can the normal probability integral table be u "P"? Why?
9. Who is responsible for the χ^2 test? For what was it first used?
10. Explain how to compute χ^2 for goodness of fit.
11. What precautions are necessary in the application of the χ^2 test fo fit to genetic ratios? Why?
12. In the fitting of observed data to that expected on the basis of th how many constants are used? Which ones?
13. Under what conditions may it be desirable to partition χ^2 into its
14. What does "P" = 0.01 indicate when obtained from a contingency tabl
15. Explain how the probability is calculated for a cell in a contingen
16. How does the χ^2 test for independence differ from that for goodness
17. What is meant by the null hypothesis?

PROBLEMS

I. In a barley cross, Robertson (1929) tested black vs. white glumes (Bb) and hoods vs. awns (Kk) for a 9 : 3 : 3 : 1 ratio in the F_2. His data were as follows:

Classes	Observed No.	Calculated No.
Black hooded (BK)	2611	2656.7
Black awned (Bk)	920	885.5
White hooded (bK)	860	885.5
White awned (bk)	332	295.3
Totals	4723	4723

Calculate χ^2 and interpret it.* Do these data fit a 9 : 3 : 3 : 1 ratio for independent inheritance?

II. Some data on hoods and awns (Kk) and covered vs. naked (Nn) in barley were tested for a 9 : 3 : 3 : 1 ratio. The observed and calculated results were as follows: (Data from Robertson, 1929)

Classes	Observed No.	Calculated No.
Hooded covered (KN)	1969	2046
Hooded naked (Kn)	681	682
Awned covered (kN)	737	682
Awned naked (kn)	250	227
Totals	3637	3637

Apply the χ^2 test and interpret it.

III. An F_2 segregation of a barley cross, Colsess x Minnesota 84-7, gave these results: (Data from Robertson, 1929)

Classes		Observed No.
Hooded green	(KF)	931
Hooded chlorina	(Kf)	326
Awned green	(kF)	326
Awned chlorina	(kf)	119

(a) What ratio fits these data? (b) Apply χ^2 test and interpret it. (c) Calculate the probability both from the table by Fisher and from the table of Elderton.

IV. In the F_2 of a certain barley cross there were 249 plants with high fertility of the lateral spikelets and 67 with low fertility. Test these data for a 3:1 ratio by the χ^2 test for goodness of fit.

V. A second generation segretation in a barley dihybrid for high and low fertility (Hh) and for black and white glume color (Bb) gave counts as follows:

HB	Hb	hB	hb
1547	568	478	184

*Note: Statement for P: "A worse result might be expected on the basis of random sampling ________ times in ________ trials."

When these data were tested for a calculated 9: 3: 3: 1 ratio, χ^2 was 8.5718 with P = 0.0365. Partition χ^2 into its components and determine whether the discrepancy is due to the individual 3 : 1 ratios or to linkage.

VI. Some F_2 oat plants were classified on the basis of crown rust and stem rust resistance as follows:

		Stem Rust Reaction		
		Resistant	Susceptible	
Crown	Resistant	66	43	109
Rust	Susceptible	75	24	99
Reaction	Intermediate	17	5	22
	Totals	158	72	230

Use the χ^2 test for independence to determine whether or not there is an association between the reaction to stem rust and crown rust.

CHAPTER IX

SIMPLE LINEAR CORRELATION

I. Nature of Correlation

So far, statistical analysis has dealt with a single set of observations to measure a single character. It is now desirable to consider two such sets of observations that measure two different characters. These observations are such that, to any observation in one set, there is naturally paired a corresponding observation of the other. One naturally inquires as to whether there exists any association or connection between the measured characters. Such association exists when an abnormality[1] in one character tends to be accompanied by an abnormality in the other. The characters are said to be correlated when such is the case. For example, height and weight in human beings are said to be correlated. In the aggregate, tall persons are heavier than short persons.

To condense what has been said into a precise definition, it may be stated that two characters are correlated when, to a selected set of values of one, there correspond sets of values of the other whose means are functions of those selected values.

II. Description of Correlation

A graphical representation of the totality of paired observations can be obtained by the treatment of each pair of measurements as the rectangular coordinates of a point. Such a diagram of scattered points is called a scatter diagram. To illustrate, one may consider 20 pairs of observations that relate length (in inches) to weight (in ounces) of ears of corn:

Length (x)	Weight (y)	Length (x)	Weight (y)
2.5	3.5	6.5	6.5
2.5	3.0	7.5	10.0
3.0	5.0	8.0	8.0
4.0	7.0	8.0	10.0
4.5	5.5	8.0	12.0
5.0	8.0	8.5	13.0
5.5	8.0	9.0	12.0
6.0	10.0	9.0	14.0
6.0	7.0	9.5	13.0
6.5	10.5	10.5	14.0

Mean length ($\bar{x}$) = 6.5 inches. Mean weight ($\bar{y}$) = 9.0 ounces

From these pairs of measurements a scatter diagram can be made as follows:

[1]Abnormality refers to ± deviations from the mean.

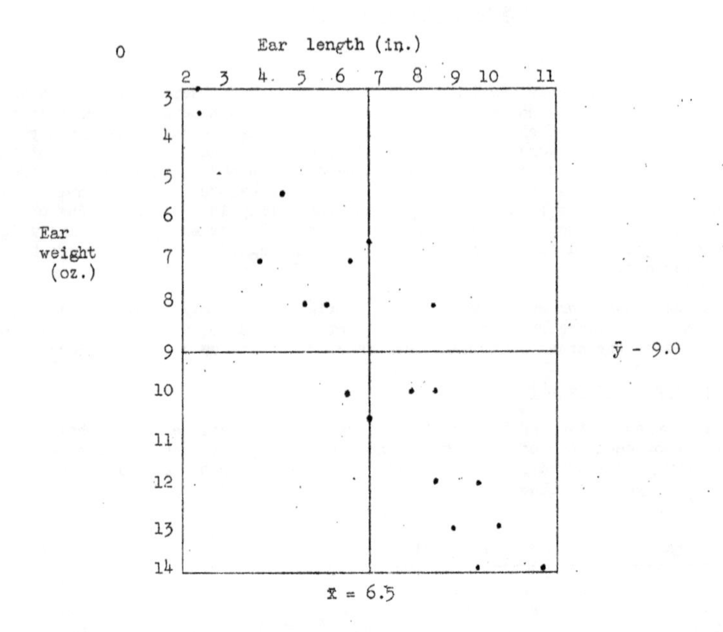

From the diagram, it is clear that the horizontal and vertical lines that represent the mean length and weight of the ears in the sample separate the plane, in which the points are plotted, into four regions or quadrants. It is also evident that most of the points fall into two of these regions, i.e., those which describe the abnormalities in regard to the characters to be of the same type above the average and below the average. Thus, there appears to exist a direct or positive correlation between the characters.

The totality of points that form the scatter very often possess the rough geometrical form of an ellipse. The position of the ellipse indicates the type of association, i.e., whether positive (direct) or negative (inverse). The shape of the ellipse roughly estimates the degree of correlation. The characters are closely related when the ellipse is narrow. A diagramatic representation of correlation is given in figures A, B, and C.[1]

---

[1]Some statisticians use the first quadrant in correlation analysis while others use the fourth.

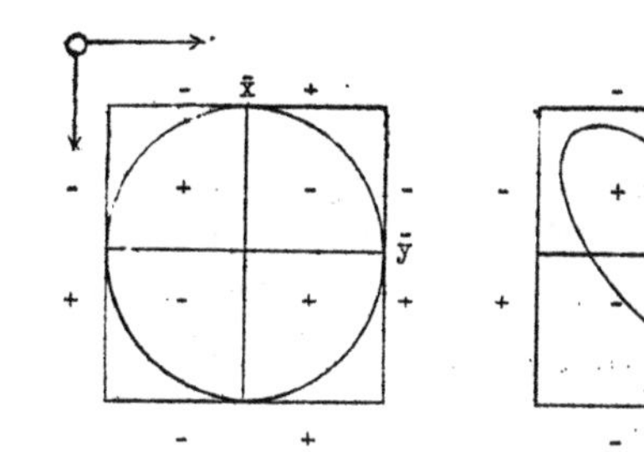

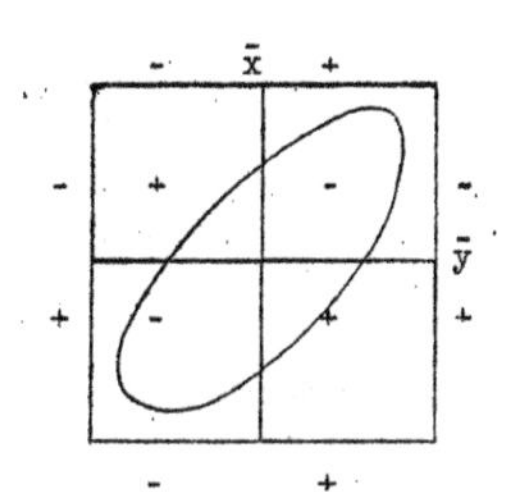

Figure A
Low Correlation

Figure B
Positive Correlation

Figure C
Negative Correlation

The signs for the quadrants are depicted in figure A. It is noted that the values of x above the mean ($\bar{x}$) are positive, while those below the mean are negative. The same applies for the y values. The sign for the quadrant is the product of the corresponding marginal signs.

There are two methods employed to describe correlation, i.e., the correlation surface method and the regression method. For an account of the correlation surface method, a text on mathematical statistics should be consulted.

A -- The Correlation Coefficient

III. Measurement of Correlation

A precise mathematical measure of the degree of association between two characters is desirable. In any case, it must be based on an assumption in regard to the mathematical functional relationship that exists between the variables. The most important measure is called the coefficient of correlation, symbolized as r. In the discussion that follows it is assumed that the association is linear, i.e., that the variables x and y are related by an equation, $y = ax + b$, where a and b are constants.

Suppose one considers each pair of measurements of the two characters as an argument, either strong or weak, for one or the other of two opposite theories of association between the two characters. These theories are that the two characters are related, either positively or negatively. A linear relationship is said to exist between two characters when the means of the values of one character are plotted with the selected values of the other character that correspond to them so that the resulting points are well-fitted by a straight line. To measure the contribution of any given pair of measurements (x, y) to one theory of association or the other, one measures the amount of abnormality exhibited by the pair of measurements with respect to each character in units of the respective standard deviation of the samples provided by the 2 sets of variates, i.e., $(x - \bar{x})/s'_y$. When these measures of the abnormalities of the pairs of observations are multiplied, i.e.,

$\frac{(x - \bar{x})}{s'_x} \frac{(y - \bar{y})}{s'_y}$ the result gives a numerical measure of the argument presented by (x, y) toward a theory of correlation. The product of both abnormalities will be positive when both are of the same type, either positive or negative. Their product will be negative when the abnormalities are opposite in type. A numerical measure of correlation between the characters under investigation is obtained when the procedure is repeated for every pair of measurements in the sample and the arithmetic mean of the several products is found. The formula for the correlation coefficient (r) is as follows:

$$r = \frac{1}{N} S \left(\frac{x - \bar{x}}{s'_x}\right)\left(\frac{y - \bar{y}}{s'_y}\right) \quad \text{- - - - - - - - - - - - - - (1)}$$

It is obvious that "r" can be plus or minus, thus depicting a positive or negative correlation. It will be shown later that "r" is numerically equal to or less than 1.0. Thus, the association that exists between two characters may be strong, as evidenced by a value of "r" numerically close to 1.0, or weak when "r" is close to 0.

The above statement must not be construed too literally but in the light of sampling theory.

IV. Computation of "r" for Ungrouped Data

The relation, $r = \frac{1}{N} S \left(\frac{x - \bar{x}}{s'_x}\right)\left(\frac{y - \bar{y}}{s'_y}\right)$ may be transformed to many different arbitrary forms for computation. Formulas which are useful for small samples are as follows:

$$r^{1} = \frac{S(xy) - N\bar{x}\bar{y}}{\sqrt{(Sx^2 - N\bar{x}^2)(Sy^2 - N\bar{y}^2)}} \quad \text{- - - - - - - - - - - - - - - - - (2)}$$

$$r - \frac{S(xy)/N - \bar{x}\bar{y}}{\sqrt{[S(x^2)/N-\bar{x}^2]\,[S(y^2)/N-\bar{y}^2]}} \quad \text{- - - - - - - - - - - - - - - - - (3)}$$

$$r = \frac{NS(xy) - (Sx)(Sy)}{\sqrt{[NS(x^2) - (Sx)^2]\,[NS(y^2) - (Sy)^2]}} \quad \text{- - - - - - - - - - - - - - (4)}$$

Formula (3) is the one given by J. Arthur Harris, which is direct, but not suited so well to machine calculation as (2) or (4).

The computation may be illustrated with these data on the length of corn ears in centimeters and their weight in ounces.

$$\text{1/}\ (s'_x)^2 = \frac{Sx^2 - N\bar{x}^2}{N} \quad \text{and} \quad (s'_y)^2 = \frac{Sy^2 - N\bar{y}^2}{N}$$

$$r = \frac{1}{N}\,\frac{S(x - \bar{x})(y - \bar{y})}{s'_x\, s'_y} = \frac{1/N\ \left[Sxy - \bar{x}S(y) - \bar{y}S(x) + N\bar{x}\bar{y}\right]}{\sqrt{\left(\frac{Sx^2 - N\bar{x}^2}{N}\right)\left(\frac{Sy^2 - N\bar{y}^2}{N}\right)}}$$

$$= \frac{1/N\ (Sxy - N\bar{x}\bar{y} - N\bar{x}\bar{y} + N\bar{x}\bar{y})}{1/N\sqrt{(Sx^2 - N\bar{x}^2)(Sy^2 - N\bar{y}^2)}} = \frac{Sxy - N\bar{x}\bar{y}}{\sqrt{(Sx^2 - N\bar{x}^2)(Sy^2 - N\bar{y}^2)}}$$

Length (x)	Weight (y)	x^2	y^2	xy
2.5	3.5	6.25	12.25	8.75
2.5	3.0	6.25	9.00	7.50
3.0	5.0	9.00	25.00	15.00
4.0	7.0	16.00	49.00	28.00
4.5	5.5	20.25	30.25	24.75
5.0	8.0	25.00	64.00	40.00
5.5	8.0	30.25	64.00	44.00
6.0	10.0	36.00	100.00	60.00
6.0	7.0	36.00	49.00	42.00
6.5	10.5	42.25	107.62	68.25
6.5	6.5	42.25	42.25	42.25
7.5	10.0	56.25	100.00	75.00
8.0	8.0	64.00	64.00	64.00
8.0	10.0	64.00	100.00	80.00
8.0	12.0	64.00	144.00	96.00
8.5	13.0	72.25	169.00	110.50
9.0	12.0	81.00	144.00	108.00
9.0	14.0	81.00	196.00	126.00
9.5	13.0	90.25	169.00	123.50
10.5	14.0	107.62	196.00	147.00
S(x) =130.5	S(y) = 180.0	$s(x^2)$ = 949.87	$S(y^2)$ = 1834.37	S(xy)= 1310.50
$\bar{x}$ = 6.5	$\bar{y}$ = 9.0	--	--	

The symbols $\bar{x}$ and $\bar{y}$ are the means of the x and y arrays. The values, $S(x^2)$ and $S(y^2)$, are the squared values for each separate entry of x and y, respectively, and the summation of the same. The value, S (xy), is the summation of the product of each value of x by the corresponding value of y. In practice, only the sums of the various values are recorded in machine calculation.

The values may be substituted in (2) as follows:

$$r = \frac{S(xy) - N\bar{x}\bar{y}}{\sqrt{(Sx^2 - N\bar{x}^2)(Sy^2 - N\bar{y}^2)}} = \frac{1310.50 - (20)(6.5)(9.0)}{\sqrt{(949.87 - 845.00)(1834.37 - 1620.00)}}$$

$$= \frac{1310.50 - 1170.00}{\sqrt{(104.87)(214.37)}} = \frac{140.50}{\sqrt{22480.9819}} = 0.937$$

Those who use this formula for the computation of r are warned that a serious error may be introduced by dropping decimals. The means should be carried out to twice the number of decimal places as appear in the original data. The formulae given above are particularly valuable when N is small, i.e., less than 50.

V. Calculation from a Correlation Surface

The correlation coefficient may be calculated from a correlation surface with the deviations from the assumed means taken on an arbitrary scale. It is necessary to apply corrections for the means, standard deviations, and class intervals. Fisher (1934) has made a contribution to simplicity in the mechanical computation of the correlation coefficient, his method[1] being used below. The data are for the correlation of total grain weight (x) in grams and culm length (y) in centimeters in wheat plants.

[1]In the determination of standard deviations where Sheppard's Correction has been used, the uncorrected standard deviations should be used in computing r.

Table 1. Correlation Table for Grain Weight and Culm Length in Wheat.

Culm length (y)	Gr. Wt. (x)	9.5	29.5	49.5	69.5	89.5	109.5	129.5	149.5	169.5	189.5	209.5	
(y)	Y \ X	1	2	3	4	5	6	7	8	9	10	11	f_y
62	1		1		2								3
67	2		4	1	2								7
72	3	1	4	3	2	2	2						14
77	4		2	8		4	2	2					18
82	5		2	7	4	7	8	5	1				34
87	6		2	6	12	5	8	17	6	1		4	61
92	7			8	28	2	2	16	20	5	2		83
97	8			7	21	24	14	3	7	4	2	1	83
102	9			2	22	34	13	1	1	8	6	1	88
107	10			1	3	32	26	6		1	2		71
112	11					4	15	6		1			26
117	12							5		1			6
f_x		1	15	43	96	114	90	61	35	21	12	6	494

The data may be arranged as follows:

Table 2. Computation of the Correlation Coefficient

(1)	(2)	(3)	(4)	(5)	(6)	(7)	(8)	(9)	(10)	(11)	(12)	(13)	(14)
Av. length culms (y)					Total for Wt.	Product	Av. grain weight (x)					Total for length	Product
Class Center	Y	f_y	Yf_y	Y^2f_y	S'Xf	YS'Xf	Class Center	X	f_x	Xf_x	X^2f_x	S'Yf	XS'Yf
62	1	3	3	3	10	10	9.5	1	1	1	1	3	3
67	2	7	14	28	19	38	29.5	2	15	30	60	51	102
72	3	14	42	126	48	144	49.5	3	43	129	387	254	762
77	4	18	72	288	74	296	69.5	4	96	384	1536	696	2784
82	5	34	170	850	167	835	89.5	5	114	570	2850	963	4815
87	6	61	366	2196	363	2178	109.5	6	90	540	3240	770	4620
92	7	83	581	4067	495	3465	129.5	7	61	427	2989	466	3262
97	8	83	664	5312	453	3624	149.5	8	35	280	2240	246	1968
102	9	88	792	7128	500	4500	169.5	9	21	189	1701	178	1602
107	10	71	710	7100	402	4020	189.5	10	12	120	1200	104	1040
112	11	26	286	3146	161	1771	209.5	11	6	66	726	41	451
117	12	6	72	864	44	528							
Totals		494	3772	31108	2736	21409			494	2736	16930	3772	21409
			SY	SY^2		SXY				SX	SX^2		SXY

$$\bar{Y} = \frac{3772}{494} = 7.6356 \qquad \bar{X} = \frac{2736}{494} = 5.5385$$

The details of computation are explained as follows:

1. To simplify the arithmetic, the variables X and Y are used in place of x and y, respectively. They are related by:

$$X = (x - x_1)/C_x + 1$$

$$Y = (y - y_1)/C_y + 1$$

where x_1 and y_1 represent the class centers of the first classes, and C_x, C_y, are the class intervals of the x and y distributions, respectively.

2. The values for Yf_y (in column 4) are the products of the class values, Y, and their respective frequencies. The values for column 11 are computed in a similar manner.

3. The values for Y^2f_y (in column 5) are the products of columns 2 and 4 for the respective values of Y. The X^2f_x values in column 12 are computed from columns 9 and 11.

4. The total deviations in culm length (y variable) are shown in column 13 for each column for grain weight (x variable). Here the symbol (f) without subscripts indicates the frequency of one cell of the correlation table, i.e., the frequency of a particular value of X accompanied by a particular value of Y. The symbol S' denotes the total over just one array. It is necessary to refer to Table 1 (columns 2 and 3) to compute these values.

1st Y-array = (1)(3) = 3
2nd Y-array = (1)(1) + (4)(2) + (4)(3) + (2)(4) + (2)(5) + (2)(6) = 51
3rd Y-array = (1)(2) + (3)(3) + (4)(8) + (5)(7) + (6)(6) + (7)(8) + (8)(7) + (9)(2) + (10)(1) = 254 etc.

The values for the X-arrays in column 6 are computed in a similar manner.

5. For the product (XS'Yf) multiply each value of the total for length in column 13 by its respective X-value in column 9. For example, (3)(1) = 3, (51)(2) = 102, etc. The values in column 7 are computed similarly. It is noted that the ultimate result (SXY) of the computations carried out in columns (6)(7) and (13)(14) is the same. Thus, one provides a check on the other.

6. The computed values in Table 2 are then substituted in formula No. 2 above: [1]

$$r = \frac{S(XY) - N\bar{X}\bar{Y}}{\sqrt{(SX^2 - N\bar{X}^2)(SY^2 - N\bar{Y}^2)}} = \frac{21409 - (494)(7.6356)(5.5385)}{\sqrt{[16930 - (494)(5.5385)^2][31108 - (494)(7.6356)^2]}}$$

$$= \frac{21,409 - 20,891.16}{\sqrt{(16,930 - 15,153.45)(31,108 - 28,801.39)}} = \frac{517.84}{\sqrt{4,097,808.00}}$$

$$= 517.84 / 2024.30 = 0.2558$$

---

[1] The data in the problem above have been coded. Suppose a = assumed mean, and C = class interval. It can be shown that the correlation coefficient from coded data is equal to that from the natural numbers, viz., $r_{xy} = r_{XY}$.

$$x = (X - a_x)/C_x \text{ and } y = (Y - a_y)/C_y$$

$$x - \bar{x} = C_x(X - \bar{X}), \text{ and } s'_x = C_x s'_X$$

$$r_{xy} = \frac{S(x - \bar{x})(y - \bar{y})}{s'_x \; s'_y} = \frac{C_x \; C_y \; S(X - \bar{X})(Y - \bar{Y})}{C_x \; C_y \; s'_X \; s'_Y} = r_{XY}.$$

7. The true means can be computed from the above values as follows:

$$\bar{y} = (\bar{Y} - 1)C_1 + Y_1 = (7.6356 - 1)(5) + 62 \doteq 95.1780$$

$$\bar{x} = (\bar{X} - 1)C_x + X_1 = (5.5385 - 1)(20) + 9.5 = 100.2700$$

VI. Use of the Correlation Coefficient for Error of a Difference

The correlation coefficient may be used to reduce the standard error of a difference (σ_d) when there exists a correlation between the paired values of two variables. This usually enables one to obtain significance with smaller differences than is possible with the formula, $\sigma_d = \sqrt{a^2 + b^2}$, given previously (See Chapter 6). However, it is seldom worthwhile to apply the correlation formula unless "r" is large because the reduction in error is usually insufficient to justify the greater amount of calculation. The extended formula for the standard error of a difference is as follows:

$$\sigma_d = \sqrt{a^2 + b^2 - 2 r_{ab} ab} \quad \text{- -} \quad (5)$$

In this formula a and b represent the standard errors of the separate values being compared, and r, the coefficient of correlation between the separate measurements of these quantities.

The averages for the heading and blossoming stages of irrigation of spring wheat over a 9-year period[2] may be taken to show the value of the correlation coefficient in the reduction of the standard error. The average yields of grain in pounds per plot, together with their standard errors ($\sigma_{\bar{x}}$), are as follows:

Year	Stage of Irrigation Heading	Blossoming
1921	478 ± 52	430 ± 46
1922	776 ± 37	637 ± 31
1923	1114 ± 38	947 ± 49
1924	1218 ± 53	1189 ± 52
1925	555 ± 28	524 ± 27
1926	774 ± 59	645 ± 49
1927	1043 ± 39	1035 ± 39
1928	639 ± 34	614 ± 33
1929	895 ±113	839 ±107
Mean	833 ± 19	762 ± 18

The coefficient of correlation was calculated for the paired annual yields by the use of the formula for the ungrouped data as explained in paragraph IV, viz., r = + 0.406.

$$\sigma_d = \sqrt{a^2 + b^2 - 2r_{ab}ab} = \sqrt{(19)^2 + (18)^2 - (0.812)(19)(18)} = 20.17$$

$$d/\sigma_d = 71/20.17 = 3.52.$$

The standard error of the difference, calculated without the use of the correlation coefficient to reduce the error, was as follows:

$$\sigma_d = \sqrt{a^2 + b^2} = \sqrt{(19)^2 + (18)^2} = 26.17$$

$$d/\sigma_d = 71/26.17 = 2.71$$

[1] Y_1 = first true class value. C_x, C_y = class intervals.

[2] Robertson, D. W., et al. Studies on the Critical Period of Applying Water to wheat. Data from Colorado Experiment Station.

It is apparent how the test comparing the averages of the yearly means is strengthened by taking into account the correlation due to years.

VII. Significance of the Correlation Coefficient

The test for significance is to determine the probability (P) that the observed correlation could have arisen by random sampling from a population in which the correlation is zero. The t-test is more accurate for small samples while the standard error test is satisfactory for large samples.

(a) The Standard Error Test

In large samples drawn from a population in which the mean value of r is zero, the standard error of "r" is given by:

$$\sigma_r = \frac{1 - r^2}{\sqrt{N - 1}} \quad (6)$$

From the standard error, r/σ_r is computed to determine significance. When r/σ_r is less than 2.0, the relation is probably due to chance rather than to correlation between the variables compared. Fisher (1934) states that, in the use of the above test, the value of r itself introduces an error which is magnified when r is squared. Only in the case of large samples (greater than 100 pairs of observations) can the standard error test be used safely. Further, the distribution of r, at least for the stronger values, is so skewed that it is unwise to make any interpretation of difference in terms of σ_r based on probabilities related to the normal curve.

(b) The t-test for Significance

For small samples, the distribution of r is not sufficiently close to normal to justify the ordinary standard error test. Fisher (1934) has developed the "t" test as a more accurate test for significance. This test measures the probability of obtaining a given value of r from a sample of paired values of a given size due to chance alone. A value of this probability of less than P = 0.05 indicates that the association of the characters is not due to chance, therefore being significant. The formula for "t" for a correlation coefficient is as follows:

$$t = \frac{r\sqrt{N-2}}{\sqrt{1 - r^2}} \quad (7)$$

In this formula N = the number of pairs of observations. The degrees of freedom for the estimation of a correlation coefficient are N - 2 due to the fact that two statistics are calculated from the sample.

The use of "t" may be illustrated with the correlation of ear length (x) and weight (y) in corn (Par. IV).

$$t = \frac{r\sqrt{N - 2}}{\sqrt{1 - r^2}} = \frac{0.937\sqrt{20-2}}{\sqrt{1 - (0.937)^2}} = 11.38$$

In the "t" table it is noted that for 18 degrees of freedom, the value of t required for P = 0.05 is 2.101. Thus, the above value is judged to be highly significant. The same result can be obtained from Table VA in Fisher, (1934).

(c) Difference between Correlation Coefficients

A test for the significance of differences between correlation coefficients has been suggested by Fisher (1934) as follows:

$$z' = 1/2 \left[\log_e (1 + r) - \log_e (1-r)\right] \quad (8)$$

The standard error would be as follows:

$$\sigma_{z'} = 1/\sqrt{N - 3} \quad \text{- (9)}$$

The method may be illustrated from an example given by Goulden (1937) who studied the relation between the carotene content of wheat flour and the color of bread for 139 wheat varieties. The correlation coefficients were as follows:

Carotene in whole wheat with crumb color, $r_1 = -0.4951$

Carotene in flour with crumb color, $r_2 = -0.5791$.

The z' test would be applied as follows:

$$z'_1 = 1/2 \left[\log_e (1 + 0.4951) - \log_e (1 - 0.4951)\right]$$

$$= 1/2 \left[\log_e 1.4951 - \log_e 0.5049\right]$$

$$= 1/2 \log_e \left[\frac{1.4951}{0.5049}\right] = 1/2 \log_e 2.9612 = 0.5428$$

$$z'_2 = 1/2 \left[\log_e (1 + 0.5791) - \log_e (1 - 0.5791)\right]$$

$$= 1/2 \log_e \left[\frac{1.5791}{0.4209}\right] = 1/2 \log_e 3.7517 = 0.6612$$

$$z'_2 = z'_1 = 0.6612 - 0.5428 = 0.1184$$

$$\sigma_{z'2 - z'1} = \sqrt{\frac{1}{136} + \frac{1}{136}} = 0.1213$$

$$dz'/\sigma_{z'} = 0.1184/0.1213 = 0.9761$$

Since the difference is less than its standard error, it is not significant.

The formula for z' deals only with the numerical value of r, no attention being paid to algebraic signs. It may be noted that the z test for significance of r is superior to the devices heretofore described.

VIII. Interpretation of the Correlation Coefficient

Certain precautions are necessary in correlation analysis. First of all, the characters of the individuals under consideration must be paired for some logical reason. The sample should also be representative of the population. Ordinarily it is inadvisable to calculate correlations on numbers where N is less than 30. Caution should be used in the application of correlation statistics where N is less than 50.

Spurious correlation is a condition where the things compared are not causally related, but which are related to a third cause. Frequently there is a tendency to assume that a significant correlation coefficient is proof of a causal relation between two variables. This may not be true. Extreme caution should be used in inferring cause from a correlation coefficient.

B -- Linear Regression

IX. Theory of Regression

A regression is said to be linear when the means of the sets of values of one character which correspond to given values of the other character can be well-fitted graphically by a straight line. Under such conditions the coefficient of correlation (r) is a valid measure of association. From the definition, it is evident that there must be two regression lines. They are termed the lines of regression of x on y, and y on x.

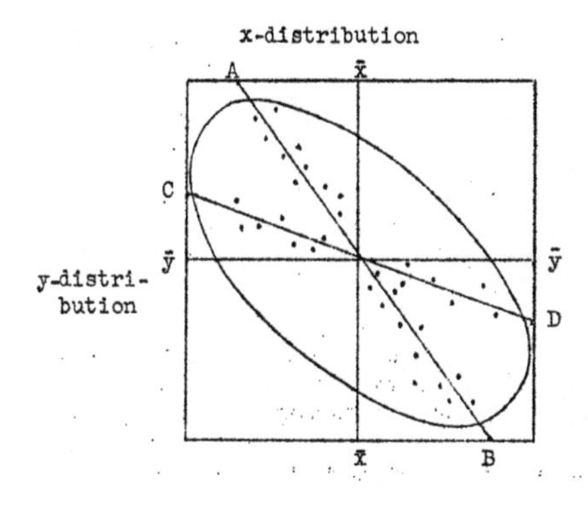

This diagram should give a clear conception of what is meant by regression. The elliptical nature of the scatter is shown with the dots which indicate the means of the individuals in each array, both horizontal and vertical. The means of all the rows fall approximately in a straight line, as well as those for columns. These lines, called the regression lines, intersect at a point which indicates the means of the two general distributions, x and y. The mathematical equations of these lines can be obtained by the method of least squares. The line AB is the regression of x on y. Its equation is as follows:

$$x - \bar{x} = r \frac{s'_x}{s'_y} (y - \bar{y}) \qquad (9)$$

Where x is the value estimated, say x_e.
The line CD is the regression line of y on x. Its equation is as follows:

$$y - \bar{y} = r \frac{s'_y}{s'_x} (x - \bar{x}) \qquad (10)$$

Where y is the value estimated, say y_e.

These equations may be used to predict or estimate the most probable value of one character to accompany or be associated with a given value of the other character. When a certain value is given y in the equation $x_e - \bar{x} = s'_x/s'_y (y - \bar{y})$, one can solve for the predicted value of x that corresponds to it. Likewise when a value is given to x, in $y_e - \bar{y} = s'_y/s'_x (x - \bar{x})$, the most likely value for the y that accompanies it can be found. Predicted values given by the regression equations are conservative. Actually, the term "regression" is a result of this tendency. These equations have little or no value for prognosis unless r is quite strong. The two diagrams below depict predicted values given by regression equations in two cases, i.e., where the correlation is strong and where it is weak.

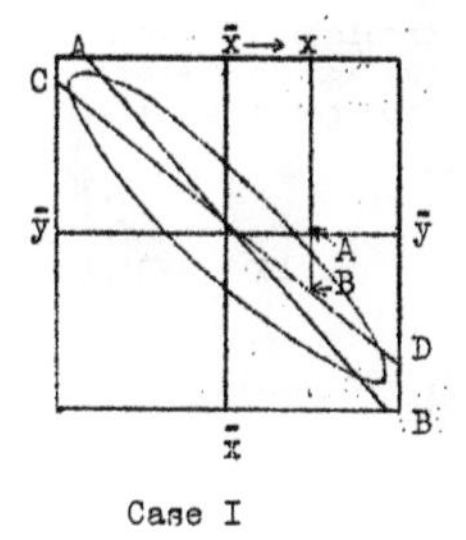

Case I

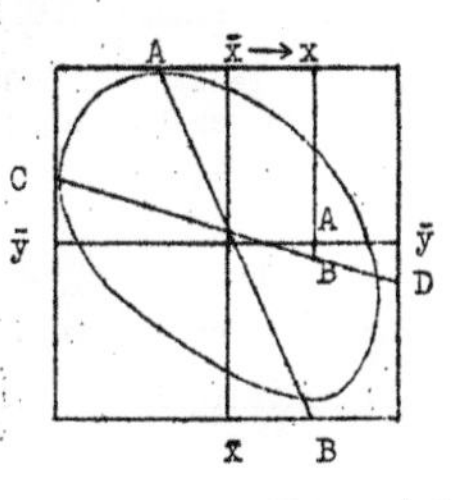

Case II

In each case, observe the same given value of x indicated by the point x, on the upper line of each diagram. The vertical line from the point, x, to the line CD measures the predicted value of y. The portion (A B) illustrates the amount of abnormality predicted. This is seen to be much smaller in case II where the correlation is weak. Moreover, the standard error of an estimated value is so large that, unless "r" is high, the reliability of an estimated value is small. Although a single predicted value is of little avail unless a very high degree of association exists between two characters, the regression measured by the coefficients $r\ s_y / c_x$ and $r\ s_y / s_x$ may be quite appreciable in one case or the other, even when r is small. This is due to the fact that variation in one character may be quite low. For instance, the association between the yield of a crop obtained from several plots and a certain treatment given in various degrees of intensity to the plots may be quite low. The first reaction would be that the treatment is not justified. However, the regression might be appreciable, so that the treatment might be very worth while for the crop as a whole.

The more important interpretation of a predicted value from a regression equation lies in the fact that it may be considered as a mean estimated value of the variable which may be expected to result in connection with a number of identical values of the second variable. Such an estimated mean would have a standard error = $\frac{s_e}{\sqrt{m - 1}}$ where m = number of repeated cases of the second variable, and s_e is the standard error of regression (See Section X (c) below).

X. Computation of Regression Equations for Grouped Data

The equation for the regression coefficient is as follows:

$$b_{yx} = \frac{SY(X - \bar{X})}{S(X - \bar{X})^2} \qquad (11)$$

The most convenient formulae for machine computation are as follows:

$$b_{yx} = \frac{S(XY) - (SX)(SY)/N}{S(X^2) - (SX)^2/N} \qquad (12)$$

or

$$b_{yx} = \frac{NS(XY) - (SX)(SY)}{NS(X^2) - (SX)^2} \qquad (13)$$

Where the transformed variables (X,Y) are not used, the same relations hold in terms of the original variables (x, y).

(a) Computation of Regression Coefficients

The computation may be illustrated for the correlation between total grain weight and average lengths of culms in wheat plants (Paragraph V above). Calculations from Table 2, which can be used here, are as follows for coded data:

SXY = 21,409	SX = 2736	SY = 3772
	$\bar{X}$ = 5.5385	$\bar{Y}$ = 7.6356
N = 494	SX^2 = 16,930	SY^2 = 31,108

By substitution in formula 13:

$$b_{yx} = \frac{NS(XY) - (SX)(SY)}{NS(X^2) - (SX)^2} = \frac{(494)(21409) - (2736)(3772)}{(494)(16,930) - (2736)^2}$$

$$= \frac{10,576,046 - 10,320,192}{8,363,420 - 7,485,696} = \frac{255,854}{877,724} = 0.2915$$

$$b_{xy} = \frac{NS(XY) - (SX)(SY)}{NS\ (Y^2) - (SY)^2} = \frac{(494)(21409) - (2736)(3772)}{(494)(31108) - (3772)^2}$$

$$= \frac{255,854}{15,367,352 - 14,227,984} = \frac{255,854}{1,139,368} = 0.2246$$

(b) Substitution in Regression Equation

The equation for the regression of Y on X is as follows:

$$Y_e = \bar{Y} - b_{yx}\ \bar{X} + b_{yx}\ X$$

$$= 7.6356 - (0.2915)\ (5.5385) + (0.2915)\ X$$

$$= 6.0211 - 0.2915X$$

The equation for the regression of X on Y is calculated in a similar manner.

$$X_e = \bar{X} - b_{xy}\ \bar{Y} + b_{xy}\ Y$$

$$= 5.5385 - (0.2246)(7.6356) + (0.2246)\ Y$$

$$= 3.8235 + 0.2246\ Y$$

(c) Significance of Regression Coefficients

The "t" test for significance of the regression coefficient, $b_{yx} = 0.2915$ (coded basis) can be determined as follows:

$$S\ (Y - Y_e)^2 = S\ (Y - \bar{Y})^2 - b_{yx}^2\ S\ (X - \bar{X})^2$$

$$= SY^2 - N\bar{Y}^2 - b_{yx}^2\ (SX^2 - N\bar{X}^2)$$

$$= 31,108 - (494)(7.6356)^2 - (0.2915)^2 \left[16,930 - (494)(5.5385)^2\right]$$

$$= 31,108 - 28,801.3856 - 0.0850\ (16,930 - 15,153.4500)$$

$$= 2,306.6144 - (0.0850)(1,776.55) = 2155.6076$$

$$s_e = \sqrt{\frac{S(Y - \bar{Y})^2 - b_{yx}^2\ S(X - \bar{X})^2}{N - 2}} = \sqrt{\frac{2155.6076}{492}} = 2.0932$$

$$t = \frac{b_{yx}\sqrt{S\ (X - \bar{X})^2}}{s_e} = \frac{0.2915\ \sqrt{1776.55}}{2.0932} = 5.87$$

This indicates that the regression coefficient is highly significant. The coefficient, $b_{xy} = 0.2246$, can be tested in a similar manner.

[1]The coded values are changed into actual values by the conversion of $\bar{Y}$ to $\bar{y}$, $\bar{X}$ to $\bar{x}$, and the multiplication of b_{yx} by C_y/C_x as follows:

$$\bar{y} = (\bar{Y} - 1)\ C_y + y_1 = (7.6356 - 1)(5) + 62 = 95.1780$$

$$\bar{x} = (\bar{X} - 1)\ C_x + x_1 = (5.5385 - 1)(20) + 9.5 = 100.2700$$

$$b_{yx} = (0.2915)(5/20) = 0.0729$$

$$y_e = \bar{y} - b_{yx}\ \bar{x} + b_{yx}x = 95.1780 - (0.0729)(100.27) + (0.0729)x$$

$$= 87.8683 + 0.0729x.$$

References

1. Elderton, W. P. Frequency Curves and Correlation. Leyton, 1906.
2. Ezekiel, M. Methods of Correlation Analysis. John Wiley & Sons. 1930.
3. Fisher, R. A. Statistical Methods for Research Workers (5th edition) Oliver and Boyd. pp. 160-197. 1934.
4. Gouldon, C. H. Methods of Statistical Analysis. Wiley. pp. 52-77. 1939.
5. Hayes, H. K., and Garber, R. J. Breeding Crop Plants. McGraw-Hill. pp. 43-55. 1927.
6. Snedecor, G. W. Statistical Methods. Collegiate Press. pp. 89-133. 1937.
7. Tippett, L. H. C. The Methods of Statistics (2nd edition). Williams and Norgate. pp. 140-188. 1937.
8. Treloar, A. E. Outlines of Biometric Analysis. Burgess, pp. 40-63. 1935.
9. Wallace, H. A., and Snedecor, G. W. Correlation and Machine Calculation. Collegiate Press (Ames). 1931.

Questions for Discussion

1. Define correlation.
2. What is a scatter diagram? How is it influenced by high correlation? Low correlation?
3. When are two variables said to be correlated? Not correlated?
4. What is the generally accepted method for the measurement of correlation? Its limitations?
5. What is meant by r = + 1, r = - 1, and r = 0?
6. How can the standard error of the difference be reduced by the use of correlation?
7. Why is the "t" test preferable to the standard error test for testing the significance of r?
8. What precautions must be exercised in the interpretation of the correlation coefficient? Why?
9. Under what conditions is "r" a valid measure of paired relationships?
10. What is regression? Its use?
11. Explain what is meant by the regression of y on x. Regression of x on y.

Problems

1. These data were collected to study the relationship between the soil moisture content and the yield of wheat (Data from Salmon):

Moisture(x)	Yield(y)	Moisture(x)	Yield(y)	Moisture(x)	Yield(y)	Moisture(x)	Yield(y)
21	1	25	35	18	10	18	3
17	1	23	24	21	28	15	4
17	3	26	39	21	28	18	8
18	8	18	0	22	25	17	12
21	21	18	0	23	29	17	13
20	24	18	0	17	0	17	16
19	20	18	0	16	7	16	15
19	7	18	3	15	6	19	11
17	19	19	3	19	9	19	10
16	21	15	11	18	23	18	4
16	21	15	9	18	23	21	4
16	20	15	9	18	27	18	36
23	32	15	9	18	28	26	47
24	37	18	13	19	5		

(a) Calculate the coefficient of correlation (r) by the machine method for ungrouped data. (b) Test the significance of "r" by the "t" test.

2. The average length of culms and the average diameter of culms was measured on 496 wheat plants at the Colorado Experiment Station. The data follows:

Average length of culms (cm.)

	y \ x	60 64	65 69	70 74	75 79	80 84	85 89	90 94	95 99	100 104	105 109	110 114	115 119
	91-100		1		1								
	101-110		1	2									
Av.	111-120					1	1		1				
Diameter	121-130	1	4	2	1	2	2	1					
culms	131-140			1	6	3	3	2					
(mm.)	141-150	2		3	1	6	4	2		1			
	151-160			2	5	5	16	27	8	7	1		
	161-170			1	3	8	13	23	25	26	6	2	1
	171-180			3	1	9	10	15	26	13	8	8	
	181-190		1				9	10	10	23	32	4	3
	191-200						2	2	8	13	20	12	1
	201-210						2	2	3	5	4		1
	211-220								2				

Calculate r and test it for significance with "t" test.

3. The correlation between the reaction to Helminthosporium in F_3 and F_5 barley lines was studied at the Minnesota Experiment Station. The reactions are given in percentage infection for 1921 and 1922. The data follow:

Percentage in 1921

Percentage in 1922		9	12	15	18	21
	12	2				
	15	2	2	1		
	18	1	3	2	3	1
	21	1	2	3	2	1
	24	1	1	4	2	
	27	1		1		

Calculate the coefficient of correlation and the regression lines. Plot the regression lines.

4. The 9-year average yields of wheat for the period 1921-29 were as follows when irrigated at the germination and filling stages. Five plots wore averaged each year. Tho data follow in grams per plot:

Year Grown	Germination (gm.) ($\sigma_{\bar{x}}$)	Filling (gm.) ($\sigma_{\bar{x}}$)
1921	511 ± 37	321 ± 23
1922	655 ± 21	518 ± 17
1923	914 ± 32	733 ± 26
1924	952 ± 28	1125 ± 34
1925	470 ± 16	538 ± 18
1926	557 ± 29	601 ± 31
1927	783 ± 20	812 ± 21
1928	566 ± 20	5[illegible] ± 18
1929	758 ± 65	733 ± 63
Means	685 ± 11	657 ± 11

Determine whether or not the wheat irrigated at germination differs significantly in yield from that irrigated at jointing and tillering. Calculate σ_d of an average of a difference by the formula $\sigma_d = \sqrt{a^2 + b^2}$, and by the extended formula for use of r.

Compare d/σ_d for both formulas.

CHAPTER X

THE ANALYSIS OF VARIANCE

I. Generalized Standard Error Methods

The basis and purpose of all statistical methods is to analyze and measure variability. Variation between observations may be due to one or more recognizable causitive factors. In addition, in all statistical work, there occur variations between observations that result from the coalition of a large aggregate of chance factors which defy control. This latter type of variation between observations results in various types of statistical distributions when it is attempted to describe homogeneous populations.

Suppose one considers a population wherein the variability may be due to both the combinations of innumerable chance factors and the non-homogeneity of the population. In other words, the population naturally and logically submits to sub-division into several homogeneous groups or sub-populations. Such a situation is common in variety tests in field experimentation. Generalized standard error methods have been devised for data of this kind.

The purpose of the generalized standard error methods is to compute the standard error of an entire experiment in order to increase the accuracy of the estimate of error. In a variety test where each variety or treatment is replicated, say four times, the reliability of the results would be very low were one to compute the standard error for each variety separately. However, the estimate of error would be much more reliable when computed on 10 different varieties, each replicated say four times, in the same experiment. In this case a total of 40 plots would contribute to the estimate of error instead of four.

The analysis of variance, developed by R. A. Fisher, has proved to be the most precise, flexible, and readily usable method available for the analysis of the results from field and many other biological experiments. It consists essentially in the partition and apportionment of the total variation to its known causes with a residual portion ascribable to unknown or uncontrolled variation and therefore called experimental error. When the variability is measured in suitable terms, i.e., sums of squares of deviations about the means, the variability ascribed to the various causes will be strictly additive. The calculations are therefore extremely simple. The mean value of the sums of squares (mean square, variance, or standard error squared) is found by division of the sums of squares by the appropriate number of degrees of freedom.

The literature on the analysis of variance has become very extensive during the past 15 years. It was first set forth in its complete form by Fisher and MacKenzie in 1923. For its application to field experiments, Fisher (1934), and Fisher and Wishart (1930) have given excellent discussions. Among other sources of information on the application of the analysis of variance to field experiments may be mentioned the books by Snedecor (1934 and 1937), and Tippett (1937), and papers by Eden and Fisher (1929), Goulden (1931), Immer, et al (1934), and Wishart, (1931). For a summary of the mathematical theorems involved in the analysis of variance, the work of Irwin (1931) is recommended.

A -- One Criterion of Classification

II. Theory of First Special Case

Suppose a sample is formed from the general population with random samples of equal size taken from each of the sub-populations. In case m sub-samples contain n measurements each, the total sample will contain $N = nm$ measurements. It is now proposed to analyze the total variance, i.e.,

$$s^2 = \frac{S(x - \bar{x})^2}{N - 1} \quad \text{- -} \quad (1)$$

where $\bar{x}$ is the mean of the total sample. Let x represent an individual measure of any (i-th) sub-sample. Then,

$$x - \bar{x} \equiv (x - \bar{x}_i) + (\bar{x}_i - \bar{x}) \quad \text{- -} \quad (2)$$

where $\bar{x}_i$ is the mean of the i-th sample.

First, the above identity should be squared and summed for all the n individuals which form the i-th sub-sample. The symbol S' will be used for this summation.

$$S'(x - \bar{x})^2 \equiv S'(x - \bar{x}_i)^2 + 2(\bar{x}_i - \bar{x})\, S'(x - \bar{x}_i) + n(\bar{x}_i - \bar{x})^2 \quad \text{- - - -} \quad (3)$$

Since $S'(x - \bar{x}_i) = 0$, it is evident that the second term on the right vanishes.

Now suppose one sums over all the m different sub-groups by use of the symbol, $\overset{m}{\underset{1}{S}}$. The combination, $\overset{m}{\underset{1}{S}}S'$ is simply S, or summation for all individuals of the total sample.

$$S(x - \bar{x})^2 = \overset{m}{\underset{1}{S}}S'(x - \bar{x})^2 = \overset{m}{\underset{1}{S}}S'(x - \bar{x}_i)^2 + n\overset{m}{\underset{1}{S}}(\bar{x}_i - \bar{x})^2 \quad \text{- - - - -} \quad (4)$$

The term on the left is the sum of the squares of the deviations of the individual observations from the means of the sub-samples. The second term on the right is n times the sum of squares of the deviations of the means of the sub-samples from the mean of the total sample.

The computation of these three terms is most easily accomplished as follows:

(1) Compute the term on the left,

$$S(x - \bar{x})^2 = Sx^2 - \frac{(Sx)^2}{N} \quad \text{- -} \quad (5)$$

(2) Next, the second term on the right,

$$n\overset{m}{\underset{1}{S}}(\bar{x}_i - \bar{x})^2 = n\overset{m}{\underset{1}{S}}\bar{x}_i^2 - \frac{(Sx)^2}{N} = \frac{S(x_a^2)}{n} - \frac{(Sx)^2}{N} \quad \text{- - - - - - - - - - -} \quad (6)$$

where x_a is an abbreviation for S'(x), the total of the variates in a single sub-sample.

(3) The other term may be found by mere subtraction.

The difficult thing to explain comes at this point. It would be easy to merely apportion, for the sample in question, the total variance into the variance within sub-samples and into that between sub-samples. These two respective variances could be obtained by division of the first and second terms on the right of the identity by N. However, the real desire is to obtain the best estimate to the variance of the population as exhibited within the sub-populations on the one hand, and between the sub-populations on the other.

The best estimate of the total variance of the general population is given by:

$$s^2 \Downarrow = \frac{S(x - \bar{x})^2}{N - 1} \quad \text{- -} \quad (7)$$

Where N - 1 (or nm - 1) is the number of degrees of freedom.

Likewise, the best estimate of the variance within sub-populations (replicates agronomically) will be:

$$s_r^2 \Downarrow = \overset{m}{\underset{1}{S}}S'\frac{(x - \bar{x}_i)^2}{N - m} \quad \text{- -} \quad (8)$$

$\Downarrow$ "is estimated by" in this sense.

Where $N - m = m(n - 1)$ is the number of degrees of freedom. This is true because $\underline{m}$ separate means of the m sub-samples were used in the computation.

This expression, s^2_r = the variance within sub-samples, is often called the residual variance.

The last term, $n \overset{m}{\underset{1}{S}} (\bar{x}_i - \bar{x})^2$, (Equation No. 4 above), must now be considered.

At first glance, it would seem that the sum of squares of the deviations of the means of the sub-samples from the mean of the total sample when divided by $m - 1$, the number of degrees of freedom, would give a proper estimate of the variance between sub-populations. However, $\bar{x}_i$ does not represent the mean of the i-th sub-population, but rather the mean of the i-th sub-sample. Therefore, the difference, $\bar{x}_i - \bar{x}$, is due to the combination of (1) the inherent nature of the i-th sub-population and (2) sampling fluctuations within the i-th sub-sample. Thus, $n \frac{\overset{m}{\underset{1}{S}}(\bar{x}_i - \bar{x})^2}{m - 1}$ can be interpreted to estimate the sum of ns_t^2, n times the variance of the means of the sub-populations, and s_r^2, the variance within sub-populations.

It should be remarked that the expression,

$$s_r^2 = \frac{\overset{m}{\underset{1}{S}}S' (x - \bar{x}_i)^2}{N - m},$$ called the "variance within sub-samples", is simply one estimate of the variance of the total population.

The term, $n \frac{\overset{m}{\underset{1}{S}} (\bar{x}_i - \bar{x})^2}{m - 1} = n\, s_t^2 + s_r^2$, called the "variance between sub-samples", is n times the variance of the sub-sample (treatment) means about the total sample mean with s_r^2 added. On the assumption (null hypothesis) that the true variance of the sub-population means about the total population mean is zero, it then becomes clear that "variance within sub-samples" and "variance between sub-samples" are both independent estimates of the same concept, i.e., variance of the total population.

The material may be placed in tabular form for clarity:

Source of Variation	Sums of Squares	Degrees of Freedom	Estimated Mean Variances
Between Sub-samples	$n\overset{m}{\underset{1}{S}}(\bar{x}_i - \bar{x})^2$	$m - 1$	$n\, s_t^2 + s_r^2 = V_t^2$
Within sub-samples	$\overset{m}{\underset{1}{S}}S'(x - \bar{x}_i)^2$	$N - m$	$s_r^2 = V_r^2$
Total	$S(x - \bar{x})^2$	$N - 1$	$s^2 = V^2$

The first two entries in the last column may be examined, i.e., the estimates given by the sample. These are $ns_t^2 + s_r^2$. It is obvious that the first should exceed the second unless s_t^2 is zero.[2]

It is now desired to determine whether or not there is a significant variation between the sub-population, i.e., whether σ_t^2 is significantly different from zero. An estimate of σ_t^2 may be made by subtraction of the estimate of σ_r^2 from that of $n\sigma_t^2$. This result is then divided by n.

---

[2] Occasionally the reverse is true. This apparent contradiction is explained by the fact that the results are merely estimates which may be distorted to whatever extent sampling fluctuations may account.

So far it is obvious that the first step in the analysis of variance serves two purposes: (1) It gives a method to test the homogeneity of a population; (2) It gives a convenient method to test the differences between several means as a whole.

Probably the best method to test for association between sub-populations is through the use of the "z" index devised by Fisher (1934). This affords a test of significance between two variances, e. g., s_1^2 and s_2^2.

$$z = \frac{1}{2} \log_e s_1^2 - \frac{1}{2} \log_e s_2^2 = \frac{1}{2} \log_e \frac{s_1^2}{s_2^2} \qquad (9)$$

The "z" table devised by Fisher (1934) may be used to test these values for significance through use of the number of degrees of freedom pertinent to each computed variance. In this case, $n\sigma_t^2 + \sigma_r^2$ takes the place of s_1^2, while σ_r^2 takes the place of s_2^2.

Tests of significance may also be made by means of the "F" test derived by Mahalonobis (1932) and by Snedecor (1934)(1937). The table by Snedecor is the more extensive The value "F" is the quotient obtained by division of the larger by the smaller variance. The "F" and "z" tests are equivalent since $z = \frac{1}{2} \log_e F$.

III. Computation for Single Criterion of Classification

This case may be illustrated with some data for the yields of two barley varieties, (See Chapter 6). The yields in bushels per acre for the Glabron and Velvet varieties grown in single plots on 12 Minnesota farms were as follows: (Data from F. R. Immer)

Farm No.	Glabron (x_1)	Velvet(x_2)	Total
1	49	42	91
2	47	47	94
3	39	38	77
4	37	32	69
5	46	41	87
6	52	41	93
7	51	45	96
8	57	56	113
9	45	42	87
10	45	39	84
11	48	47	95
12	64	39	103
Totals (Sx)	580	509	1089
Means ($\bar{x}$)	48.3333	42.4167	45.3750

$S(x^2) = 30{,}599.00 \qquad \frac{(Sx)^2}{N} = 49{,}413.38$

Suppose that the total variability is separated into two components, viz., that "due to varieties" and that due to variation between plots of the same variety. The expression "due to varieties" simply means that it is proposed to make varieties the criterion for the break-down of the total sample into sub-samples.

The sum of squares for total variation is found by summation of the squares of the 24 plot yields, e.g. $(49)^2 + (47)^2 + \ldots\ldots + (39)^2 = 30{,}599$, and the subtraction of the correction factor $(Sx)^2/N$ from this value.

Algebraically, this is given by:

$$\text{Total} = S^{1} (x - \bar{x})^2 = S(x^2) - (Sx)^2/N = 50{,}599.00 - 49{,}413.38 = 1185.62$$

The sum of squares for varieties is obtained by the summation of the squares for the two totals for varieties, dividing by number of plots or values contained in each variety total, and subtracting the correction factor, e.g.

$$\text{Between Varieties} = n\overset{m}{\underset{1}{S}} (\bar{x}_i - \bar{x})^2 = \frac{Sx_v^2}{n} - \frac{(Sx)^2}{N}$$

$$= \frac{595{,}481.00}{12} - 49{,}413.38$$

$$= 49{,}623.42 - 49{,}413.38 = 210.04$$

The sum of squares for within varieties, here used as error, is the remainder after subtraction of the sums of squares for varieties from the total, e.g. 1185.62 - 210.04 = 975.58.

The analysis of variance follows:

Variation due to	D.F.	Sum Squares	Mean Square	Standard Error(s)	z-value obtained	5 pct. point	F
Between Varieties	1	210.04	210.0400		0.7777	0.7294	4.737
Within Varieties	22	975.58	44.3446	6.6592			
Total	23	1185.62					

The degrees of freedom for "between varieties" and "total" are one less than the number of varieties and total number of plots, respectively. The degrees of freedom for within varieties are those for a single variety (11 in this case) multiplied by the number of varieties (2 in this case), or (2)(11) = 22.

The mean squares are obtained by division of the sums of squares by their respective degrees of freedom. The standard error of a single determination (s) is the square root of the mean square for error (or variance).

The z-test may be used to test the significance for variance "between varieties" and that "within varieties". The value, z, is 1/2 $\log_e$ of the difference of the variances to be compared. The values of the logarithms needed in computing z are found in a table of natural logarithms (See Table 4, appendix).

$$z^{2} = 1/2 \log_e 210.04 - 1/2 \log_e 44.3446$$

$$= 1/2 \log_e \left(\frac{210.0400}{44.3446}\right) = 1/2 \log_e 4.737 = 0.7777$$

---

[1] $S(x - \bar{x})^2 = Sx^2 - 2\bar{x}\,S(x) + N\bar{x}^2$

$= Sx^2 - 2\bar{x}\cdot N\bar{x} + N\bar{x}^2$

$= Sx^2 - N\bar{x}^2$

Since $N\bar{x} = S(x)$

$Sx^2 - N\bar{x}^2 = Sx^2 - S(x)\bar{x} = Sx^2 - (Sx)^2/N$

[2] The decimal point may be moved to the left on the mean square values to shorten the work, so long as the resultant numbers are greater than 1.0. The true loge values will not be obtained but the difference of z-value is unaffected. A shift of decimals is particularly desirable, when any of the mean squares are less than 1.0 to avoid taking a negative loge.

The theoretical z-value is looked up in the table given by Fisher (1934), where N_1 is the number of degrees of freedom for the larger and N_2 the degrees of freedom for the smaller variance. In this case z = 0.7294 for the theoretical value for the 5 per cent point. However, the interpretation is made more easily by using the "F" value. The "F" value is the quotient of the larger by the smaller variance, e.g., F = 210.04/44.3446 = 4.74. In Snedecor's table (Table 2, Appendix) for N_1 = 1 D.F. and N_2 = 22 D.F., it is found that the observed "F" lies between the 5.0 per cent and 1.0 per cent points. The theoretical value for the 5.0 per cent point is 4.30. It may be noted that $F = t^2$ for one degree of freedom.

IV. The More General Case

Suppose that the number of observations in each sub-sample varies, and that they are represented by $n_1, n_2 \ldots\ldots n_m$. Then $N = \overset{m}{\underset{1}{S}} n_i$. The equation for the sample is as follows:

$$S(x - \bar{x})^2 = \overset{m}{\underset{1}{S}}S'(x - \bar{x})^2 = \overset{m}{\underset{1}{S}}S'(x - \bar{x}_i)^2 + \overset{m}{\underset{1}{S}}n_i(\bar{x}_i - \bar{x})^2 \qquad (10)$$

Again, $\overset{m}{\underset{1}{S}}S'(x - \bar{x}_i)^2$ divided by N - m, the degrees of freedom, will give the variance within a group. However, it is now impossible to arrive at an estimate of s_t^2 because $\overset{m}{\underset{1}{S}}n_i(\bar{x}_i - \bar{x})^2$ is affected by the different number of observations in each sub-sample. Therefore suppose that s_t^2 is truly zero so that $\overset{m}{\underset{1}{S}} \frac{n_i(\bar{x}_i - \bar{x})^2}{m - 1}$ will estimate s_r^2.

This assumption may be tested for the existence of an association between sub-populations. To do this, the values $\overset{m}{\underset{1}{S}}S' \frac{(x - \bar{x}_i)^2}{N - m}$ and $\overset{m}{\underset{1}{S}} \frac{n_i(\bar{x}_i - \bar{x})^2}{m - 1}$ are compared for a significant difference.

In the field of agronomic experimentation this situation is rarely found because the experiments are designed to permit a simpler set-up for the computation of the statistical constants.

B -- Two or More Criteria of Classification

V. Theory of the Extended Case of the Analysis of Variance.

Frequently, the complexity of the experiment that affords the data makes it necessary to analyze the total variance into more than two parts in order to make the most of the possibilities. First, re-examine the tabular arrangement for the first special case, (Paragraph II). Suppose the classification of the total population into sub-populations, which forms the basis of the above analysis, be termed classification "A". Now suppose the total population lends itself to an independent classification, "B", which contains "m'" classes. For simplicity, assume that the sample sub-divides evenly for this classification with "n'" observations in each class. Thus, N = nm = n'm'.

Previously, the heterogeneity in the total population for classification "A" was tested by a comparison of V_t^2 with V_r^2. It was necessary to tacitly assume that each sub-population for classification "A" was homogenous. Now, if the population submits to a new classification "B", it is quite likely that the original sub-populations were not homogenous if classification "B" has any logical basis. Lack of homogeneity in the sub-populations increases the variance therein. The residual variance, V_R^2, may be so affected in the comparison between V_R^2 and V_t^2 that the differences between groups for classification "A" may appear to be insignificant when the opposite is true.

Therefore, it is proposed to remove from the squared residual errors, $\overset{m}{S}S'(x - \bar{x}_i)^2$, the sum of the squared errors between groups for classification "B". [1]This amount will be termed $n' \sum_{j=1}^{m'} (\bar{x}_j - \bar{x})^2$. It represents m' - 1 degrees of freedom, while $\bar{x}_j$ indicates the mean of the j-th class of classification "B". The mean variance between groups for classification "B" will be designated as $V_{t'}^2$.

At first, one might expect the mean residual variance (V_R^2) to be definitely reduced in this manner, regardless of any justification for classification "B". This is not true because the reduced sum of the residual squared errors now represents only N-m-m'+ 1 degrees of freedom where N - m degrees of freedom were represented before. Thus, it is apparent that the new V_R^2 will not differ sensibly from its former value, should the differences between groups be insignificant for classification "B". However, the greater the significance of the differences between the groups for classification "B", the more markedly V_R^2 will be reduced. Then the ratio V_t/V_R will be sensibly increased, with the result that the test for significance of differences between groups for classification "A" is strengthened. The new tabular arrangement of the analysis is as follows:

Source of Variation	Sum of Squares	Degrees of Freedom	Mean Variance
Between groups (A)	$n \overset{m}{\underset{1}{S}} (\bar{x}_i - \bar{x})^2$	m - 1	V_t^2
Between groups (B)	$n' \overset{m'}{\underset{1}{S}} (\bar{x}_j - \bar{x})^2$	m' - 1	V_t^2
Residual	- - - - - -	N - m - m' + 1	V_R^2
Total	$S(x - \bar{x})^2$	N - 1	V^2(or s^2)

The entry for the sum of squares of the residual errors is left blank because, in computation, it would be found by subtraction.

This process may be extended in the same manner to take into account other possible classifications which might contribute to the heterogenous character of the original population. The object is for the residual variance to represent variance due to chance alone as nearly as possible. Furthermore, an increase in the scope of an experiment will proportionately increase, to within differences due to sampling fluctuations, all the sums of squared deviations incorporated into the analysis. Hence, V_t^2 and $V_{t'}^2$ will be increased proportionately since the number of degrees of freedom they respectively represent are unchanged. The value V_R^2 will be increased to a lesser extent due to the fact that the number of degrees of freedom represented will be more than proportionately increased. Thus, V_t/V_R and $V_{t'}/V_R$ will be increased which, together with the fact that a smaller z value is required to prove significance, make it more likely that positive conclusions can be drawn from the analysis of variance.

VI. Computation for Two Criteria of Classification

The same data on the yields of Glabron and Velvet barley varieties are used to illustrate this case. It is desired to determine whether or not there is a significant variation from farm to farm as well as between varieties. Hence, the computations will be for total variance, that due to farms, and that due to varieties. The residual variance will be obtained by subtraction.

$$S(x - \bar{x})^2 = S(x^2) - (Sx)^2/N = 50,599.00 - 49,413.38 = 1185.62$$

$$n \overset{m}{\underset{1}{S}}(\bar{x}_i - \bar{x})^2 = S(x_v^2) - \frac{(Sx)^2}{n} = \frac{595,481.00}{12} - 49,413.38 = 210.04$$

$$n' \overset{m'}{\underset{1}{S}}(\bar{x}_j - \bar{x})^2 = S(x_f^2) - \frac{(Sx)^2}{N} = \frac{100,269.00}{2} - 49,413.38 = 721.12$$

The subscripts, v and f evidently indicate "varieties" and "farms".

The new tabular arrangement now becomes:

Variation due to	D.F.	Sums Squares	Mean Square	Standard Error (s)	F-value
Farms	11	721.12			
Varieties	1	210.04	210.0400		9.08
Error	11	254.46	23.1327	4.8096	
Total	23	1185.62			

When the F-table is consulted it is found that an F-value of 4.84 is required for the 5 per cent point. Thus, the added refinement through the removal of the variation between farms greatly increased the significance of the difference due to varieties.

VII. Introduction to Analysis of Variance in Agricultural Experiments

The principal difficulty to contend with in field experiments is the variation in soil fertility over the area used in experimentation. The natural fertility usually varies continuously. The art of planning an experiment lies in the arrangement of the varieties, treatments or conditions under investigation in nearby plots. They are usually placed within as small a land area as is practically feasible. The entire arrangement is then replicated over a larger area so that the variations caused by regional changes in fertility may be removed from the comparison. The randomized block arrangement and its more restricted form, the latin square arrangement, are commonly used to make possible the removal of the general effect of soil heterogeneity by means of the analysis of variance. (See Chapter on "Design of Simple Field Experiments".)

In the use of the analysis of variance in field experiments it is assumed that the distribution of the plot yields is normal, i.e., that it fits the normal curve. The agronomist is familiar with the fact that the variability between plots of the same variety grown on land of high fertility is often less than between similar plots of low fertility. The variability among plots of high fertility may be considered as restricted by what may be termed "ceiling effect" which imparts an abnormal distribution to the population. Fisher and others (1932) found evidence of negative skewness in heights of barley plants selected at random from plots that received various nitrogen treatments. Eden and Yates (1933) obtained similar results with height measurements of wheat plants. They made a practical test on these data to determine whether the validity of the z-test would be destroyed by such non-normal data. They concluded that the z-test could be safely applied.

References

1. Eden, T. and Fisher, R. A. Studies in Crop Variation VI. Experiments on the Response of the potato to potash and nitrogen. Jour. Agr. Sci. 19:201-213. 1929.

2. Eden, T. and Yates, F. On the validity of Fisher's Z test when applied to an actual example of non-normal data. Jour. Agr. Sci. 23:6-16. 1933.
3. Fisher, R. A. Statistical methods for research workers. Oliver and Boyd, Edinburgh, Ed. 5. pp. 199-231. 1934.
4. Fisher, R. A., Immer, F. R., and Tedin, Olof. The genetical interpretation of statistics of the third degree in the study of quantitative inheritance. Genetics 17:107-124. 1932.
5. Fisher, R. A. and MacKenzie, W. A. Studies in crop variation. Jour. Agr. Sci. 13:311-320. 1923.
6. Fisher, R. A. and Wishart, J. The arrangement of field experiments and the statistical reduction of the results. Imperial Bureau of Soil Science. Tech. Comm. No. 10. 1930.
7. Goulden, C. H. Modern methods of field experimentation. Sci. Agr. 11:681-701. 1931.
8. Immer, F. R., Hayes, H. K., and Powers, LeRoy. Statistical determination of barley varietal adaptation. Jour. Am. Soc. Agron. 26:403-419. 1934.
9. Irwin, J. O. Mathematical theorems involved in the analysis of variance. Jour. Royal Stat. Soc. 94:284-300. 1931.
10. Mahalanobis, P. C. Auxilliary tables for Fisher's Z test in analysis of variance. Indian Jour. Agr. Sci. 2:679-693. 1932.
11. Snedecor, George W. Calculation and interpretation of analysis of variance and covariance. Collegiate Press, Inc., Ames, Iowa. 1934.
12. Snedecor, G. W. Statistical Methods. Collegiate Press, Ames. pp. 171-218. 1937.
13. Tippett, L. H. C. The Methods of Statistics. Williams and Norgate, London. (2nd edition). pp. 125-139. 1937.
14. Wishart, John. The analysis of variance illustrated in its application to a complex agricultural experiment on sugar beets. Wissenschaftliches Archiv für Landwirtschaft, 5:561-584. 1931.

Questions for Discussion

1. What is meant by generalized standard error methods? Why are they useful in agronomic experiments?
2. What are the general features of the analysis of variance?
3. What is the basis of sub-division of the sample for one criterion of classification?
4. Why is it logical to use the variance for within varieties to compare with that between varieties?
5. What is the "z" test for significance? "F" test?
6. How may the sub-division of the total sample into two criteria of classification strengthen the experiment?
7. What assumptions are made in the use of the analysis of variance for plot yield data?

Problems

1. Yield data in bushels per acre for 5 wheat varieties are given on the following page:

Variety	Replications 1	2	3	Total
A	32.4	34.3	37.3	104.0
B	20.2	27.5	25.9	73.6
C	29.2	27.8	30.2	87.2
D	12.8	12.3	14.8	39.9
E	21.7	24.5	23.4	69.6
Totals	116.3	126.4	131.6	374.3

(a) Calculate the analysis of variance for one criterion of classification, i.e., between and within varieties.

(b) Obtain the "F" value and determine whether or not the varieties differ significantly in yield.

(c) Use the "z" test to determine significance.

2. Calculate the data in problem 1 for 2 criteria of classification, i.e., replicates and varieties. Determine whether or not the varieties differ significantly in yield by use of the "F" test.

CHAPTER XI

COVARIANCE WITH SPECIAL REFERENCE TO REGRESSION

I. Relationship of Covariance, Correlation, and Regression

The concepts of covariance, correlation, and regression are interwoven, being fundamentally equivalent. Suppose one considers N pairs of measures that relate to two characters represented by the variables x and y. In the chapter on correlation, it was seen that the basis for the measurement of correlation and regression was the product sum, $S(x - \bar{x})(y - \bar{y})$. The entire subject can well be treated by the analysis of variance principle.

II. Analysis of Covariance

Suppose the sample of N pairs of measures is divided into m sub-samples that contain n pairs of variates each. Let $\bar{x}_i$ and $\bar{y}_i$ represent the pair of means that correspond to the i-th sub-sample. Then, for any pair of variates in the i-th sub-sample this equation can be formed:

$$(x - \bar{x})(y - \bar{y}) = \left[(x - \bar{x}_i) - (\bar{x}_i - \bar{x})\right] \left[(y - \bar{y}_i) + (\bar{y}_i - \bar{y})\right] \quad (1)$$

By an analogous procedure to the first treatment of the analysis of variance, the right side of this expression may be expanded and summed for all the pairs of variates in the i-th sub-sample, viz.,

$$S'(x - \bar{x})(y - \bar{y}) = S'(x - \bar{x}_i)(y - \bar{y}_i) + n(\bar{x}_i - \bar{x})(\bar{y}_i - \bar{y}).$$

It is noticed that the two middle terms of the expansion become zero for the summation. The summation is taken again to include all the sub-samples, viz.,

$$\overset{m}{\underset{1}{S}}S'(x - \bar{x})(y - \bar{y}) = S(x - \bar{x})(y - \bar{y}) = \overset{m}{\underset{1}{S}}S'(x - \bar{x}_i)(y - \bar{y}_i) + n\overset{m}{\underset{1}{S}}(\bar{x}_i - \bar{x})(\bar{y}_i - \bar{y}) \quad (2)$$

The total covariance or correlation, $S(x - \bar{x})(y - \bar{y})$, may be most easily computed by this formula:

$$S(x - \bar{x})(y - \bar{y}) = Sxy - \frac{(Sx)(Sy)}{N} \quad (3)$$

The term, $n\overset{m}{\underset{1}{S}}(\bar{x}_i - \bar{x})(\bar{y}_i - \bar{y})$, which measures the covariance between sub-sample means, can be computed as follows:

$$n\overset{m}{\underset{1}{S}}(\bar{x}_i - \bar{x})(\bar{y}_i - \bar{y}) = n\overset{m}{\underset{1}{S}}\bar{x}_i\bar{y}_i - \frac{(Sx)(Sy)}{N} = \overset{m}{\underset{1}{S}}\frac{x_a^2\, y_a^2}{n} - \frac{(Sx)(Sy)}{N} \quad (4)$$

where x_a and y_a are abbreviations for S'x and S'y the sums of the variates in a single sub-samplo.

The term, $\overset{m}{\underset{1}{S}}S'(x - \bar{x}_i)(y - \bar{y}_i)$, which measures the covariance within the sub-samples, can be found by subtraction.

The computation is analogous to the ordinary case of analysis of variance. In fact, it should be incorporated with it for each variable separately. An illustrative example will make the computation and analysis clear.

III. Computation of Covariance[1]

The data used to illustrate this problem involve height measurements of 5 plants from each of 13 inbred lines of sweet clover together with a determination of the percentage of leaves (by weight) on each of these plants. The data are given in table 1 for height in inches (x) and per cent leaves (y) for each of the 65 plants.

Table 1. Data on Height and Percent Leaves of 5 Plants from each of 13 Lines of Sweet Clover

Line No.	Plant 1 x (In.)	Plant 1 y (%)	Plant 2 x (In.)	Plant 2 y (%)	Plant 3 x (In.)	Plant 3 y (%)	Plant 4 x (In.)	Plant 4 y (%)	Plant 5 x (In.)	Plant 5 y (%)	Total Height Sx	Total Leaves Sy
1	63	43	66	38	59	42	62	39	69	40	319	202
2	70	38	77	37	64	39	53	38	61	40	325	192
3	54	37	51	50	56	49	61	35	56	49	278	220
4	40	50	39	44	44	42	38	43	45	45	206	224
5	36	49	40	50	44	42	39	45	40	44	199	230
6	44	48	50	50	54	44	54	42	56	44	258	228
7	58	38	60	38	58	42	60	40	64	40	300	198
8	54	42	52	48	46	40	52	48	46	47	250	225
9	52	41	56	39	52	42	46	43	48	42	254	207
10	58	40	48	41	52	39	52	40	60	39	270	199
11	68	41	54	42	58	37	58	40	54	40	292	200
12	48	50	45	53	45	52	41	49	41	55	220	259
13	43	47	31	41	45	56	41	45	44	46	204	235
Tot.											3375	2819

Since there was no replication of these lines the total variability will be divided into only two components: (1) between lines and (2) between plants within lines. Let the height of plants be designated as (x) and the per cent leaves be designated as (y).

The sum of squares for total variation in height of plants will be:

$$S(x^2) - (Sx)^2/N = 180{,}831.0 - 175{,}240.4 \doteq 5590.6$$

The sum of squares for variation between lines is calculated from the sums of five plants per line as follows:

$$\frac{S(x^2_h)}{5} - \frac{(Sx)^2}{N} = \frac{898{,}467}{5} - 175{,}240.4 = 4453.0$$

In like manner the total sum of squares for per cent leaves will be:

$$S(y^2) - \frac{(Sy)^2}{N} = 123{,}747 - 122{,}257 = 1489.1$$

The sum of squares for the 13 lines in per cent leaves will be:

$$\frac{S(y^2_1)}{5} - \frac{(Sy)^2}{N} = \frac{615{,}713}{5} - 122{,}257.9 = 884.7$$

[1] This illustrative example is one prepared by Dr. F. R. Immer, with minor modifications.

The sum of products for total variation will be obtained by multiplication of each plant height by the per cent leaves on that plant. The results are then summed. This will be:

$$S(xy) - \frac{(Sx)(Sy)}{N} = 114,697 - 146,371.2 = -1674.2$$

The sum of products for variation between lines is obtained by a similar process, viz.,

$$\frac{S(x_h y_1)}{5} - \frac{(Sx)(Sy)}{N} = \frac{724,014}{5} - 146,371.2 = -1568.4$$

The analysis of variance and co-variance table can now be constructed as given in table 2.

Table 2. Analysis of variance and co-variance

Variation due to:	D.F.	Sum of Squares due to: x^2	xy	y^2	Mean Sq. due to: x^2	y^2
Lines (Between)	12	4453.0	-1568.4	884.7	371.08**	73.72**
Within Lines (Error)	52	1137.6	-105.8	604.4	21.88	11.62
Total	64	5590.6	-1674.2	1489.1		

**Exceeds the 1 per cent points.

The sums of squares and sum of products for variation between plants within lines is obtained by subtraction.

Differences between lines with regard to height of plants (x), and percentage of leaves (y), may now be tested separately for significance in the ordinary manner. It is noted that these lines were significantly different in both height of plants (x) and per cent leaves (y), the mean square for lines compared with error being greater than the 1 per cent point.

However, there is no method to determine the significance of co-variance itself (xy). That is determined by tests of significance performed on correlation or regression coefficients calculated from it. This problem will be considered next.

IV. Calculation of Correlation and Regression Coefficients

The coefficients of correlation can be calculated directly from the sums of squares, since

$$r = \frac{S(x - \bar{x})(y - \bar{y})}{\sqrt{S(x - \bar{x})^2}\sqrt{S(y - \bar{y})^2}} \quad \text{- (5)}$$

By substitution of the sums of squares and products for variation between lines, given in table 2, one obtains:

$$r = \frac{-1568.4}{\sqrt{4453.0}\sqrt{884.7}} = -.790$$

The other correlation coefficients can be calculated in like manner from table 2, merely by substitution of the sums of squares and products found in the appropriate row in the table, for the source of variability to be considered.

The coefficient of regression of y on x will be given by $b = \frac{S(x - \bar{x})(y - \bar{y})}{S(x - \bar{x})^2}$ i.e., prediction of y from x.

By substitution of the sum of products and sum of squares for "lines" from table 2, $b = \frac{-1568.4}{4453.0} = -.3522$.

The correlation and regression coefficients are given in table 3.

Table 3. Coefficients of correlation and regression

Variation due to:	D.F.	Correlation between height (x) and per cent leaves (y)	Regression of per cent leaves on height (y on x)
Lines (Between)	11	-.790**	-.3522
Within Lines	51	-.114	-.0930
Total	62	-.580**	-.2995

**Exceeds the 1 per cent point of Fisher's table V.A. $\sigma_z = \frac{1}{\sqrt{N-3}}$

From Fisher's table V.A. it is seen that r = .790 is greater than the expected value of r for P = .01. The chances are, therefore, in excess of 99:1 against the occurrence of so large a correlation coefficient thru errors of random sampling from uncorrelated material. The degrees of freedom for Fisher's table V.A. are 2 less than the number of pairs in the sample and would, therefore, be one less than the degrees of freedom in table 2.

The correlation coefficient within lines, r = -.114 is not significant. The degrees of freedom are 51 in this case.

V. Tests for Significance for Regression Coefficients

The regression coefficients can be tested for significance by means of an analysis of variance or by means of a "t" test. The former method will be illustrated first.

(a) Test by Analysis of Variance

Suppose there exists a linear regression of percentage of leaves (y) on plant height (x). Then y_e, the estimated percentage of leaves from a sample of N pairs of values of y and x, is given by the regression equation:

$$y_e = a + b(x - \bar{x}) \qquad (6)$$

In this equation, $a = \bar{y}$ and $b = \frac{S(y - \bar{y})(x - \bar{x})}{S(x - \bar{x})^2}$ are estimates of the true mean percentage of leaves and the true regression coefficient, respectively.

Since the regression equation can be written as $\bar{y} = y_e - b(x - \bar{x})$, it is apparent that:

$$S(y - \bar{y})^2 = S\left\{y - \left[y_e - b(x - \bar{x})\right]\right\}^2$$
$$= S(y - \bar{y}_e)^2 + 2bS(y - y_e)(x - \bar{x}) + b^2S(x - \bar{x})^2$$

Due to the fact that the middle term on the right is zero, [1]

---

[1] Consider $S(y - y_e)(x - \bar{x})$

Since $y_e = \bar{y} + b(x - \bar{x})$, we have:

$S\left[y - \bar{y} - b(x - \bar{x})\right](x - \bar{x})$ or $S(y - \bar{y})(x - \bar{x}) - bS(x - \bar{x})^2$.

It is clear that the whole expression is zero due to the fact that $b = \frac{S(y - \bar{y})(x - \bar{x})}{S(x - \bar{x})^2}$.

$$S(y - \bar{y})^2 = S(y - y_e)^2 + b^2S(x - \bar{x})^2 \quad \text{- -} \quad (7)$$

Thus, the sum of the squares of the deviations of the percentage of leaves has been analyzed into two components, one dependent on b, and therefore ascribable to regression, and the other a sum of squares that represents deviation from regression or residual.

Since $b = \frac{S(y - \bar{y})(x - \bar{x})}{S(x - \bar{x})^2}$, it is obvious that the value of $b^2S(x - \bar{x})^2$, the component ascribable to regression will be:

$$b^2S(x - \bar{x})^2 = \frac{[S(x - \bar{x})(y_e - \bar{y})]^2}{S(x - \bar{x})^2} \quad \text{- -} \quad (8)$$

This procedure is now applied to the illustrative problem. The values from table 2 will be used to compute total regression:

$$S(y - \bar{y})^2 = 1489.10$$

$$\frac{[S(x - \bar{x})(y - \bar{y})]^2}{S(x - x)^2} = \frac{(-1674.2)^2}{5590.6} = 501.37$$

The analysis to test the significance of total regression follows in table 4.

Table 4. Analysis of Variance to Test the Significance of Total Regression

Variation	D.F.	Sum of Products	Mean Product	F-value
Due to Regression	1	501.37	501.37	31.98**
Deviations from Regress.	63	987.73	15.68	
Total	64	1489.10		

The total sum of squares for y (leaf percentage) is taken directly from table 2. Here y is used as the dependent variable, i.e., y (leaf percentage) is predicted from x (plant height) which is known. The sum of squares due to deviations from regression is obtained by subtraction, i.e., 1489.10 - 501.37 = 987.73. There will be one degree of freedom due to linear regression with a remainder of N-2 degrees of freedom for deviations from regression. It is also to be noted that N-2 is the number of degrees of freedom used to test the significance of r (Fisher, Table V.A). It is obvious from table 4 that the regression coefficient is highly significant, since the "F" value exceeds the one per cent point. The same conclusion was obtained when r was tested for significance. In fact, the two tests for significance are equivalent. When the correlation coefficient is significant, the regression coefficient must be significant, and vice versa.

To test for the significance of regression between lines, the values already computed for that source of variation in table 2 are used:

$$S(y - \bar{y})^2 = 884.7$$

$$\frac{[S(x - \bar{x})(y - \bar{y})]^2}{S(x - \bar{x})^2} = \frac{(-1568.4)^2}{4453.0} = 552.4$$

The values are summarized in table 5:

Table 5. Analysis of Variance to Test the Significance of Regression Between Lines

Variation	D.F.	Sums of Products	Mean Product	F-value
Due to regression	1	552.4	552.40	18.29**
Deviations from regress.	11	332.3	30.21	
Total	12	884.7		

It is thus evident that the regression between lines is extremely significant. The regression within lines will not be tested for significance since r is not significant (See table 3).

(b) The "t" Test of Significance

Regression coefficients may be tested by means of the "t" test also (See Fisher, 1934, pp. 126-137). As an illustration, the significance of the regression of y on x between lines may be tested. From table 3, b = -0.3522, $S(x - \bar{x})^2$ = 4453.0, and $S(y - \bar{y})^2$ = 884.7.

Then,

$$t = \frac{b\sqrt{S(x - \bar{x})^2}}{s} \qquad (9)$$

where

$$s^2 = \frac{S(y - \bar{y})^2 - b^2 S(x - \bar{x})^2}{m - 2} \qquad (10)$$

Then,

$$s^2 = \frac{884.7 - (0.3522)^2(4453.0)}{13-2} = \frac{332.3}{11} = 30.21$$

$$s = 5.496$$

$$t = \frac{0.3522\sqrt{4453.0}}{5.496} = 4.276 \text{ for 11 D.F.}$$

From the "t" table it is obvious that the observed t-value exceeds the 1.0 per cent point.

Since $\sqrt{F}$ = t for one degree of freedom, it is noted that $\sqrt{F} = \sqrt{18.39}$ = 4.277 (from table 5). Thus, it is apparent that tests of significance of regression coefficients by means of the analysis of variance and the "t" test are equivalent. Moreover, they give the same result as tests of significance of the correlation coefficient (Table V A, Fisher, 1934).

VI. Substitution in Regression Equation

The regression equation is usually expressed as $y_e = \bar{y} + b(x - \bar{x})$.

For such a regression between lines one may substitute $\bar{y}$ = 43.37, $\bar{x}$ = 51.92 and b = -.3522. The mean values of y and x are obtained directly from table 1 by division of the totals by 65. The value of b is taken from table 3. Numerically y_e = 43.37 - 0.3522 (x - 51.92). This regression equation can be simplified to y_e = 61.66 - 0.3522 x, where x is any value of plant height. In table 6 is given the mean height of each line, the mean leaf percentage of each line and the leaf percentage predicted from plant height by means of the equation above.

Table 6. Observed Mean Height, Mean Leaf Percentage and Predicted Leaf Percentage of the 13 Lines of Sweet Clover.

Line No.	Observed mean height (x)	Observed mean % leaves (y)	Predicted mean % leaves (y_e)	Line No.	Observed mean height (x)	Observed mean % leaves (y)	Predicted mean % leaves (y_e)
1	63.8	40.4	39.2	8	50.0	45.0	44.0
2	65.0	38.4	38.8	9	50.8	41.2	43.8
3	55.6	44.0	42.1	10	54.0	39.8	42.6
4	41.2	44.8	47.1	11	58.4	40.0	41.1
5	39.8	46.0	47.6	12	44.0	51.8	46.2
6	51.6	45.6	43.5	13	40.8	47.0	47.3
7	60.0	39.6	40.5				

The differences between observed mean leaf percentage and predicted leaf percentage in table 6 represent errors in prediction. The sum of squares of these differences would be given by $S(y - y_e)^2$ where y represents the observed mean leaf percentage and y_e the predicted value. This quantity can be computed from table 5 by subtraction of the observed and predicted leaf percentage, these values being squared and added to give $S(y - y_e)^2 = 66.54$. Now this sum of squares is based on means of 5 plants per line while the analysis of variance in table 4 was on a single plant basis. Therefore, multiplication of 66.54 by 5 to place it on a single plant basis gives 332.7. This agrees with the sum of squares due to deviation from regression, i.e., 332.3 considering that the predicted leaf porcentages have been computed to only one place of decimals.

It may be noted also that s^2 used in the "t" test could be written $s^2 = \frac{S(y - y_e)^2}{m - 2}$, since $S(y - y_e)^2 = S(y - \bar{y})^2 - b^2S(x - \bar{x})^2$, the latter form being simpler for computation purposes.

While the application of analysis of variance and co-variance to correlation and regression problems has been illustrated here with data from a very simple experiment, it is evident that it is equally applicable to problems of any degree of complexity.[1] The analysis of variance and co-variance are keyed out for the particular problem under investigation after which the correlation and regression coefficients are calculated for any component of the total variability. The tests of significance are made in a manner similar to the ones illustrated.

VII. Use of Covariance

The analysis of covariance is often successfully applied in an artificial reduction of experimental error in certain types of experiments where preliminary or uniformity trial data are available. There may be factors which it is impossible to equalize satisfactorily between the different treatments, and yet there may be reason to suppose that greater accuracy would arise from their equalization, were that possible. Availability of preliminary data may provide the basis for such equalization.

The possible use of data from a previous uniformity trial to reduce errors due to soil heterogeneity in the experimental years has been given considerable attention in field trials in recent years. The assumption is that soil fertility is constant from year to year. Thus, a significant correlation between the same plots in successive years may be used to reduce the error in the experimental year. The regression equation is applied to predict the yields in the experimental year from the yields of the same plots grown under uniform treatment in the previous year. The deviations from the predicted yields should then contribute to the error of the experiment. Methods to utilize information from previous crop records have been outlined by Fisher (1934), Sanders (1930), Eden (1931), and by Wishart and Sanders (1935). With annual crops, Summerby (1934) found that it was not worthwhile to sacrifice a year to a uniformity trial in order to obtain information to reduce the error in the experimental year. The method seems to have the greatest possibilities with perennial crops. (See Fisher, 1937).

Another possible application of covariance arises where stand counts are available in addition to yield. Mahoney and Baten (1939) have made such an application. Stand counts may furnish a good index of plot variability provided they have been unaffected by treatment. Correction for stand, which can be made from the regression relation,

---

[1] For a consideration of curvilinear regression and its treatment by the analysis of variance, the reader is referred to more advanced works on the subject.

provides an adjustment of the data to what they would be if all plots had the same number of plants (proportionality assumed). When yield is related to plant number it is obvious that the experimental error will be decreased when this factor is taken into account and a correction made for it. It is first necessary to determine whether or not such a relationship exists.

The simpler aspects of covariance as applied to between and within groups have already been considered. The method will be used here for ordinary field experiments where the total variation is sub-divided into more than two parts. An illustrative example used by Fisher (1934) will be followed.

VIII. Use of Preliminary Trial Data for Error Reduction

Some data collected by Eden (1931)[1] on tea will be used to illustrate the calculations for covariance for preliminary and experimental yields. Four "dummy" treatments for yields of tea expressed in per cent of the mean in a randomized block experiment are given in table 7.

Table 7. Preliminary and Experimental Yields of Tea Plants

Treatment	Preliminary (x) or Experimental (y)	Blocks 1	2	3	4	Treatment Total	Mean
A	x	91	118	109	102	420	105.00
	y	85	121	114	107	427	106.75
B	x	88	94	105	91	378	94.50
	y	81	93	106	92	372	93.00
C	x	88	110	115	96	409	102.25
	y	90	106	111	102	409	102.25
D	x	102	109	94	88	393	98.25
	y	93	114	93	92	392	98.00
Block Total	x	369	431	423	377	1600	
	y	349	434	424	393	1600	

The preliminary yields will be designated as x and the experimental yields as y in the subsequent calculations.

(a) Analysis of Variance and Covariance for Preliminary and Experimental Yields.

The sums of squares for preliminary yields:

Total: $Sx^2 - \frac{(Sx)^2}{N} = 1526.0$

Treatments: $\frac{Sx_t}{n_t} - \frac{(Sx)^2}{N} = 253.5$

Blocks: $\frac{Sx_b^2}{n_b} \quad \frac{(Sx)^2}{N} = 745.0$

Sums of squares for experimental yields:

Total: $Sy^2 - \frac{(Sy)^2}{N} = 2040.0$

Blocks: $\frac{Sy_b^2}{n_b} - \frac{(Sy)^2}{N} = 1095.5$

Treatments: $\frac{Sy_t^2}{n_t} - \frac{(Sy)^2}{N} = 414.5$

[1]Cited by R. A. Fisher (1934).

Sums of Products:

Total: $Sxy - \frac{(Sx)(Sy)}{N} = 1612.00$

Treatments: $\frac{Sx_t y_t}{n_t} - \frac{(Sx)(Sy)}{N} = 323.25$

Blocks: $\frac{Sx_b y_b}{n_b} - \frac{(Sx)(Sy)}{N} = 837.0$

The above results are incorporated in table 8.

Table 8. Analysis of Variance and Covariance

Variation due to	D.F.	Sums of Squares (x)	(y)	Sums of Products (xy)	Mean Squares (x)	(y)	F-Value (x)	(y)
Blocks	3	745.0	1095.5	837.00	248.33	365.17	4.24*	6.20*
Treatments	3	253.5	414.5	323.25	84.50	138.17	1.44	2.35
Error	9	527.5	530.0	451.75	58.61	58.89		
Total	15	1526.0	2040.0	1612.00				

From this analysis it is clear that no significance resulted between yields in case of the "dummy" treatments, while a considerable degree of soil heterogeneity evidently exists because the variation between blocks proved significant for both the preliminary and experimental data.

It is now proposed to test the covariance as a basis to provide a correction for the mean experimental yields in an effort to reduce the soil heterogeneity effect further. The analysis of covariance is given in table 9.

Table 9. Analysis of Covariance and Test of Significance of Adjusted Experimental Means

Variation due to	D.F.	Sum of Squares (x)	Sum of Products (x,y)	Sum of Squares (y)	Errors of Estimate: Sums Squares	D.F.	Mean Squares
Blocks	3	745.0	837.00	1095.5			
Treatments	3	253.5	323.25	414.5			
Error	9	527.5	431.75	530.0	143.12	8	17.89
Total	15	1526.0	1612.00	2040.0			
Tr. + Error	12	781.0	775.00	944.5	175.45	11	15.95
Test of significance for adjusted treatment means					32.33	3	10.78[1]

[1] F = 17.89/10.78 = 1.66 non-significant

Since the total has been broken down into more than two parts, it is necessary to form a new total which contains only the two effects under study, viz., treatment and error. This new total is in the line, treatment + error, in table 9. The degrees of freedom, sums of squares, and products are added to obtain the appropriate numbers.

The sums of squares for errors of estimate, $S(y - y_e)^2$, are calculated by use of the principle of subtraction, viz.,

$$\left[Sy^2 - (Sy)^2/N\right] - \frac{[Sxy - (Sx)(Sy)/N]^2}{Sx^2 - (Sx)^2/N}$$, in the lines for error, and treatment + error. These computations are as follows:

(1) Error: $530.0 - (451.75)^2/527.5 = 143.12$

(2) Treatments + Error: $944.5 - (775)^2/781 = 175.45$

The sums of squares for error is subtracted from that for treatment + error to yield the sum of squares appropriate for the test of significance for the adjusted treatment means, viz., 175.45 - 143.12 = 32.33.

In this particular case, the mean square for adjusted treatment means is not significantly different from error since "dummy" treatments were used.

(b) Calculation of the Regression Coefficient

The regression coefficient (b) is calculated from the values in table 9. The regression required is the regression of y on x in the row designated error. Since the regression coefficient is the ratio of the products to the sums of squares of the independent variable,

$$b = \frac{Sxy - (Sx)(Sy)/N}{Sx^2 - (Sx)^2/N} = \frac{451.75}{527.50} = 0.8564$$

The significance of the error regression, b = 0.8564, should be tested at this point. Unless it is significant, there will be little advantage to use it to reduce the error for the experimental year. The sum of squares due to linear regression will be:

$$\frac{[Sxy - (Sx)(Sy)/N]^2}{Sx^2 - (Sx)^2/N} = \frac{(451.75)^2}{527.50} = 386.88$$

The test for significance is summarized in table 10.

Table 10. Test of Significance for Error Regression

Variation due to	Formulas	D.F.	Sums Squares	Mean Square	F
Regression	$\frac{[Sxy - (Sx)(Sy)/N]^2}{Sx^2 - (Sx)^2/N}$	1	386.88	386.88	21.63
Deviations from regression[1]	$\left[Sy^2 - \frac{(Sy)^2}{N}\right] - \frac{[Sxy - (Sx)(Sy)/N]^2}{Sx^2 - (Sx)^2/N}$	8	143.12	17.86	
Total for Error	$Sy^2 - (Sy)^2/N$	9	530.00	58.89	

[1]Error for adjusted yields.

The observed F-value is highly significant. It indicates that it will be worth while to proceed with the correction of the experimental test data on the basis of their regression on the preliminary yields.

(c) The Adjusted Treatment Means

The adjusted treatment means can be calculated and compared with the unadjusted. The formula for the adjusted values is, $y_e - bx$, where y is the individual treatment in the experimental year. The computations are given in table 11. The mean yields per treatment of the original data, x and y, are taken directly from table 8.

Table 11. Calculation of Mean Yields of Treatments in Experimental Test corrected for Yields in Preliminary Test.

Treatment	Mean Yield Preliminary Test (x)	Deviations from Mean $(x - \bar{x})$	Product[1] $b(x - \bar{x})$	Mean Yield Experimental Year (y)	Corrected Yields for Experimental Test $y_e - b(x - \bar{x})$
A	105.00	5.00	4.28	106.75	102.47
B	94.50	-5.50	-4.71	93.00	97.71
C	102.25	2.25	1.93	102.25	100.32
D	98.25	-1.75	-1.50	98.00	99.50
Gen. Mean	100.00	0.00	0.00	100.00	100.00

$$^{1}b = \frac{Sxy - (Sx)(Sy)/N}{Sx^2 - (Sx)^2/N} = 0.8564$$

The regression equation for error for x on y is as follows:

$$y_e = \bar{y} + b(x - \bar{x}) = 100.0 + 0.8564(x - 100.00) = 0.8564x + 14.36$$

The graphical representation is shown below, the points for the determination of the line determined by substitution in the regression equation.

Let $x = 92.5, y_e = 93.58$. Let $x = 107.5, y_e = 106.42$.

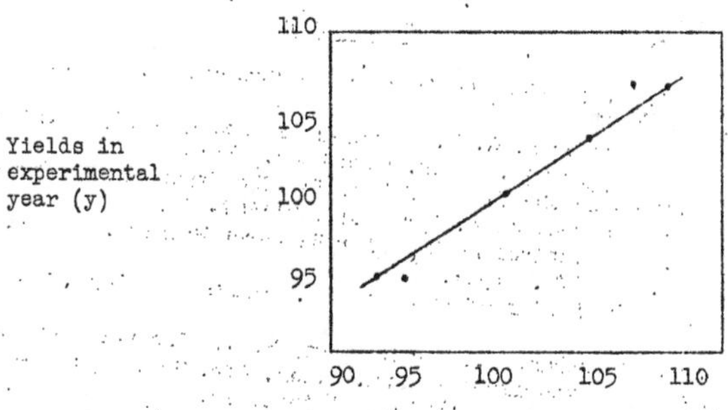

(d) Standard Error of a Difference

The standard error of a given difference between the corrected mean yields is given by Wishart and Sanders (1936) as follows:

$$\sigma_{c\bar{x}} = \sqrt{\frac{2s^2}{n} + \frac{s^2(\bar{x}_1 - \bar{x}_2)^2}{A'}} \quad (11)$$

where s^2 = the variance of the corrected yields (17.89), n = the number of plots per treatment (4), A' = the sum of squares for error in the original preliminary trial (527.5), and $\bar{x}_1$ and $\bar{x}_2$ = the means of the preliminary treatment plots being compared (105.00 - 94.50 = 10.50).

For treatment A and B, the mean difference of the corrected yields (table 11) is 102.47 - 97.71 = 4.76. The standard error is computed as follows:

$$\sigma_{c\bar{x}} = \sqrt{\frac{2(17.89)}{4} + \frac{(17.89)(10.50)^2}{527.5}} = 3.56$$

$d/\sigma_{c\bar{x}}$ = 4.76/3.56 = 1.34, a non-significant value.

(e) Factors in Use of Independent Variable

The investigator usually wishes to know when it is worthwhile to introduce the independent variable into the experiment. This question is answered by Snedecor (1937) who states that three items will aid him. First, the list of actual and adjusted means. Sometimes the rank order of adjusted means is quite different from that of the unadjusted and the shifts may be interpreted. Second, a comparison of the sum of squares of errors of estimate (table 8) used to test treatment significance, 32.33, with $Sy^2 - (Sy)^2/N$ = 414.5. The latter is far greater than the former. Third, the change in precision of the experiment due to the adjustment of the error sums of squares. This is indicated in table 8. The sum of squares, $Sy^2 - (Sy)^2/N$ = 530.00 with 9 degrees of freedom, is analyzed into two parts, one with a single degree of freedom that measures the variation attributable to regression, the other 8 degrees of freedom being assigned to error. The mean square for error is reduced from 58.59 to 17.89, which is highly significant. These factors will enable the investigator to decide whether to retain the independent variable in similar experiments. It has already been mentioned that the use of preliminary uniformity data to reduce the error in the subsequent experimental test may be useful in perennial crops, but probably is not worth while for annual crops.

References

1. Bartlett, M. S. A Note on the Analysis of Covariance. Jour. Agr. Sci. 26:488-491. 1936.
2. Cox, G. M., and Snedecor, G. W. Covariance used to Analyze the Relation between Corn Yields and Acreage. Jour. Farm Econ., 18:597-607, 1936.
3. Eden, T. Studies in the Yield of Tea. I: The Experimental Errors of Field Experiments with Tea. Jour. Agr. Sci., 21:547-573. 1931.
4. Fisher, R. A. Statistical Methods for Research Workers. Oliver and Boyd, 5th edition, pp. 257-272. 1934.
5. ____________ The Design of Experiments. Oliver and Boyd. 2nd Ed., pp. 172-189. 1937.
6. Garner, F. H. Grantham, J., and Sanders, H. G. The Value of Covariance in Analyzing Field Experimental Data. Jour. Agr. Sci., 24:250-259. 1934.
7. Goulden, C. H. Methods of Statistical Analysis. Burgess Publ, Co., pp. 151-157, and 185-196. 1937.
8. Immer, F. R. A Study of Sampling Technic with Sugar Beets. Jour. Agr. Res., 44:663-647. 1932.
9. ____________, and Raleigh, S. M. Further Studies of Size and Shape of Plot in Relation to Field Experiments with Sugar Beets. Jour. Agr. Res., 47:591-598. 1933.
10. Mahoney, C. H., and Baten, W. D. The Use of the Analysis of Covariance and its Limitation in the adjustment of Yields Based upon Stand Irregularities. Jour. Agr. Res., 58:317-328. 1939.
11. Sanders, H. G. A Note on the Value of Uniformity Trials for Subsequent Experiments. Jour. Agr. Sci., 20:63-73. 1930.
12. Snedecor, G. W. Statistical Methods. Collegiate Press, pp. 219-241. 1937.

13. Summerby, R. The Value of Preliminary Uniformity Trials in Increasing the Precision of Field Experiments. McDonald Col. Tech. Bul. 15. 1934.
14. Wishart, J. and Sanders, H. G. Principles and Practices of Field Experimentation. Emp. Cotton Growing Corp., pp. 45-56. 1935.

Questions for Discussion

1. What is covariance? Where useful in experimental work?
2. Interpret the use of the analysis of variance for the determination of significance of the regression coefficient.
3. How is the error for linear regression computed?
4. How can covariance be used on preliminary trial data to reduce the error in the experimental year?
5. Discuss conditions in field experimentation where it might be useful to use preliminary trial data to reduce the error in the subsequent test.
6. What assumption is made in the correction of stand by covariance? What precautions are necessary?
7. Upon what is the error of estimate based? Explain.
8. What does it mean when the difference to adjust treatment means actually is less than the mean square for the error of estimate for error?
9. Name 3 types of agronomic tests where covariance might prove useful. Give the reason in each case.

Problems

1. The yields of soybeans in a randomized block experiment with split plots are given below. Let x represent the yield of hay in tons per acre and y represent the yield of seed in bushels per acre.

   The total yields of the 4 plots of each spacing are assembled below.

Bu. of seed per acre (y)						Tons of hay per acre (x)					
Width of rows	Spacing within rows					Width of rows	Spacing within rows				
	1/2"	1"	2"	3"	Sum		1/2"	1"	2"	3"	Sum
16"	89.8	91.8	79.6	88.6	349.8	16"	11.40	10.72	9.63	9.68	41.43
20"	92.7	85.6	87.2	87.1	352.6	20"	11.31	10.06	9.73	9.31	40.41
24"	90.6	82.3	84.3	80.7	337.9	24"	10.02	9.21	9.00	8.41	36.64
28"	86.0	83.0	82.4	78.3	329.7	28"	9.62	9.41	9.09	8.28	36.43
32"	85.1	78.4	74.6	72.9	311.0	32"	9.58	8.72	8.45	7.77	34.52
40"	78.4	70.7	71.7	69.2	290.0	40"	8.81	8.19	7.34	7.59	31.93
Sum	522.6	491.8	479.8	476.8	1971.0	Sum	60.74	56.34	53.24	51.04	221.36

The analyzis of variance for x and y are given below

Variation due to:	D.F.	$(y-\bar{y})^2$	$(x-\bar{x})(y-\bar{y})$	$(x-\bar{x})^2$	Correlation of x and y	Regression of y on x
Blocks	3	10.4370		.1438		
Width of rows	5	182.0500		3.9684		
Error (a)	15	32.5542		.3513		
Rows	23	225.0412		4.4655		
Spacings	3	54.7512		2.2108		
Width x spacing	15	30.4300		.2389		
Error (b)	54	268.6038		1.5984		
Total	95	578.8262		8.5115		

(a) Calculate an analysis of co-variance.
(b) Calculate the correlation coefficient for the different lines in the analysis of variance and co-variance.
(c) Do the same for the regression coefficient of y on x.
(d) Test the significance of the correlation coefficients by means of Fisher's table V.A. Mark the coefficients which exceed the 5 per cent point with one asterisk(*) and those which exceed the 1 per cent point with two astericks (**).
(e) Test the significance of the regression for error (b) by means of an analysis of variance. Get the $\sqrt{F}$ also.
(f) Test the significance of the regression for error (b) by means of "t" test.
(g) Test the significance of the correlation coefficient for error (b) by means of the "t" test, given by Fisher in section 34 of his book.
(h) Calculate the mean yield of seed in bushels for the six different width of rows. Calculate the predicted mean yields for each width, using the regression for width of rows.

2. Some data on number of sugar beet plants per plot and yield in tons per acre are given by Snedecor (1937) for a fertilizer experiment conducted in a randomized block test. The data for 3 replications are given below:

Fertilizer Applied	No. (x) or Yield (y)	Block 1	2	3	Treatment Sum	Sums Squares	Sums Products
None	x	183	176	291			
	y	2.45	2.25	4.38			
P	x	356	300	301			
	y	6.71	5.44	4.92			
K	x	224	258	244			
	y	3.22	4.14	2.32			
PK	x	329	283	305			
	y	6.34	5.44	5.22			
PN	x	371	354	352			
	y	6.48	7.11	5.88			
KN	x	230	221	237			
	y	3.70	3.24	2.82			
NPK	x	322	367	400			
	y	6.10	7.68	7.37			
Block Sums	x						
	y						
Sum Squares	x						
	y						
Sums Products	-						

Due to the fact that the number of plants varies it is necessary to examine the effect of the variable stand and to estimate the yields on the basis of equal numbers of plants. Calculate as follows:

(a) Yield in a simple randomized block experiment.
(b) The analysis of covariance of stand and yield.
(c) Calculate the test for significance.
(d) Give the conclusions for the test.

3. A sugar beet variety test was conducted at Rocky Ford as a randomized block experiment in which the number of plants differed in each plot. The yields were taken on the basis of competitive plants per plot, and also on the basis of all the beets in the plot. The object of the experiment was to determine the yields of the different varieties. Since the number of beets varies from plot to plot, it is desired to examine the effect of the variable stand and to estimate the yields on the basis of equal numbers of beets. The data follow (unpublished data from G. W. Deming):

Variety No.	No.(x) or Yield (y)	Block 1	2	3	4	5	6	Treatment Totals
1	x	248	217	227	210	218	215	
	y	17.49	18.65	14.39	16.33	11.28	15.75	
2	x	245	217	239	210	205	219	
	y	17.99	19.22	18.21	14.11	16.84	13.76	
3	x	238	228	205	191	224	211	
	y	14.13	16.62	15.99	13.81	14.80	13.00	
4	x	254	223	189	180	216	209	
	y	20.19	21.45	14.01	12.54	14.20	17.65	
5	x	249	221	226	242	246	216	
	y	20.08	17.04	14.04	16.05	13.86	9.75	
6	x	225	212	194	211	202	215	
	y	17.49	19.63	17.55	16.02	15.46	14.05	
Block Totals	x							
	y							

Calculate the analysis of covariance and adjust the yields to a uniform stand basis.

FIELD PLOT TECHNIQUE

PART III

Field and Other Agronomic Experiments

CHAPTER XII

SOIL HETEROGENEITY AND ITS MEASUREMENT

I. Universality of Soil Heterogeneity

One of the difficulties in yield tests is the fact that uniform soil conditions rarely exist, even over a small portion of any field. Soil variability has been noted by many investigators, but it was J. Arthur Harris (1915)(1920) who first presented data to show its extreme importance in field experimentation. Lyon (1911) states that it is "quite likely that productivity of plots change from year to year even with the same treatment", altho the work of Harris and Schofield (1920) (1928) and of Garber, et al. (1926)(1930) indicates a tendency for the differences in plot yields to be permanent.

A soil with differences so slight as to escape the most observant eye may have very great effects on plants which grow in it. Parker (1931) is authority for the statement that two plots of the same crop variety grown in "an apparently uniform soil and treated alike in every respect may differ from one another in yield by 20 per cent or more solely as a result of differences in soil conditions." Small plots have generally replaced large ones to correct for this condition, because it is obvious that two plants of the same variety grown one yard apart are more likely to yield alike than when 200 feet apart as probably would be true of one-acre plots. It is impossible to avoid variation even under such conditions. Davenport and Frazer (1896) report results with 77 varieties of wheat grown on plots two rods square. Nine check plots of the same variety were systematically distributed over the area. The variation in the check plots was so great that only 8 varieties yielded more than the highest check, and but 3 lower than the lowest check.

Soils vary in texture, depth, drainage, moisture, and available plant nutrients from yard to yard. After the analyses of large amounts of data from all over the world, Harris (1920) concluded that soil heterogeneity was practically universal. He estimated it to be the most potent cause of variation in plot yields and the chief difficulty in their interpretation. In 1915 he stated: "It is obviously idle to conclude from a given experiment that variety 'A' yields higher than variety 'B', or that fertilizer 'X' is more effective than fertilizer 'Y', unless the differences found are greater than those which might be expected from differences in the productive capacity of the plots of soils upon which they are grown." Even earlier than this, Piper and Stevenson (1910) remarked that soil variability was so great that "doubt was cast on the greater portion of published field experiments where yield was primarily involved." The yield differences must be large enough to overshadow soil variation, or the experiment designed so as to remove its effect.

Much of the improvement in experimental methods for field experiments in recent years has been brought about thru special devices to measure much of the soil fertility variation and essentially eliminate it from the actual comparisons being made.

II. Uniformity Trial Data

Uniformity trial data have been used for the measurement of soil heterogenity as well as for many other purposes in field experimentation. The usual procedure is to plant a bulk crop, the area being later partitioned into small plots, usually of the same dimensions. The same cultural operations are carried out over the entire area. The yield of each plot is recorded separately at harvest. The usefulness of the uniformity trial lies in the fact that the small units can be combined into larger plots of various sizes and shapes in order to study variability. The variation in yield over the field is due to soil heterogeneity, as well as to plant variation, errors in weighing, etc., (generally summed up as experimental error). The most obvious use for the data is to provide information on the optimum size and shape of plot. Uniformity trial data can also be used to compare the relative efficiencies of different experimental designs, particularly in relation to a certain crop. Data from previous uniformity trials may also be used to reduce the error of subsequent experiments laid down on the same plots.

The method offers promise for perennial crops where the same plants are concerned, but offers little or no advantage for annual crops. A catalogue of uniformity trial data has been published by Cochran (1937).

Some agronomists conduct so-called blank trials (planted to a bulk crop) to observe soil heterogeneity as a preliminary step in experimentation on a new field. Love (1928) advocates such trials, especially as a preliminary to long-time experiments. They afford an opportunity for the investigator to detect good and poor spots on a field so that unsatisfactory areas may be eliminated. One objection to the blank trial used in this manner is that it takes time. Time may be an important element in an experiment.

III. Criteria for the Measurement of Soil Variability

Some accurate measure of soil heterogeneity may be desirable preliminary to steps for its correction.

(a) Correlation Coefficient

Harris (1915) supplied the first quantitative measure based on correlation, his heterogeneity coefficient being an intra-class correlation coefficient. For use of the formula, the field must be planted uniformly to the same crop and harvested in small units. Harris grouped nearby plots. The number in a group was arbitrary, it being common to use 2 by 1, 2 by 2, and 2 by 3-fold groupings. The size of the heterogeneity coefficient is influenced by the size of group. The more plots that are put together, the greater is the correlation coefficient. The heterogeneity coefficient is expressed on a relative scale from 0.0 ± 1.0 so that comparisons from field to field can be made directly. This coefficient measures the degree to which nearby plots are similar in productivity. Should the correlation be sensibly zero, the irregularities of the field are not so great as to influence in the same direction the yields of nearby small plots. The higher the correlation, the greater the soil heterogeneity. One may grasp the significance when he remembers that the correlation coefficient multiplied by 100 gives the most probable percentage deviation of the yield of an associated plot when the deviation of one plot of the group from the general average is known. Hayes and Garber (1927), in explanation, state that in "patchy" fields certain contiguous units tend to yield high while others show a tendency in the opposite direction. Under these conditions a high correlation coefficient results. "Where variability is due only to random sampling the correspondence between contiguous plots will be counter-balanced by lack of correspondence in others." The same result can be obtained with the ordinary inter-class correlation coefficient as with the heterogeneity coefficient when a 2 by 1-fold arrangement is used.

The analysis of variance can be used to obtain the same result as with the heterogeneity coefficient as indicated by Fisher (1934). Intra-class correlation merely measures the relative importance of two groups of factors that cause variation. In the calculation it is necessary to obtain these equalities:

$$\overset{nm}{\underset{1}{S}} (x - \bar{x})^2 = (m - 1)n\, s^2 \quad \text{-------------------} \quad (1)$$

$$n\overset{m}{\underset{1}{S}} (\bar{x}_b - \bar{x})^2 = (m - 1)s^2 \left[1 + (n - 1)\, r\right] \quad \text{-----------} \quad (2)$$

where m = the number of arbitrary blocks and n = the number of ultimate units within a block

The principal value of the correlation coefficient, either inter-class or intra-class, is to demonstrate that the fertilities of adjacent areas are correlated and that variability exists in the field.

(b) Fertility Diagram

The suitability of a particular lay-out adopted in an experiment can be judged to a considerable extent by a fertility diagram constructed from the individual plot yields. This is possible from uniformity trial data. An example taken from Crowther and Bartlett (1938) is given in Figure 1.

Figure 1
Variation in natural fertility at Bahtim, 1934 (yield in kantars per feddan).

IV. Computation of Heterogeneity by the Analysis of Variance

Suppose a field is divided into N small plots, all sown to the same variety. Some uniformity trial data from Mercer and Hall (1911) on the grain yields of one acre of wheat when harvested in 1/500-acre plots will be used to illustrate the method of computation. The yields in pounds per plot for the 24 plots in the northwest corner are as follows:

3.63		4.15	4.06		5.13
	m_1			m_4	
4.07		4.21	4.15		4.64
4.51		4.29	4.40		4.69
	m_2			m_5	
3.90		4.64	4.05		4.04
3.63		4.27	4.92		4.64
	m_3			m_6	
3.16		3.55	4.08		4.73

It is noted that the area is divided into an arbitrary number of blocks all equal in size, i.e., there are 6 blocks with 4 ultimate plots in each.

Let x = the value of an ultimate plot unit.
N = the total number of plots = 24.
S(x) = sum of all the ultimate plots = 101.54.
$\bar{x}$ = mean yield of the ultimate plots = 4.2308.
$S(x^2)$ = sum of squares of yields of the ultimate plots = 434.5582.

Then, the sums of squares are computed as follows:

$$\text{Total} = S(x^2) - \frac{(Sx)^2}{N} = 434.5582 - 429.5988 = 4.9594$$

$$\text{Between blocks} = \frac{S(x_b^2)}{n} - \frac{(Sx)^2}{N} = 1727.9410 - 429.5988 = 2.3864$$

$$\text{Within blocks} = 4.9594 - 2.3864 = 2.5730$$

The analysis of variance is as follows:

Variation	D.F.	Sums Squares
Between blocks	5	$2.3864 = (m-1)s^2[1 + (n-1)r]$
Within blocks	18	$2.5730 = (m-1)s^2(n-1)(1-r)$
Total	23	$4.9594 = (m-1)ns^2$

Now, let m = the number of blocks, m-1 = the degrees of freedom for blocks, n = the number of plots per block, and s^2 = the estimated variance.

Then m = 6, m - 1 = 5, and n = 4.

Since $(m-1)ns^2 = 4.9594$, $20s^2 = 4.9594$, and $s^2 = 0.2480$

From the formula for the sum of squares between blocks,

$$(m - 1)s^2 \left[1 + (n - 1) r\right] = 2.3864$$
$$\text{or } (5)(0.2480)(1 + 3r) = 2.3864$$
Then $r = 0.3082$

V. Amount of Soil Heterogeneity

In his studies of soil heterogeneity, Harris (1915)(1920) used fields planted to the same crop, but harvested in separate small plot units. The relative productivity of contiguous plots was determined.

(a) Variations in Yield in Same Season

Some of the results obtained by Harris are given by Hayes and Garber (1927):

Crop	Characters	Plot Size	Investigator	r
Wheat	Grain Yield	5.5 by 5.5 ft.	Montgomery	0.603 ± 0.029
	N-Content			0.11[illegible] ± 0.044
Oats	Grain Yield	1/30 acre	Kiesselbach	0.495 ± 0.035
Mangels	Roots	1/200 acre	Mercer & Hall	0.346 ± 0.037
	Leaves			0.466 ± 0.043
Potatoes	Tuber Yield	12-foot row	Lyon	0.311 ± 0.043
Corn	Grain Yield	0.085 acre	Smith	0.830 ± 0.019

The amount of soil heterogeneity in rod-row trials was measured by Hayes (1925) at the Minnesota Station in connection with a variety test. Four systematically distributed plots were used. To obtain the heterogeneity coefficient the average yield of each strain in the trial was considered as 100. The yielding ability of each plot was obtained by dividing its actual yield by the average yield of all four replicates and expressing the result in percentage. By the ordinary method, correlations in yielding ability of adjacent plots or of plots at any distance apart were determined. The results were as follows for oats, spring wheat, and winter wheat:

	Correlation Coefficients (r)		
Factors Correlated	Oats	Spring Wheat	Winter Wheat
Adjacent plots	0.572 ± 0.025	0.618 ± 0.023	0.552 ± 0.068
Separated by one plot	0.490 ± 0.029	0.518 ± 0.028	0.293 ± 0.028
Separated by four plots	0.264 ± 0.041	0.449 ± 0.034	0.114 ± 0.118
Separated by ten plots	0.275 ± 0.057	0.429 ± 0.060	---------

The correlation coefficient explains very little unless one knows the factors involved. However, it affords the best means to consider the amount of replication that should be practiced.

Similar results were obtained by Garber, Hoover, and McIlvaine (1926) in West Virginia experiments. They found a marked correlation between the yields of oat hay in contiguous plots. The correlation for the yields of replicated plots was sensibly zero.

(b) Permanence of Differences

It is important to know whether or not there is a tendency for plots that produce low yields one season to produce low yields the next season, etc. The results of Harris and Scofield (1920) indicate a tendency for plots to yield in a similar manner from year to year, altho there are some exceptions. Their data for inter-annual correlations for hop yields are as follows:

Series	1st and 2nd Years	1st and 3rd Years	1st and 4th Years	1st and 5th Years	1st and 6th Years
1909	0.768 ± 0.051	0.662 ± 0.075	0.380 ± 0.105	0.259 ±0.115	0.061 ± 0.125
1910	0.577 ± 0.082	0.447 ± 0.099	0.451 ± 0.098	0.274 ±0.114	
1911	0.062 ± 0.123	0.313 ± 0.111	0.126 ± 0.121		
1912	0.311 ± 0.111	0.705 ± 0.062			
1913	0.597 ± 0.079				

In a later paper, Harris and Schofield (1928) give the results of a 15-year study on a uniform cropping experiment at Huntley, Montana. In general, a positive correlation between the yields of a series of plots was found thruout a period of years. The plots which show a heavier yield one year will in general show heavier yields in other years during the perior under investigation. Under some conditions negative correlations were found which were interpreted as indicating the importance of a preceding crop in determining the characteristics of an experimental field.

Garber, et al. (1926) found some tendency for plots which produced relatively high yields of oat hay in 1923 to produce relatively high yields of wheat grain in 1924. The correlation coefficient for the two years was 0.304 ± 0.036. The study was continued by Garber and Hoover (1930) to determine whether or not the natural variation in soil productivity among plots as revealed by a crop uniformity test persisted after an experiment is started that involves different crops and different soil treatments. They correlated the relative yields from duplicate oat plots in 1923 and the relative average yields from the same duplicate plots of other crops in a rotation experiment from 1924 to 1929 (incl.). The data were as follows:

"r" value between yields in 1923 and	N	r
1924	130	0.38 ± 0.05
1925	130	0.35 ± 0.05
1926	126	0.48 ± 0.05
1927	128	0.41 ± 0.05
1928	120	0.42 ± 0.05
1929	136	0.27 ± 0.05
Average 1925 to 1929 (incl.)	115	0.60 ± 0.04

These correlation coefficients were all statistically significant, and indicate that the differences in natural productivity may persist over a period of years (five, in this case) even though the soil be subjected to different treatments.

VI. Causes of Soil Heterogeneity

Many factors may contribute to soil variation. Yields of crops from foot to foot may vary as a result of soil topography, soil moisture, and soil fertility. Harris (1920 has demonstrated by the use of the correlation coefficient that substratum heterogeneity is sufficient to influence experimental results.

(a) Soil Topography

The topography of the soil may directly or indirectly influence the variation in soil productivity. Steep hillsides are unadapted to experimentation because heavy rains gully the field and carry the fertilizers from plot to plot. Moreover, water is apt to pond on certain areas and influence crop yields. Sometimes errors are introduced by variation in the subsoil. For example, there are gravel pockets in the subsoil on the Judith Basin (Montana) field station.

(b) Soil Moisture

The water content of soil was studied by Harris (1915) on the U.S.D.A. Experimental Farm at San Antonio (Texas). He took borings 6 feet deep at 20-foot intervals on a field 150 by 264 feet in size. The coefficients ranged from r = +0.32 to 0.70, being statistically significant for each foot section of the upper 6 feet of soil.

(c) Fertility Elements

The carbon and nitrogen content of soils was studied by Harris (1915) at Davis, California. The heterogeneity coefficient for carbon was 0.417 ± 0.063, while that for nitrogen was 0.498 ± 0.057. On blow sands at Oakley, the r-value for carbon was 0.317 ± 0.068, and that for nitrogen was 0.230 ± 0.072. Wide fluctuations in nitrate nitrogen were reported by Blaney and Smith (1931) on 1/30 acre plots. They found that the probable error was usually greater than 5 per cent where less than 20 soil cores were considered. In fact, they recommended 50 soil samples on a 1/30-acre plot to reduce the error to approximately 5 per cent of the mean. When soils outside of Rhode Island were considered, they found that 6 to 81 samples were necessary to obtain a probable error that low. Some Colorado Station data show extremely wide fluctuations in p.p.m. nitrate nitrogen on an irrigated soil. A 13 by 10-foot plot was sampled in 5 places to a depth of 6 feet. The nitrate nitrogen varied from 5 to 35 p.p.m. on this small area. It is obvious that variations in nitrate nitrogen can cause yield differences from area to area.

VII. Corrections for Soil Variability

Once soil heterogeneity is recognized, some means must be obtained to avoid or correct its influence in field experiments. A decrease in size of plots and an increase in the number of replications (as will be shown later) has been the general practice to overcome soil variation. The repetition of plots of varieties or treatments to be tested against each other are scattered out so that they may sample the different conditions of the trial area. One variety, for instance, may be grown partly on favorable portions and partly on less favorable portions. This usually means that the variety encounters somewhere near average soil conditions. Efficient experimental designs provide for the removal of a portion of variability due to soil. Artificially constructed field plots were studied by Garber and Pierre (1933) over a 3-year period. They found that soil heterogeneity was largely removed by a thorough mixture of soil placed in 30 artificial bins. These soil bins were 9 feet 4 inches by 4 feet 8 inches (inside area) by 24 inches in height, and were 0.001-acre in area. They obtained a probable error of a single determination in per cent of the mean of 3.4 for soybean hay, and 6.2 for wheat. They found, however, that the variation in crops was still too high to make replication unnecessary.

VIII. Relation to the Experimental Field

Many early experimental fields were poorly selected because of the belief that an experimental farm should contain many different soil types, i.e., the soil should be extremely heterogenous. The Ohio Experiment Station was allowed to relocate after the first 10 years due to the poor choice of the original site. (Thorne, 1909). For all ordinary field experiments the land should be as uniform as possible in regard to topography, fertility, subsoil, and previous soil management. However, extreme uniformity may defeat the purpose of the investigator unless such soil is representative of the area for which the results are to apply.

(a) Topography

A perfectly level piece of land is as undesirable for field experiments as one with surface inequalities because water may pond on it. A slope of 1 or 2 per cent will permit water from heavy rains to flow off uniformly and completely. A slight slope is highly desirable on land to be irrigated. Some experimenters use low land or "draws" and irregular areas for bulk crops or for seed increase plots.

(b) Previous Soil Treatment

It is desirable to have soils which have had uniform previous treatment because there may be a carry-over effect of previous treatments. According to the American Society of Agronomy standards (1933): "When a field or series of plots has been occupied by varietal or cultural tests of such a nature as to seriously increase soil variability, one or more uniform croppings should intervene (or follow) before it is again used for such tests. It is frequently helpful to arrange the plots at right angles to the direction of the previous plots."

(c) Subsoil Conditions

When it is necessary to drain lands in the humid regions, the tile lines should be located so as to influence all plots alike. They should run across the plots rather than with them. In the case of soil fertility experiments, it is recommended that a soil profile be taken to a depth of 3 feet for each series of plots. Before soil treatment experiments are begun, representative samples of the soil and subsoil should be carefully taken for such analyses as may be desired for future reference.

References

1. Blaney, J. E., and Smith, J. B. Sampling Market Garden Soils for Nitrates. Soil Sci., 31:281-290. 1931.
2. Cochran, W. G. A Catalogue of Uniformity Trial Data. Suppl. Jour. Roy. Stat. Soc., 4:233-253. 1937.
3. Crowther, F., and Bartlett, M. S. Experimental and Statistical Technique of Some Complex Cotton Experiments. Emp. Jour. Exp. Agr. 6:53-68. 1938.
4. Davenport, E., and Frazer, W. J. Experiments with Wheat, 1888-1895. Ill. Agr. Exp. Sta. Bul. 41, pp. 153-157. 1896.
5. Fisher, R. A. Statistical Methods for Research Workers. Oliver and Boyd. 5th Ed. pp. 210-214. 1934.
6. Garber, R. J., et al. A Study of Soil Heterogeneity in Experiment Plots. Jour. Agr. Res., 33:255-268. 1926.
7. Garber, R. J., and Hoover, M. M. Persistence of Soil Differences with Respect to Productivity. Jour. Am. Soc. Agron., 22:883-890. 1930.
8. ______ and Pierre, W. H. Variation of Yields Obtained in Small Artificially Constructed Field Plots. Jour. Am. Soc. Agron., 25:98-105. 1933.
9. Harris, J. Arthur. On a Criterion of Substratum Homogeneity (or Heterogeneity) in Field Experiments. Am. Nat., 49:430-454. 1915.
10. ______. Practical Universality of Field Heterogeneity as a Factor Influencing Plot Yields. Jour. Agr. Res., 19:279-314. 1920.
11. ______, and Schofield, C. S. Permanence of Differences in the Plots of an Experimental Field. Jour. Agr. Res., 20:335-356. 1920.
12. ______. Further Studies on the Permanence of Differences in the Plots of an Experimental Field. Jour. Agr. Res., 36:15-41. 1928.
13. Hayes, H. K., and Garber, R. J. Breeding Crop Plants. McGraw-Hill, pp. 56-69. 1927.
14. ______. Control of Soil Heterogeneity and Use of the Probable Error Concept in Plant Breeding Studies. Minn. Agr. Exp. Sta. Tech. Bul. 30. 1925.
15. Love, H. H. Planning the Plat Experiment. Jour. Am. Soc. Agron., 20:426-432. 1928.
16. Lyon, T. L. Some Experiments to Estimate Errors in Field Plot Tests. Proc. Am. Soc. Agron., 3:89-114. 1911.

17. Mercer, W. B., and Hall, A. D. Experimental Error of Field Trials. Jour. Agr. Sci., 4:107-127. 1911.
18. Parker, W. H. Methods Employed in Variety Trials by the National Institute of Agricultural Botany. Jour. Natl. Inst. Agr. Bot., 3:5-22. 1931.
19. Piper, C. V., and Stevenson, W. H. Standardization of Field Experimental Methods in Agronomy. Proc. Am. Soc. Agron., 2:70-76. 1910.
20. Standards for the Conduct and Interpretation of Field and Lysimeter Experiments. Jour. Am. Soc. Agron., 25:803-828. 1933.
21. Thorne, C. E. Essentials of Successful Field Experimentation. Ohio Agr. Exp. Sta. Cir. 96. 1909.
22. Wishart, J., and Sanders, H. G. Principles and Practice of Field Experimentation. Emp. Cotton Growing Corp., pp. 7-8, and 60-65. 1935.

Questions for Discussion

1. Discuss how soil heterogeneity might influence yield trials.
2. Why and how may small plots overcome the influence of soil variation?
3. What is a uniformity trial? How conducted?
4. What uses can be made of uniformity trial data?
5. How can correlation be used to measure soil heterogeneity?
6. Fundamentally, what is the so-called "heterogeneity coefficient" used by J. Arthur Harris? How interpreted?
7. What evidence did Harris have that soil heterogeneity was universal?
8. What general results were obtained at the Minnesota Station when the yields of adjacent plots were correlated? Those separated by other plots?
9. Are differences in the productivity of plots constant from year to year? Explain.
10. How may soil topography, moisture, and nitrogen account for soil heterogeneity?
11. What corrections can be used for soil variability?
12. Would artificial soil bins do away with the need for replication? Explain.
13. What precautions should be taken in the selection of an experimental field?
14. Is extremely uniform soil always desirable for experimental work? Explain.
15. What is the value of a bulk crop preceding an experiment?
16. To what use would you put uneven and low land in an experimental field? Why?

Problems

1. One acre was planted uniformly to the same variety of wheat and harvested in units 1/500-acre in size. (Data from Mercer and Hall). The 16 plots in the southwest corner of the acre gave yields in pounds as follows:

3.87	4.21	3.68	4.06
3.76	3.69	3.84	3.67
3.91	4.33	4.21	4.19
3.54	3.59	3.76	3.36

(a) Calculate the correlation coefficient by the analysis of variance for a 1 by 2-fold arrangement.

(b) Calculate the simple correlation coefficient for the same paired values.

2. Some unpublished data from the Akron Field Station give the average yields of corn and oats (combined) in bushels for a particular piece of land for 20 years as follows:

	North				Totals
	43.9	39.8	40.6	39.4	163.7
	39.0	40.4	37.7	35.0	152.1
West	32.9	40.5	39.5	36.7	149.6
	38.6	41.3	41.1	35.5	156.5
	35.4	43.1	37.3	30.5	146.3
Totals	189.8	205.1	196.2	177.1	

(a) Calculate the correlation coefficient by the analysis of variance to determine the heterogeneity from north to south, i.e., for a 1 by 4 combination.

(b) What is the correlation coefficient for a west to east direction? Calculate r for 1 by 5 combinations.

(c) In what direction is the soil most variable? Why?

(d) Assume that the yield for each plot is 40 bushels in the above problem 2. Calculate the correlation coefficient by the analysis of variance.

3. Some yields of wheat plots of a single variety grown in 10 by 10-foot plots were as follows: (Data from Montgomery).

67	56	70	76	76	69	59	74	73	71
66	71	61	63	67	65	66	77	79	77
66	67	62	72	54	61	64	80	76	76
64	68	65	67	76	71	68	72	77	86
74	58	62	68	77	74	77	70	65	66
58	60	71	64	70	64	65	57	75	74
73	63	62	69	86	64	66	61	62	62
57	37	56	65	69	64	63	57	61	65
58	43	73	73	71	73	74	66	67	79
59	53	60	73	78	78	72	73	60	74

Calculate the heterogeneity coefficient (intra-class correlation) by the analysis of variance for a 2 by 1-fold combination (2 horizontal rows and 1 vertical row).

CHAPTER XIII

SIZE, SHAPE AND NATURE OF PLOTS

I. Early Use of Field Plots

Modern field experiments began in 1834 when Jean Boussingault started a series of tests on his farm near Bechelbronne in Alsace. Early agriculture investigators favored large plots because of their attempts to conduct field trials in essentially the same manner as the farmer handled his crops.

The size of plot was considered at the Virginia Experiment Station as early as 1890 by Alwood and Price (1890) who suggested that, within limits, the larger the plot the more reliable the results. However, they conceded that small plots were sufficiently accurate for preliminary trials and for obtaining information on earliness and general quality of varieties. Taylor (1908) found a wide variation in size of plots used in this country in 1908. They varied from two acres in a Georgia cotton experiment to 1/40-acre in size, with all sizes between the two extremes. The average size of plot in America at that time was 1/10-acre.

The size of plots in relation to the experimental error was first studied at the Rothamsted Experimental Station in 1910 by Mercer and Hall (1911). As a result of their work and that carried on subsequently by others, the trend has been toward smaller plots and increased replication. A questionnaire, sent out by the Committee on the Standardization of Field Experiments of the American Society of Agronomy in 1918, reflected this tendency. The plot sizes used by different agronomists varied in size from one acre to 1/200-acre, with very few using plots larger than 1/10-acre or less than 1/80-acre.

At the present time, plot sizes vary from 1/10 to 1/1000-acre in size. The basis for the smaller plots with increased replication has been data from various blank or uniformity trials conducted by Mercer and Hall (1911), Day (1920), Summerby (1925), McClelland (1926), Wiebe (1935), Smith (1938) and many others. The catalogue by Cochran (1937) should be consulted for uniformity trials with specific crops.

A -- Size and Shape of Plots

II. Factors that Influence Plot Size

There are several factors to consider in plot size aside from the accuracy of the results. Some of these are: Kind of crop, number of varieties or treatments, kind of machinery to be used on them, and the amount of land, labor, and funds available for the tests. (1) Kind of Crop: It is the general practice to use larger plots for corn, sugar beets, and the forage plants than for small grains. The plots must be large enough to carry a representative population of the crop involved. (2) Number of Varieties or Treatments: Small plots are a necessity when large numbers of varieties or strains are in various testing stages. In small grains, it is not uncommon to have from 500 to 20,000 strains in the various stages of a breeding program. (3) Amount of Seed: In the early years of selection in small grains and in many other plants, only a very small amount of seed is usually available. Obviously, the plots must not be too large for the seed supply. (4) Kind of Machinery: The area and shape of field plots should be such as to enable the operation of standard farm machinery and to reduce to a reasonable minimum the errors concerned therewith. Larger plots are necessary when the crop is planted, cultivated, and harvested with standard farm machinery than where hand methods are used. (5) Land Area: For a given area of land, the plot size varies inversely with the number of varieties or treatments to be included. This is true until the minimum practical size is reached. As a result, to quote Goulden (1929): "The general practice is to use quite small plots adequately replicated for strain tests, i.e., when there are a large number of varietal units, and larger plots when the number of varieties is small enough to permit their use with the amount of land available." (6) Funds Available: In general, it is more costly to use large plots than small plots.

III. Kinds of Experimental Plots

It is necessary to distinguish between nursery and field plots more or less arbitrarily. Nursery plots are usually small plots cared for by hand while field plots are larger and adapted to the use of standard farm machinery. The present tendency is to reduce the size of field plots and to enlarge nursery plots from single to multiple short rows (rod-rows in many cases).

(a) Nursery Plots

Nursery plots may be as small as one square yard in area, but the rod-row is probably the most common unit size. Small plots allow the preliminary testing of many strains. However, uniform soil and careful technic is vital to accuracy for small plots. Taylor (1908) points out that small mistakes on small plots may greatly modify the results. For example, an error of 5 pounds on a 1/20-acre plot would mean an error of 100 pounds on an acre basis. The rod-row unit has been widely used in this country for small grain trials while the chessboard plot has been used in England. Engledow and Yule (1926) describe the latter as being one yard square with the crop space-planted at 2 by 6 inches. The principal objection to the chessboard is the amount of detailed hand labor involved and the fact that it affords less opportunity to observe strength of straw, evenness of germination, etc. As plant individuality must be considered in row crops, there is some variation in type of nursery plots.

(b) Field Plots

For standard farm machinery, field plots usually vary from 1/10 to 1/100 acre in size. They offer more opportunity to observe crop behavior under conditions comparable to those found on the farm. Field plots are used for variety tests, crop rotation experiments, fertilizer trials, forage experiments, pasture experiments, irrigation studies, cultural trials, etc. Ordinarily, such plots are long and narrow in shape as most convenient for farm machinery.

(c) Comparison of Nursery vs. Field Plots

The use of small hand-sown nursery plots to test yields of agricultural crop varieties has been frequently criticised on the ground that such plots do not represent normal agricultural conditions. In general, small plots have been found to compare favorably with large field plots in accuracy so long as adequate precautions have been taken against competition and other errors. There is further evidence that nursery plots give results that are valid when applied to agricultural practice.

As early as 1910, Lyon (1910) reported a comparison of seven 1/10-acre plots with seven groups of 10-row plots 17 feet long. The probable errors were 5.09 and 4.49, respectively. Moreover, less land was required for the small plots. Seven 1/10-acre plots covered an area of 30,492 square feet, while 70 of the 17-foot rows required only 1,190 square feet in area.

A general correspondence of rod-rows and field plots has been shown by Klages (1933) for 11 to 14 varieties of spring wheat, 7 varieties of durum wheat, 12 to 15 varieties of oats, 13 to 29 varieties of barley, and 7 varieties of flax in each of 4 years. He calculated the correlation coefficients (r) for the two sets of plots. Hayes and others (1932) compared the yields of 16 wheat varieties sown in rod rows by hand and by a drill at different rates with those sown by a farm drill in 1/40-acre plots. The correlation coefficients indicate some agreement between the yields obtained from the small and large plots. Smith (1936) has criticized the correlation coefficient as inefficient in such comparisons: "If real differences between varieties were either small or non-existent, then the correlation coefficients would be zero or insignificant, altho the trials might agree in showing no significant differences between them. On the other hand, the correlation coefficient could not become unity unless experimental error could be entirely eliminated. Consequently, r may vary from 0 to + 1 even while the two forms of trial are in perfect agreement."

In a study of 12 timothy varieties, Smith and Myers (1934) showed that the yields from rod-rows and 1/50-acre field plots agreed to precisely the degree required by statistical theory. Smith (1936) later compared 9 wheat varieties sown by a farm drill in 1/100-acre plots and dibbed in square yard plots. Agreement of the two experiments was excellent with respect to yield of grain. Tysdal and Kiesselbach (1939) compared 2 varieties of alfalfa in 1/80-acre field plots with various 16-foot nursery plots which differed as to number of rows and spacing. They combined the forage yields into a single analysis of variance from which they concluded that the several types of nursery plots gave essentially the same yields of the two varieties as did the field plots. The interaction of varieties x type of plot was not significant.

The problem resolves itself into whether small nursery plots with more precise control of soil heterogeneity will give the same results as large field plots with less control of soil variability. The sacrifice in plot (sample) size must be balanced by more effective control of soil heterogeneity for the small nursery plot to be as satisfactory as the large field plot. This can be brought about to some extent by increased replication of small plots.

IV. Relation of Plot Size to Accuracy

In general, it has been found that the variability is decreased as the plot is increased in size up to about 1/10-acre. However, the variability is less when a unit of a certain area is made up of several distributed units than when a single large unit is used. In a theoretical discussion, Siao (1935) states "Increasing the size of plot decreases the variability of the experiment by increasing the precision of a single plot yield. On the other hand, there is an increase in the variability within the block through expanding the area included in the block. There are two opposing tendencies that affect the experimental error as the plot changes in size, the final result being due to a balance between these two tendencies. The slow rate of reduction in experimental error through increase in size of plot and, in exceptional cases, the greater variability for larger plots, may be explained by increase in variation within the block as the plot increases in size." The work of Stadler (1921) and Wiebe (1935) indicates that the total variation tends to increase as more land is added to the experimental area, provided the size and shape of the ultimate units remains the same. It should be emphasized that plot size varies with the conditions of the experiment, there being no one size best for all crops on all soils. Comparative studies on plot size have been carried out in most instances on blank or uniformity tests. After optimum plot size has been determined, the standard error per plot and the number of replications to reach a given degree of accuracy in the comparison of the mean treatment yields is usually computed. Typical investigations on plot size will be considered for small grains and for other crops separately.

(a) Small Grain Plots

Much of the earlier work was conducted with small grains. The conclusions applicable to one are generally applicable to the others. Mercer and Hall (1911) used uniformity trial data for an acre of wheat, the field being divided into 500 small plots each of which was harvested separately. Adjacent plots were grouped so as to form plots of different sizes. The standard deviations in per cent for the 1/500, 1/250, 1/125, 1/50, 1/25, and 1/10-acre plots were 11.6, 10.0, 8.9, 6.3, 5.7, and 5.1, respectively. The standard deviation was reduced as the plots were made larger, but the increase in plot size above 1/50-acre produced a relatively small decrease in variability. These investigators found that precision was increased more rapidly by replication. When five scattered 1/500-acre plots were combined so as to give a total area of 1/100-acre the standard deviation in per cent of the mean was reduced to 4.6 per cent. Olmstead (1914) found with wheat that a number of small plots ranging down to 0.0007-acre in size is much better than the same total area in one plot, and also that one large plot is more accurate than one small one. In wheat studies, Day (1920) found that the probable error decreased with an increase in plot size up to 1/20-acre. From a blank trial with wheat, Smith (1938) concluded that the reduction in variability by increasing plot size is less than could be obtained by equivalent random replication.

Hayes (1923) compared 16 and 32-foot rows of wheat, oats, and barley. He failed to find a significant difference in favor of 32-foot rows. A comparison of one and two rod-rows plots indicated little advantage for the harvest of two rod-row per plot over one. Stadler (1921) obtained data on three and five-row plots, the border rows being discarded. His results follow:

Crop	No. Plots	Coefficient of Variability	
		One Central Row	Three Central Row
Barley	21	24.80	22.18
Oats	20	24.80	22.59
Wheat	80	27.68	25.11

Summerby (1925) found very little difference in accuracy between large and small plots when eight replications were used. His oat plots were 1,2,4,8,16, and 32 rows in width, spaced one foot, and 15 feet long. Love and Craig (1938) made an analysis of data from 2 oat crops for various types of plots and various numbers of replications, and for rows 15 and 30 feet in length. The data indicate that 3-row plots with several replications (8 or 10), when all rows are harvested, give accurate results. They are preferred to single-row plots. The 15-foot rows were considered more satisfactory than those 30 feet in length. Such data as these support the widespread practice of using three rod-row plots for small grain nursery trials with the center row harvested for yield.

(b) Other Crops

Different crop plants are known to differ in variability. The coefficients of variability for different crops were compared by Smith (1938) for a standard 1/40-acre plot from the published data for 39 uniformity trials. The crops fell roughly into 3 groups: (1) wheat, mangolds, sugar beets, soybeans, and sorghums (forage) seem to be less variable; (2) corn, potatoes, cotton, and natural pasture were intermediate; and (3) fruit trees were most variable.

In the case of corn, Bryan (1933) reports that "variability of plot yields decreased as the size of plots increased from 8 to 16, to 24, and to 48 hills, but the decrease was not proportional to the size of plot. The experimental error for a given area, therefore, would be lower with larger numbers of small plots." McClelland (1926) obtained a similar reduction in error as the size of plot was increased, the error being 11.2 per cent for 1/80-acre plots, and 6.2 per cent for those 1/2 acre in size.

With sorghums, Stephens and Vinall (1928) concluded that the errors decrease with an increase in plot size up to 1/20-acre. Increasing the plot from 1/800-acre to 1/20-acre, with the same total area concerned, reduced the probable error about 60 per cent.

The standard error was found by Immer (1933) to be actually reduced in sugar beet plots when the plot size was increased from one to two rows in width, or for an increase in length from two to four rods. However, efficiency in the use of land decreased as the size of plot was increased. Some of his data for the harvest of the entire plot are as follows:

Length Plot in Rods	Percentage Efficiency of Plots of Indicated Width (Rows) 1	2	3	4	6	12
2	100.0	88.0	77.7	53.3	34.9	27.4
4	76.2	62.5	48.2	35.2	21.2	28.8
10	50.0	37.6	28.6	26.1	10.2	9.4
20	35.1	24.5	21.6	10.1	5.8	6.7

Similar results were reported by Immer and Raleigh (1933).

Uniformity trial data with soybeans, computed by Odland and Garber (1928), indicate that 16-foot plots in single rows replicated three times were the most satisfactory when both accuracy in results and land economy were taken into account.

Westover (1924) experimented with 220 rows of potatoes, 150 feet long. He harvested them in 10-foot lengths, and found a sharp reduction in probable error between row lengths of 10 and 40 feet. Beyond 60-foot lengths, there was very little reduction in error.

Ligon (1930) found no necessity for rows greater than 100 feet in length for cotton, the shorter rows being just as accurate when sufficiently replicated. Unit rows of cotton 24 feet long and spaced one foot apart were used by Siao (1935) in studies on size of plot for cotton. When combined into plot sizes of 1,2,3,4, and 8 rows, the efficiency was greatest for the smallest plot.

In plot size studies with millet, Li and others (1936) concluded that plots 15 feet long and two rows wide were the most efficient, i.e., 113.9 per cent compared to 100 per cent for 15-foot plots one row wide.

Batchelor and Reed (1918) studied the variability of orchard plot yields from the standpoint of increasing the number of adjacent trees per plot. The average reduction in variability for all fruits was 37.78 to 24.27 per cent when the plot was increased from one to eight trees, but little was gained by including 16 to 24 trees per plot.

The reasons for variability in small plots may be summarized as follows: (1) Variability in soil, (2) losses in harvest and errors in measurement have a relatively great effect, (3) in row crops, plant variability may be important because of fewer plants, (4) competition and border effects are apt to be greater on small plots.

V. Plot Sizes for Various Crops

The plot sizes depend upon the crop plant, and upon the conditions under which the test is conducted. (1) Small Grains: The majority of experiment stations use three-row plots with the center row harvested for yield, but a few use five-row plots with the center three rows used for the yield determination. A few use single-rod-row plots. (2) Corn: The Nebraska station uses four-row plots, 12 hills long, harvesting the center rows for yield. Others use single rows about 20 hills long, or three rows of the same length with only the center one harvested for yield. Bryan (1933) reports that, in a comparison of open-pollinated varieties and hybrids, equal degrees of precision were attained with about half as many plants or hills of crosses as of open-pollinated varieties. He found that 48 total hills were sufficient to represent a variety. (3) <u>Soybeans</u>: Soybeans may be grown in rows 16 feet long and 30 to 32 inches apart. Field plots are often employed. (4) <u>Sorghums</u>: The work of Stephens and Vinall (1928) indicates that "three or four replications of 1/40-acre or 1/80-acre plots will give results sufficiently reliable for the ordinary sorghum test". Slightly larger plots are advocated by Swanson (1930). When protected by borders, 2 and 4-row plots 8 rods long having an area from 1/50 to 1/25-acre, are regarded as convenient units. At Kansas, four-row plots about 100 feet long are used. The grain sorghums are thinned to eight inches in the row, while the forage sorghums are spaced four inches in the row. The rows are spaced the same distances apart as for corn. (5) <u>Alfalfa and Clovers</u>: These crops are usually grown in field plots about seven feet wide and 60 feet or longer in length, with the center five feet harvested with a mower. Tysdal and Kiesselbach (1939) state that the most serviceable types of plot for advanced nursery testing appear somewhat optional among these: (a) Solid-drilled 5 to 8 rows spaced 7 inches apart with a 12 to 14-inch alley between border rows; or (b) solid-drilled 3 to 5 rows spaced 12 inches apart with an 18-inch alley between border rows. The entire plot may be harvested since very little error due to border effect occurs. (c) Single rows spaced 18 to 24 inches apart are permissibile for preliminary nursery tests. (6) <u>Sugar Beets</u>: Immer (1932) states that four-row plots are the most efficient. The rows should be two to four rods long, spaced 20 to 22 inches apart, and the plants thinned to about 12 inches in the row.

VI. <u>Relation of Shape to Reliability</u>

Some investigators have found that long narrow plots best overcome the effects of soil heterogeneity, while others believe that plots should be approximately square. For example, Barber (1914) reported that a small square plot affords a more accurate basis for variety comparisons than a long narrow plot that has extra growth along the borders when alleys exist between the plots. On the other hand, Kiesselbach (1918) showed that the coefficient of variability for 1/10-acre oat plots 48 rods by 5.5 feet was 3.84 per cent, as compared with 5.18 per cent for plots 16 rods by 16.5 feet. Justesen (1932) found long narrow plots to be more efficient than the shorter plots of the same area. Mercer and Hall (1911) divided the plots of a single variety into plots of equal area but of different shapes. The dimensions were 20 by 12, and 50 by 5 yards. They found no significant difference in variability between them. Similar results were obtained by Stephens and Vinall (1928) with sorghums. Bryan (1930), in his work with various shapes of corn plots, concluded that shape is less important as the size of plot is reduced. With plots as small as 16 hills, either single, two or four-row plots may be expected to give similar results.

These apparent inconsistencies are explained in the work of Day (1920) who harvested, in five-foot sections, a 1/40-acre area uniformly cropped to wheat and combined the ultimate units to form plots of various shapes. He found that plots with their greatest dimensions in the direction of the least soil variation are more variable than plots having their greatest dimension in the direction of the greatest variation. He found that shape exerted no influence on accuracy where soil variation is as great in one direction as it is in the other. Some of his data follow:

Adjacent Rows (No.)	Length Rows (Ft.)	Total Length Rows in Plot. (Ft.)	Shape of Plot	C.V.
1	150	150	Long in direction of least variation	17.36
3	50	150	Long in direction of least variation	16.37
10	15	150	Rectangular	12.72
24	5	120	Long in direction of most variation	10.54

Similar conclusions were obtained by Siao (1939) for cotton and by Smith (1938) for wheat.

It is generally conceded that relatively long and narrow plots with the long dimension in the direction of the greatest soil variation best overcome the affects of soil heterogeneity. In addition, linear plots are more economical for cultural operations. However, the area occupied by a single replicate or block should approach a square in shape for the most efficient design.

VII. Practical Considerations in Plot Shape

Width of plots should be sufficient to allow for the removal of border rows when this appears desirable, or to render border effects negligible when not removed. The triple rod-row is a convenient shape for small plots, while large plots are usually rectangular in shape to accommodate farm machinery in an attempt to simulate farm conditions.

(a) Adaptation to Farm Machinery

Some multiple of seven feet provides a favorable width for field experiments as it permits convenient operations of the 3.5 and 7.0-foot farm implements. Kiesselbach (1928) calls attention to the fact that the multiple of seven feet will enable the use of the seven-foot disk, seven-foot drill, seven-foot binder, and 3.5-foot corn planter and cultivator. The standards of the American Society of Agronomy (1933) recommend 14 feet as a minimum plot width for crop rotation, fertilizer, and tillage experiments, while varietal tests with inter-tilled crops commonly should contain at least three or four rows. Extremely narrow plots, in the case of manurial or fertilizer tests, make it difficult to keep the treatments within the plot limits.

(b) Calculation of Plot Size

Plots should be made an alequot part of an acre, e.g., 1/40, 1/50, 1/80-acre plots. This sort of plan is worthwhile because of the grave possibility of error in computations made on acres expressed as decimal fractions. For instance, to calculate the dimensions of a 1/40-acre plot for a drill seven feet wide, the steps are as follows:

43,560/40 = 1089 square feet in 1/40 acre.
1,089/7 = 155.6 feet for length of the plot.

Hayes (1923) suggests for small grain nursery rows spaced 12 inches apart, that the row length be adjusted in length slightly so that gram yields per plot can be converted to bushels per acre by multiplying by a simple conversion factor. The factor 0.2 can be used for a 15-foot row of oats, the factor 0.1 for a 16-foot row of wheat, and the factor 0.1 for a 20-foot row of barley.

VIII. Calculation of Plot Efficiency

Some uniformity data on 120 rod rows of Haynes Bluestem wheat in bushels per acre, as given by Hayes and Garber (1927), will be used for the computation of plot efficiency. The method was suggested by Dr. F. R. Immer.

25.0,	22.0,	21.7,	22.0,	18.6,	23.5,	20.3,	19.9,	24.9,	22.9,	22.9,	25.0,
24.9,	25.0,	20.7,	24.5,	26.3,	25.2,	25.6,	23.1,	28.5,	27.0,	26.6,	25.9,
29.4,	24.4,	23.1,	30.7,	25.7,	21.9,	23.2,	23.2,	21.5,	25.6,	24.4,	28.2,
28.7,	26.8,	25.2,	29.6,	25.4,	26.5,	24.4,	24.1,	29.3,	24.3,	28.9,	25.5,
28.0,	23.7,	26.4,	27.8,	23.8,	23.7,	28.3,	28.8,	21.0,	23.8,	24.8,	22.8,
27.0,	23.8,	24.6,	31.7,	28.0,	28.1,	24.5,	26.7,	31.6,	23.7,	25.0,	33.0,
28.9,	27.5,	25.2,	27.9,	28.2,	25.4,	25.8,	28.1,	32.0,	29.6,	28.3,	33.7,
30.9,	27.2,	26.0,	30.0,	22.7,	21.6,	19.3,	24.0,	28.8,	25.2,	26.0,	27.6,
25.3,	27.4,	25.2,	21.6	28.3,	25.1,	21.6,	24.6,	25.9,	24.8,	26.9,	25.9,
24.1,	25.8,	22.1,	26.9,	27.6,	27.6,	27.3,	30.2,	22.4,	23.7,	23.1,	26.5.

It is assumed that 30 varieties are to be tested, and that the investigator desires to determine the relative efficiency of 1, 2, and 4-row plots. The analysis of variance will be used.

(1) One Row per Plot:

Block	$S(x_b)$	$S(x^2_b)$
I	727.2	528,819.84
II	767.6	589,209.76
III	826.7	683,432.89
IV	760.8	578,816.64
Totals	3082.3	2,380,279.13

$S(x)$ for all plots = 3,082.30 $\bar{x}$ = 25.685833.

$S(x^2)$ for all plots = 80,176.37. $S(x)^2/N$ = 79,171.4306

$S(x^2) - S(x)^2/N$ = 1,004.9394

$$\frac{S(x^2_b)}{30} - \frac{(Sx)^2}{N} = 79,342.64 - 79,171.43 = 171.21$$

Variation due to	D. F.	Sum Squares	Mean Square
Blocks	3	171.21	57.0698
Varieties and Error	116	833.73	7.1873
Total	119	1004.94	

(2) Two Rows per Plot:

Block	$S(x_b)$	$S(x^2_b)$
I	1494.8	2,234,427.04
II	1587.5	2,520,156.25
Totals	3082.3	4,754,583.29

$$\frac{S(x^2)}{2} - \frac{S(x)^2}{N} = \frac{159,661.25}{2} - 79,171.43 = 659.19$$

The total $S(x^2)$ is divided by 2 to place the results on a single plot basis so that the common correction factor $S(x)^2/N$, can be used.

$$\frac{S(x^2_b)}{60} - \frac{S(x)^2}{N} = 79,243.05 - 79,171.43 = 71.62$$

Variation due to	D. F.	Sum Squares	Mean Square
Blocks	1	71.62	71.62
Varieties and Error	58	587.58	10.13
	59	659.20	

(3) Four Rows per Plot:

$$\frac{S(x^2)}{4} - \frac{S(x)^2}{N} = \frac{318.930.59}{4} - 79,171.43 = 561.21$$

Variation due to	D. F.	Sum Squares	Mean Square
Total	29	561.21	19.3521

(4) Comparison of 1, 2, and 4 Row Plots:

No. Determinations	Rows per Plot	No. Blocks	Mean Square	Mean Square Basis One Plot	Pct. Efficiency
30	1	4	7.1873	7.1873	100.00
30	2	2	10.1307	5.0653	70.95
30	4	1	19.3521	4.8380	37.14

B -- Plot Replication

IX. Replication in Experimental Work

Replication is merely repetition. The investigator repeats a variety or treatment several times in a test in order to obtain a mean yield or value which is a more reliable estimate of the yield of the general population than that obtained from a single plot of a treatment. It also provides the mechanism for a valid estimate of the random errors in an experiment. Strictly speaking, five replications of a variety refer to six plots, i.e., the original plot and its repetition five times. For the sake of simplicity, the number of replications will be understood to mean the number of plots grown of each variety or treatment. In field experiments, a single replicate is usually planned to contain one plot of each treatment in a rather compact block. The repetition of the treatments is brought about by the repetition of the blocks. This distribution of plots over the experimental area is an effort to sample the field in an attempt to measure and, in some cases remove, the influence of soil heterogeneity. Replication in space and time is often necessary. For example, it may be desirable to repeat an experiment in other regions of the state in order to sample different soil and climatic conditions. In the same region, repetition of the experiment over a number of years may be necessary to sample the climatic conditions in different seasons.

X. History of Replication

Replication of experimental plots has been comparatively recent. In the old field tests large single plots were placed side by side. These were simple and effective for the demonstration of known facts so long as the differences to be observed were large. However, they are inadequate as soon as accurate measurements are needed because they do not take into account the tremendous variation in the soil from plot to plot.

Sir John Russell (1931) gives some of the early history of replication. The Broadbalk plots at the Rothamsted Experimental Station were split lengthwise into two halves in 1846-47 which, from that time onwards, were harvested separately. This was the first duplication of field experiments so far as can be determined. In 1847-48, and occasionally afterwards, one half of each plot was treated differently from the other with the result that they ceased to be strict duplicates. Better duplication appears to have been practiced by P. Nielsen, founder of the Danish Experiment Station about 1870, in his experiments on grass mixtures for pastures. Some Norfolk (England) experiments carried out in the later 1880's were systematically replicated as follows: ABCDDCBA. In America, some experimenters began to use replication about 1888. Some old experiments in Kansas were replicated six times. However, replication soon fell into disuse because of the demand for information and due to limited land and funds. Single plots were the rule. Nothing further was done in England until 1909 when A. D. Hall (1909) and later Wood (1911) urged the need for the estimation of experimental errors. Marked changes came about as a result. S. C. Salmon (1913) revived duplication of plots in this country in 1910. Single 1/10-acre plots were commonly used for variety and rate and date of seeding tests with small grains at that time. He split those 1/10-acre plots into 1/50-acre plots and replicated them five times for variety tests. His rate and date tests were replicated three times. Thus, the same area was required for variety tests as before and a smaller area for rate and date tests. Largely as the result of his efforts, the Office of Cereal Investigations, U. S. D. A., provided for replication in their work about 1912. In England at about the same time, Dr. E. S. Beaven designed his well-known strip method of replication which is especially suited to variety trials.

A questionnaire sent out by the Committee on Standardization of Field Experiments of the American Society of Agronomy in 1918 indicated that less than 20 per cent of the agronomic workers depended upon single plot tests even though they had been the rule 10 years previously. At present, replication is considered essential in modern field experiments.

XI. Reduction of Error by Replication

The most effective method to obtain greater accuracy in field experiments as well as in many other types of agronomic experiments, is to increase the number of replications. It can be brought about to a limited extent by an increase in plot size as shown by Summerby (1923). However, frequent replication of small plots proved to be a more efficient means to obtain a high degree of accuracy than the use of the same amount of land with less frequently replicated larger plots. Love (1936) gives some uniformity trial data with cotton that indicate the same trend. The probable error for a 2-row plot 20 feet long was 10.35 per cent, while that for two single 20-foot plots was 9.01. Further, the probable error for a single 4-row plot was 9.51 while for the same area made up of four scattered units it was 7.35 per cent. Many other investigators have obtained similar results. Since the standard error of the mean ($\sigma_{\bar{x}}$) is given by $\sigma_{\bar{x}} = \frac{s}{\sqrt{N}}$, it follows that the decrease in $\sigma_{\bar{x}}$ is proportional to the square root of the number of replications. This rule applies when the variation due to the replicates themselves is removed from the error, but not strictly otherwise. This can be illustrated with some data on 120 rod rows of bluestem wheat, cited by Hayes and Garber (1927). The value of replication was studied on the variability of yields calculated separately on the basis of 20 determinations and fo 1,2,4, and 6 systematically distributed plots. The coefficients of variabilityrwere compared with mathematical expectation as follows:

No. determinations	No. systematically distributed plots	C. V.	Mathematical expectation
20	1	9.05	9.05
20	2	6.54	$9.05/\sqrt{2}$ = 6.42
20	4	5.61	4.53
20	6	4.44	3.69

The calculated coefficient of variability decreases as a result of replication, but less rapidly than would be indicated by mathematical expectation. This is attributed to the greater land area used for several replications than can be used for single-plot trials which, on the average, brings in soils of greater difference in productivity than can be found in smaller areas. In this case, the error due to blocks has not been removed. That replication beyond a certain point may be impractical is indicated in some data compiled by Salmon (1923). He shows the relation between the number of replications and the probable error of the mean (expressed in per cent) as follows:

Number of Plots	1	2	3	4	5	6	7	8	9
Kherson oats	3.7	2.0	2.0	1.7	1.9	1.8	1.7	1.6	1.5
Alfalfa	11.2	7.5	7.1	5.0	4.8	4.7	5.5	6.0	5.8
Ear corn	9.0	5.9	5.5	5.1	4.1	3.7	4.1	4.1	3.8

It is to be noted that variability was rapidly reduced up to 4 replications, but the decrease was at a much slower rate beyond that point. Hayes and Garber (1927) questioned whether the gain in accuracy beyond three replications warranted the additional work. The relation of replication to design will be considered in a later chapter.

XII. Number of Replications

The question naturally arises as to the number of replications that should be used. Goulden (1929) states that it depends upon the degree of soil heterogeneity, the degree of precision required, and the amount of seed available. Any desired degree of precision within practical limits may be ordinarily achieved for any given set of conditions by replication. For field plots, the American Society of Agronomy (1933) recommends 3 to 6 replications, dependent upon the degree of precision required. The smaller number will suffice when average rather than annual results are stressed. From 4 to 6 replications are commonly used in corn variety trials. Nursery experiments ordinarily should be replicated 5 to 10 times to assure significant results. It is impossible to prescribe a rule for all cases. In rod-row trials with oats, Love and Craig (1938) found 8 or 10 replications more satisfactory than a smaller number, as 3 or 5. In alfalfa nursery plots, Tysdal and Kiesselbach (1939) concluded that 4 to 16 replications were necessary to make a 5 per cent difference statistically significant for plots that varied in size from 1/80-acre to a single space-planted 16-foot row. The larger plot required the 4 replications. However, little is to be gained by the use of more than 10 replications in field trials.

References

1. Alwood, W. B., and Price, R. H. Suggestions Regarding Size of Plots. Va. Agr. Exp. Bul. 6. 1890.
2. Barber, C. W. Note on the Influence of Shape and Size of Plots in Tests of Varieties of Grain. Me. Agr. Exp. Sta. Bul. 226, pp. 76-84. 1914.
3. Batchelor, L. D., and Reed, H. S. Relation of the Variability of Yields of Fruit Trees to the Accuracy of Field Trials. Jour. Agr. Res., 12:245-283. 1918.
4. Bryan, A. A. A Statistical Study of the Relation of Size and Shape of Plot and Number of Replications to Precision in Yield Comparisons with Corn. Ia. Agr. Exp. Sta. Rpt. for 1930-31, p. 67. 1931.
5. ________. Factors Affecting Experimental Error in Field Plot Tests with Corn. Ia. Agr. Exp. Sta. Bul. 163. 1933.
6. Cochran, W. G. Catalogue of Uniformity Trial Data. Suppl. Jour. Roy. Stat. Soc., 4:233-253. 1937.
7. Day, J. W. The Relation of Size, Shape, and Number of Replications of Plots to Probable Error in Field Experiments. Jour. Am. Soc. Agron., 12:100-106. 1920.
8. Engledow, F. L., and Yule, G. U. The Principles and Practices of Yield Trials. Emp. Cotton Growing Corp. 1926.
9. Goulden, C. H. Statistical Methods in Agronomic Research. Can. Seed Growers Assn. 1929.
10. Hall, A. D. The Experimental Error in Field Trials. Jour. Bd. Agr. (London), 16:365-370. 1909.
11. Hayes, H. K. Controlling Experimental Error in Nursery Trials. Jour. Am. Soc. Agron., 15:177-192. 1923.
12. Hayes, H. K., and Garber, R. J. Breeding Crop Plants. McGraw-Hill. pp. 69-77. 1927.
13. ________ et al. An Experimental Study of the Rod- Row Method with Spring Wheat. Jour. Am. Soc. Agron., 24:950-960. 1932.

14. Immer, F. R. Size and Shape of Plot in Relation to Field Experiments with Sugar Beets. Jour. Agr. Res., 44:649-668. 1932.
15. ________ and Raleigh, S. M. Further Studies of Size and Shape of Plot in Relation to Field Experiments with Sugar Beets. Jour. Agr. Res., 47:591-598. 1933.
16. Justensen, S. H. Influence of Size and Shape of Plots on the Precision of Field Experiments with Potatoes. Jour. Agr. Sci., 22:366-372. 1932.
17. Kiesselbach, T. A. The Mechanical Procedure of Field Experimentation. Jour. Am. Soc. Agron., 20:433-442. 1928.
18. ________ Studies Concerning the Elimination of Experimental Error in Comparative Crop Tests. Nebr. Agr. Exp. Sta. Res. Bul. 13. 1918.
19. Klages, K. H. W. The Reliability of Nursery Tests as Shown by Correlated Yields from Nursery Rows and Field Plots. Jour. Am. Soc. Agron., 25:464-472. 1933.
20. Li, H. W., Meng, C. J., and Liu, T. N. Field Results in a Millet Breeding Experiment. Jour. Am. Soc. Agron., 28:1-15. 1936.
21. Ligon, L. L. Size of Plot and Number of Replications in Field Experiments with Cotton. Jour. Am. Soc. Agron., 22:689-699. 1930.
22. Love, H. H. Statistical Methods Applied to Agricultural Research. Com. Press Ltd. (Shanghai). pp. 404-420. 1936.
23. Love, H. H., and Craig, W. T. Investigations in Plot Technic with Small Grains. Cornell U. Memoir, 214. 1938.
24. Lyon, T. L. A Comparison of the Error in Yield of Wheat from Plots and from Single Rows in Multiple Series. Proc. Am. Soc. Agron., 2:38-39. 1910.
25. McClelland, C. K. Some Determinations of Plot Variability. Jour. Am. Soc. Agron., 18:819-823. 1926.
26. Mercer, W. B., and Hall, A. D. The Experimental Error in Field Trials. Jour. Agr. Sci., 4:107-132. 1911.
27. Odland, T. E., and Garber, R. J. Size of Plat and Number of Replications in Field Experiments with Soybeans. Jour. Am. Soc. Agron., 20:93-108. 1928.
28. Olmsted, L. B. Some Applications of the Method of Least Squares to Agricultural Experiments. Jour. Am. Soc. Agron., 6:190-204. 1914.
29. Report of the Committee on Standardization of Field Experiments. Jour. Am. Soc. Agron., 10:345-354. 1918.
30. Report of the Committee on Standardization of Field Experiments. Jour. Am. Soc. Agron., 25:803-828. 1933.
31. Russell, E. J. The Technique of Field Experiments (Forword). Rothamsted Conf., 13, pp. 5-8. 1931.
32. Salmon, S. C. A Practical Method of Reducing Experimental Error in Varietal Tests. Jour. Am. Soc. Agron., 5:182-184. 1913.
33. ________ Some Limitations in the Application of the Method of Least Squares to Field Experiments. Jour. Am. Soc. Agron., 15:225-229. 1923.
34. Siao, Fu. Uniformity Trials with Cotton. Jour. Am. Soc. Agron., 27:974-979. 1935.
35. Smith, H. Fairfield. Comparison of Agricultural and Nursery Plots in Variety Experiments. Jour. Counc. Sci. and Ind. Res., 9:207-210. 1936.
36. ________ An Empirical Law Describing Heterogeneity in the Yields of Agricultural Crops. Jour. Agr. Sci., 28:1-23. 1938.
37. ________, and Myers, C. H. A Biometrical Analysis of Yield Trials with Timothy Varieties using Rod Rows. Jour. Am. Soc. Agron., 26:117-128. 1934.
38. Stadler, L. J. Experiments in Field Plot Technic for the Preliminary Determination of Comparative Yields in Small Grains. Mo. Agr. Exp. Sta. Res. Bul. 49. 1921.
39. Stephens, J. C., and Vinall, H. N. Experimental Methods and the Probable Error in Field Experiments with Sorghum. Jour. Agr. Res., 37:629-646. 1929.
40. Summerby, R. Replication in Relation to Accuracy in Comparative Crop Tests. Jour. Am. Soc. Agron., 15:192-199. 1923.

41. Summerby, R. A Study of Size of Plats, Number of Replications, and the Frequency and Methods of Using Check Plats, in Relation to Accuracy in Field Experiments. Jour. Am. Soc. Agron., 14:140-149. 1925.
42. Swanson, A. F. Variability of Grain Sorghum Yields as Influenced by Size, Shape, and Number of Plots. Jour. Am. Soc. Agron., 22:833-838. 1930.
43. Taylor. F. W. The Size of Experiment Plots for Field Crops. Proc. Am. Soc. Agron., 1:56-58. 1908.
44. Tysdal, H. M., and Kiesselbach, T. A. Alfalfa Nursery Technic. Jour. Am. Soc. Agron. 31:83-98. 1939.
45. Westover, K. C. The Influence of Plat Size and Replication on Experimental Error in Field Trials with Potatoes. W. Va. Agr. Exp. Sta. Bul. 189. 1924.
46. Wiebe, G. A. Variation and Correlation in Grain Yield Among 1500 Wheat Nursery Plots. Jour. Agr. Res., 50:331-357. 1935.
47. Wood, T. B. The Interpretation of Experimental Results. Jour. Bd. Agr. (London) Supplement, pp. 15-37. 1911.

Questions for Discussion

1. What was the early history of the use of field plots?
2. What type of experiment is used to compare different sizes and shapes of plots? Why?
3. What practical considerations usually determine the size of plots used in field experiments?
4. What is the general objective of nursery tests and what general relation should they have to field plots?
5. Distinguish between nursery and field plots.
6. What are the common sizes of nursery plots? Field plots?
7. Compare nursery plots and field plots as to accuracy.
8. What is the relation between size of plots and the standard error? Between size of plot and border effect?
9. How may increased size of plot increase the amount of variability?
10. What size of plot has shown the lowest variability for practical purposes with wheat? Corn? Soybeans? Millet? Cotton?
11. What is meant by efficiency in plot size?
12. What reasons can be given for the variability in results with small plots?
13. What is a common size of plots for small grain nurseries? Corn trials? Sorghums? Alfalfa? Sugar beets?
14. In general, what relation is found between shape of plots and the standard error?
15. What relation, if any, is found between the direction in which plots extend and the standard error? Why?
16. What recommendations would you make on width of field plots for the use of farm machinery? Why?
17. What relation is found between shape of plots and border effect?
18. What modifications can be made in length of nursery rows for wheat, oats, and barley, for rapid conversion of yields to bushels per acre?
19. What is replication? Why used?
20. What has been the general practice regarding replication of plots? What is the practice now?
21. Trace the early history of plot replication.
22. What serious results may result from single (unreplicated) plot trials? Why?
23. What is the theoretical relation between the number of plots and the standard error? Actual relation? Why do they not always agree?
24. What class of errors does plot replication tend to reduce or eliminate? On what class does it have no effect?
25. Discuss the statement: "Precision can be increased indefinitely by replication."

26. How does replication furnish an estimate of error?
27. Give a general rule or rules for plot replication.

Problems

1. It is desired to use 1/80-acre plots in a crop rotation experiment and to make them 14 feet wide. Calculate the plot length.

2. Some data reported by Wiebe (1935) are given below for 15-foot rows of wheat, one-foot apart. The yields are reported in grams. Assume 15 varieties, and compute the efficiency for 1, 2, and 4-row plots.

Series 1	Series 2	Series 3	Series 4
715	595	580	580
770	710	655	675
760	715	690	690
665	615	685	555
755	730	670	580
745	670	585	560
645	690	550	520
585	495	455	470
560	540	450	500
685	730	610	500
755	810	665	570
640	635	585	465
725	655	530	455
715	775	615	545
700	705	555	440

3. Calculate the number of replications required to make a 5 per cent difference in yield statistically significant for these sizes of plots:

Kind of Plot	$\sigma_{\bar{x}}$ (in per cent)
Field plot	3.3
Single-row plot (18-inch spacing)	5.2
Single-row plot (24-inch spacing) hand-planted	7.0

Use the formula, $\frac{\sigma_{\bar{x}}\,(2)\sqrt{2}}{\sqrt{n}} = 5$ (percent difference in yield), where "n" = the number of replications.

CHAPTER XIV

COMPETITION AND OTHER PLANT ERRORS

I. Plants in Relation to Error

That soil heterogeneity will contribute to experimental error has already been seen. There are also many errors due to plants that may contribute to the experimental error. These may be caused by differences in genetic constitution or variations due to environmental conditions. Variations in plant stand within plots may introduce differential responses due to intra-plot competition, while the effect of one plot on the adjacent one may bring about differences due to inter-plot competition. Other errors related to plants include such effects as differences in the moisture content of the harvested crop, differences in adaptation, etc.

II. Acclimatization

A serious systematic error may be introduced thru differences in acclimatization of the crops under test, unless acclimatization itself is the factor under consideration. Varieties in crops like corn, alfalfa, and red clover may vary widely in their climatic adaptation. Variety tests in corn may be a common source of error in this respect. In Nebraska, Kiesselbach (1922) compared Reid Yellow Dent corn grown 100 miles farther north with that grown and adapted at Lincoln. He obtained large differences in yield, plant height, date of maturity, length of ear, etc., within the same variety when originally grown under different conditions. Lyon (1911) reported similar results for corn and also for strains of Turkey wheat from other states included in winter-hardiness tests. Differences in varieties may be brought about in a very few years which may introduce either a slight or a very large error. Reliable tests are impossible when varieties are collected from different climatic regions. Each variety should be grown for a year or two in the region where it is to be tested until it has undergone the changes incidental to adaptation to the new environment.

III. Plant Individuality

Plant individuality varies with different crops. It is more marked in cross than in self-fertilized crops, e.g., it would be more important in rye than in barley. The size of plot necessary is influenced by the number of plants grown per plot, as well as by the kind of plant. For instance, it is easily possible to have 1,000,000 wheat plants on one acre, while the number of corn plants is only about 10,000 per acre. Plant individuality would be negligible in the case of small grains, but quite important in crops like corn and sorghums where the number of plants per plot may be quite low. Lyon (1911) found that quite a large error may be introduced by yield determinations from a small number of plants due to the variations in growth of certain individuals. For maize, he showed that the effect of plant individuality was practically none when each plot was composed of 100 plants.

IV. Variation in Moisture Content of Harvested Crop

In forage and cereal crops the variation in moisture content of the harvested crop may be an important source of error in yield determinations. For precise experimental results, this condition should be recognized and a remedy provided for it.

(a) Moisture in Forage Crops

Obviously, the most accurate method to determine the water or dry-matter content of the forage grown on a plot is to dry all the material to a water-free basis. Since it is impossible to do this, dry matter determinations are based on small shrinkage samples.

The problem of moisture determination is rather simple under semi-arid conditions where forage is readily field-cured. Forage weights are usually taken after the material is dry enough to stack, taking a shrinkage sample at that time. The sample is weighed immediately and allowed to air-dry for 2 or 3 weeks after which it is re-weighed for the air-dry weight. Yields corrected on this basis are found to be reliable. In case moisture-free determinations are necessary, the samples of each variety or treatment may be composited, ground, and dried in a vacuum oven for 12 to 24 hours.

Under humid conditions, reliable comparisons from the weights of field-cured forage cannot be made, except on rare occasions that cannot be predicted. As a result of the work of Farrell (1914), McKee (1914), Vinall and McKee (1916), and Arny (1916), the general practice has been to weigh the forage as soon as cut, and sampling it for air-dry or water-free determinations at that time. These green samples are usually placed in a drier at once to avoid the loss of dry matter thru oxidation, fermentation, etc. The investigations of Wilkins and Hyland (1938) indicate that the samples should be taken and weighed within 4 to 6 minutes after the forage is cut to avoid error due to moisture loss. These workers also found that the error introduced thru the use of green weights of alfalfa and red clover for plot yields without dry matter determinations was negligible so long as the weights were taken quickly. Yield determinations on the basis of green weights proved to be as accurate as where 2 or 3 samples per plot served as a basis for moisture determination and subsequent yield correction.

(b) Moisture in Cereals

The moisture content of small grains is usually of little consequence since the bundles are usually air-dry before threshing is attempted. The threshed grain may be weighed at the threshing machine and re-weighed a week later to be sure it has reached a uniform moisture content. The determination of moisture in shelled corn is regarded as an essential practice for obtaining precise yields. Moisture determinations can be made on each plot of each variety, or a composite determination for the variety on all replications. A common practice is to report yields on the basis of shelled corn with 13.5 per cent moisture, the maximum moisture permitted for U.S. No. 2 corn. Moisture determinations for corn or small grains can be made quickly with the Brown-Duvel moisture tester described by Coleman and Boerner (1927). Recently, the Tag-Heppenstall moisture meter, an electrical device, has been widely used for moisture determinations. This meter is calibrated for wheat, corn, oats, barley, rye, sorghums, rice, soybeans, and vetch. The electrical moisture meter has certain advantages for practical work: (1) It is unnecessary to clean after each sample; (2) the sample is not weighed; (3) a single determination can be made in less than one minute; (4) it will duplicate results within a tolerance that cannot be met in a single determination by other methods, and (5) the operation and maintenance cost is low. Cook, et al (1934) have made a study of rapid moisture determination devices. When determinations are made on each variety in each replicate, single rather than duplicate determinations should be sufficient.

V. Competition Concept in Plants

A "struggle for existence" results when plants are grouped or occur in communities in such a way that the demands for an essential factor are in excess of the supply. This is true in many field trials. Competition always occurs when two or more plants make demands for light, nutrients or water in excess of the supply. It is greatest between individuals of the same species which make similar demands upon the supply at the same time. This is generally the case in cultivated crops where an area is planted to the same species or variety. A detailed discussion on the nature of plant competition is given by Clements and others (1929)

A number of investigations have been conducted to determine the importance of plant competition in experimental plots. There is apt to be an effect when varieties that differ considerably in growth habit, time of maturity, and other characters are grown in adjacent plots. The principal contention is whether or not the yield of a poorer variety growing next to a high yielding variety will be adversely affected so that the yield will be actually lower than when the variety is grown next to a plot of its own kind.

Competition may or may not influence plot yields. Two distinct schools of opinion have arisen as to its importance. In areas of limited moisture supply, competition has been generally found to be a source of error in comparative crop tests. Kiesselbach (1918) obtained errors of 24 and 56 per cents due to plant competition in two different years. Hayes and Arny (1917) found errors in small grain yield trials where varieties competed with each other. In Missouri, Stadler (1921) reported errors of 50 to 100 per cent due to plant competition. Some workers in the more humid regions, where moisture is often sufficient throughout the season for ordinary stands, consider competition effects unimportant. For example, Stringfield (1927) found only occasional disturbances in Ohio, while Garber and Odland (1925) failed to find evidences of competition in adjacent soybean rows on the West Virginia Station. Love and Craig (1938) concluded that the effect of competition is not serious enough to influence the yields of wheat and oats under New York conditions.

The influence of plant competition depends upon the test being conducted, but the possible error from this source should be kept in mind constantly. It is a safer procedure to eliminate or provide for this source of error than to be led to erroneous conclusions by overlooking it.

A -- Intra-plot Competition

VI. Uneven Plant Distribution in Plots

Plants within a plot are in competition with each other when some factor such as moisture is present in insufficient quantities. Uneven plant distribution, with a normal number of plants per plot, was studied in corn by Kiesselbach and Weihing (1933) to determine whether or not this condition would alter acre yields. Corn was planted in hills 3.5 feet apart so as to average three plants per hill. The three systematically uneven distributions were planted so as to have 2-4, 1-3-5, and 1-2-3-4-5 plants in alternate hills. Essentially uniform stands of three plants per hill were grown for comparison. During a 14-year period the systematically variable stands of 2-4, 1-3-5-, and 1-2-3-4-5 plants per hill averaged 50.6, 49.3, and 50.0 bushels per acre, respectively. The three variable stands averaged 50.0 bushels while the uniform 3-plant stand averaged 49.9 bushels per acre. In another trial, these investigators tested a random variable stand by planting corn that germinated 100, 75, 60, and 50 per cents at adjusted rates to average three viable kernels per hill. The plot yields for a single season were 24.96, 25.50, 25.34, and 25.12 bushels per acre for the respective germination per cents. From these data, it was concluded that systematically and randomly variable stands did not affect the yields so long as the same number of plants occurred on a plot. The authors caution that "experience has indicated that stand irregularities materially greater than those herein considered, such as are sometimes caused by rodents, worms, birds, and soil washing, would undoubtedly increase plant variability and lower the yield."

A similar type of study was conducted by Smith (1937) on 2 Australian wheat fields planted by a farm drill in which short lengths of drill row were harvested separately. Variability of plant density as found in a drill-sown field did not by itself cause a decrease in yield of grain as compared to even spacing of seed. The correlation of yield and plant number per foot of drill row, which is invariably observed in small grain fields, was said to be due to the effects of competition between near-by densities. He makes this statement: "The true correlation between yield and plant density per area may be positive, negative, or zero according to circumstances. The yield from variable seeding may be less than, equal to, or even slightly greater than the yield from even seeding, according to how near the even seeding may be optimum for the given conditions and how far the variable seeding may fall within or overlap the optimum range within which plant density is of little importance."

VII. Differences in Stand in Plots

The ideal condition for yield trials is a perfect stand on all plots, but this is not always attained. Some allowance must be made for the lost area where the stand is injured by outside influences, particularly with some crops. When the loss in stand is due to the treatment, i.e., it injures germination, destroys part of the plants, or in any manner is directly responsible for stand, the use of a perfect-stand basis for yield calculations eliminates the effect of the treatment. The lethal effect of the treatment may be a definite part of the results obtainable and should be given consideration. The effect of stand within plots is more of a problem in crops like corn, sorghums, certain legumes, and sugar beets where the plants are large, variable *inter se*, and subject to the influences of plant competition. Less difficulty is experienced in small grains because the plants tend to tiller and fully utilize the extra space.

(a) Competition between Unlike Hills in Corn

Relative yields of one, two, and three-plant corn hills uniformly surrounded by three-plant hills was studied by Kiesselbach (1918) (1923). He found the yields to be 61, 82, and 100 per cents for the one, two, and three plants per hill. It is obvious that the fewer plants per hill made some use of the additional space. In another test, the relative yields of three-plant hills were compared when adjacent to hills with various numbers of plants.

3-plant hills surrounded by 3-plant hills except as indicated below:	Average Grain Yield per Hill Actual (lbs.)	Relative (pct.)
Surrounded by 3-plant hills	1.075	100
Adjacent to one hill with 2 plants	1.098	102
Adjacent to one hill with 1 plant	1.151	107
Adjacent to one blank hill	1.224	114

It is obvious that 3-plant hills adjacent to blank or 1 or 2-plant hills tend to yield higher than when surrounded by 3-plant hills. In comparisons of inbred lines and F_1 hybrids, Brewbaker and Immer (1931) found that a rather large error may be introduced in yields where hills have reduced stands or are adjacent to hills which lack in stand. However, under a 3-plant rate in corn it is generally conceded that 10 to 15 per cent of the stand may be lost before the yield is measurably reduced or the experimental accuracy affected in ordinary yield trials.

(b) Competitive vs. Non-Competitive Yields in Sugar Beets

Sugar beet tonnages are usually reported as (1) total weight of all beets on a unit area, or as (2) a calculated yield from "normally competitive" beets. The beets which serve as the basis for calculation are those grown surrounded by neighbors on all sides at appropriate distances for the conditions imposed in the experiment. A study of the response of sugar beets to increased space allotment was made by Brewbaker and Deming (1935). Their plants were grown in 20-inch rows and thinned to 12 inches between plants in the row. Beets around a single blank space were found to increase in weight sufficient to compensate for 96.2 per cent of the loss of a single beet. They obtained increases of 28.7, 55.2, and 95.0 per cents for beets adjacent to one blank space in the same row, between two blank spaces in the same row, and with blanks on four sides, respectively. It is evident that the beet weight was greatly influenced by the relative area available for its development. The regression of weight of beets upon stand was essentially linear for stands between 25 and 75 per cent. For each 10 per cent increase in stand there was an increase from 0.76 to 2.10 tons beets per acre for the regression within blocks. There may be situations where yields based on competitive beets would be in error, particularly in poor stands and in spacing tests. Such instances have been pointed out by Nuckols (1936). He harvested actual and competitive beets on 254 plots where the stand varied from 50 to 100 per cent. A greater difference in competitive and actual yields was obtained for poor stands than for good stands. In fact, the mathematical possibilities showed that there are only 35 per cent of competitive beets in a 90 per cent stand, 10 per cent in an 80 per cent stand, and 5 per cent in a 70 per cent stand. This indicates the greater possibility for error when competitive beets are taken from poor stands. Nuckols also found an indication that there is a greater difference between competitive and actual yields where the beets are closely spaced than where more widely spaced in the row. It is obvious that in rate of spacing tests, the method of selection of competitive beets is not the same for all plants.

(c) Stand Effects in Other Crops

In potatoes, Livermore (1927) reports that the yield of the two hills adjacent to a blank may be 40 per cent more than that for hills surrounded by hills. Werner and Kiesselbach (1929) found 62.3 per cent of the loss in yield was recovered in potatoes adjacent to one-hill blanks. In alfalfa nursery plots, Tysdal and Kiesselbach (1939) found for variable seed rates that stands tended to equalize after 4 years. Considerable latitude in the amount of seed sown per row was possible without serious effects on comparative varietal performance.

VIII. Corrections for Uneven Stands

A great deal of attention has been given to possible corrections for loss of stand. It should be emphasized that there is no entirely satisfactory method to correct for uneven stands, it being better practice to prevent them so far as possible. One method to avoid poor stands is to plant thick and thin the young plants to the desired rate. For example, where a stand of 3 plants per hill is desired in corn, the experimenter may plant 6 kernels per hill and subsequently thin the seedling plants to 3 per hill. Most empirical methods for the correction of yields on a stand basis are based upon plants surrounded by the normal stand, i.e., competitive plants. Stewart (1919) (1921) gives a formula for the correction of stand errors in potatoes where the stand is relatively satisfactory. The practice in corn experiments is to harvest the entire plot without stand corrections when the stand is 90 per cent of the theoretical or better. For less than that, it is usually harvested on a perfect-stand basis. Kiesselbach (1918) (1923) selects only perfect-stand hills surrounded by hills with the same stand and computes the yields from those. Bryan (1933) found that 26 per cent fewer hills were required to obtain any given degree of precision with only perfect-stand hills than with all hills regardless of stand. Adjustment of the yields of perfect stand hills further reduced the number of hills required for any degree of precision by 18.9 per cent. The procedure in the U. S. Department of Agriculture for the uniform corn hybrid tests is to adjust yields for missing hills but not for minor variations in stand.

Probably the most satisfactory method for the adjustment of yields on the basis of stand is by covariance in which the regression coefficients are calculated. Mahoney and Baten (1939) have outlined its use for this purpose. When there is a fairly high variation due to soil heterogeneity and no appreciable differences in stand, usually nothing is gained by adjustment.

B -- Inter-plot or Border Effect Competition

IX. Types of Inter-plot Competition

Many studies have been conducted to determine the border effects of adjacent plots. The committee on the standards for the conduct of field experiments for the American Society of Agronomy (1933) makes this statement: "In a majority of soil experiments and in many cultural and variety tests, plot yields may be modified by contiguity to other treatments, crops, or interspaces. Border competition in adjacent unlike plots often raises some yields and lowers others." A vigorous variety may benefit when grown next to a poor one, particularly in single-row plots. The same type of error may be introduced in rate and date of planting tests. As a result, multiple-row plots are often used in experimental work with the border rows discarded. This procedure is justified on the basis of experimental data which indicate that the yield order may be changed when border rows are included in the plot yields, according to Arny (1921). In some fertilizer and cultural experiments alleys between plots are necessary because the treatment may spread to the next plot through faulty application.

X. Effect in Variety Tests

Most tests to determine the amount of inter-plot competition have been on the basis of single-row vs. multiple row plots with the borders discarded.

(a) Small Grains

It is concluded by Hayes and Arny (1917) that there is considerable competition between rod rows of small grains when grown one-foot apart. This led to the adoption of three-row plots for small grain variety tests at the Minnesota station. Comparisons of three-row plot yields with the central rows showed that the latter are as accurate for yield determinations as attained by the use of all three rows. Kiesselbach (1918) found that competition caused Big Frame wheat to yield 10.3 and 12.4 per cent too high in 1913 and 1914, respectively, when grown in alternate rows with Turkey. Burt oats yielded 16 and 38 per cent too high for these years when grown in alternate rows with Kherson. Stadler (1921) found competition in small grains to be more extreme between different varieties than between different commercial strains of the same variety. As a result, it is almost the universal practice to grow small grains in multiple-row plots and discard at least one border row from each side at harvest for small grain nursery plots. The use of single-row or 3-row plots with all rows harvested appears possible under humid conditions where competition appears to be slight. (See Love and Craig, 1938)

(b) Corn

As early as 1909, Smith (1909) found one-row plots too narrow for fair tests in corn when varieties of diverse characteristics were planted in adjacent rows. A variety with short stalks was at a disadvantage when grown next to a taller one because of shading; or a variety with "strong foraging powers" may compete more successfully for moisture and plant food over a weaker or slower growing neighbor. Kiesselbach (1922) (1923) found that where large and small varieties of corn were grown in alternate rows, the smaller variety yielded 66 per cent as much as the larger one, and only 47 per cent as much when both were planted in the same hill. The smaller variety yielded 85 per cent as much when planted in alternate 5-row plots and the three center rows harvested for yield. That the smaller variety was being robbed of light, water, and nutrients was shown by the yields where each variety was surrounded by its own kind.

(c) Other Crops

Competition between soybean varieties was studied by Brown (1922) in Connecticut. Twenty-five single-row check plots of a small and early soybean variety averaged

26.9 bushels of seed per acre. When the check was adjacent to larger and later varieties like Mammoth Yellow, the checks averaged only 17.1 bushels or 63.6 per cent as much as the average of all checks. In potatoes, he concluded that yields were not influenced by competition between single-row plots.

In alfalfa, solid-drilled plots with a 7-inch row spacing has been shown to be definitely subject to serious interplot varietal competition. The work of Tysdal and Kiesselbach (1939) indicates that the effects could be overcome when the border rows were discarded at harvest. When the alley space between plots was widened to 12 inches a significant interaction between varieties was also prevented. The relative yields from single or multiple-row plots with either 18 or 24-inch row spacing likewise exhibited no significant differential interaction.

Immer (1934) made a study of the effect of competition between adjacent rows of different varieties of sugar beets, i.e., "Old Type" and "Extreme Pioneer". These were grown in alternate single-row plots and also in 4-row plots with the border rows removed for yield determinations. When grown in single-row plots the "Old Type" brand yielded 3.78 ± 0.44 tons more per acre than "Extreme Pioneer." In 4-row plots, with the central two rows alone being harvested, the increase of "Old Type" over "Extreme Pioneer" was only 1.78 ± 0.31 tons per acre. The difference between these two differences was 2.00 ± 0.54 tons, a value that is significant. Thus, "Old Type," the higher yielding sort, profited at the expense of "Extreme Pioneer" when these two brands were grown side by side in single-row plots.

In cotton variety tests, Christidis (1937) found that competition may cause a definite bias in the estimation of comparative yields of cotton varieties. Hancock (1936) tested two cotton varieties with diverse growth characteristics. The varieties were: Acala, a late tall variety, and Delfos, an early semi-dwarf type. The varieties were arranged in these combinations with the series alternated: DDDDAD and AAAADA. He observed that Delfos with Acala on only one side (DDA) showed very small differences when compared with themselves between their own border rows (DDD). For instance, DDD as an average for four years produced only 1.4 per cent more seed than DDA, while AAA produced 4.01 per cent less than AAD. Where two rows of the same variety are planted, only one row would be affected by a different variety. Since he found this effect to be small, two-row plots were advocated with both harvested for yield. Such a procedure may be satisfactory under conditions of abundant moisture, but would be questionable where habitat factors are severely limited.

XI. Rate and Date Tests

Under most environmental conditions competition will exist between plots in rates and dates of planting tests. Hulbert (1931) presents data to show that border effect on outside rows increases as the rate of seeding is increased. The border effect on Red Bobs wheat was 147.85 per cent when seeded at the rate of three pecks per acre, 175.41 per cent for five pecks, and 173.01 for seven pecks. Kiesselbach (1918) tested two rates of planting for Turkey wheat, a thin and a thick rate. The thin rate yielded 68 per cent as much as the thick rate when grown in alternate single-row plots, and 90 per cent as much when grown in alternate five-row plots. Competition between alternate single-row plots for two rates for Kherson oats caused the thin rate to yield 20 per cent too low in 1913 and 34.3 per cent too low in 1914. Nebraska White Prize corn was planted in alternate rows so as to obtain two and four plants per hill. Due to competition the thin rate yielded relatively 29.0 and 9.0 per cents too low in different years. Similar results would be expected in date of planting tests. Klages (1928) found a marked degree of competition in spacing tests with sorghums. Yields of rows with dense stands profited at the expense of the yields of adjacent rows with thinner stands. The degree of competition was influenced by environmental conditions.

XII. Border Effect

Plants that grow along the sides and ends of plots are often more thrifty and vigorous than those in the interior. This is particularly true when the plots are surrounded by alleys. Border effect is considered here to mean the effect of blank alleys on the border rows. The amount and extent of this border effect is important in comparative crop tests.

(a) Small Grains

Arny and Hayes (1918) and Arny (1921) (1922) studied (1) the distance alley effect is operative within plots, (2) the increase in yield due to alley effect, and (3) the influence of additional alley space on variety response. They used small grain plots composed of 16 drill rows six inches apart. The yields of the border rows were compared with those of the center rows. Arny (1921) gives some typical data:

	Oats		Wheat		Barley	
Description	Bu.	Pct.	Bu.	Pct.	Bu.	Pct.
Outside border rows	65.58	199.9	30.56	153.6	48.93	213.5
Middle border rows	58.53	170.4	25.75	127.4	42.74	186.5
Inside border rows	49.95	152.3	22.23	111.7	33.56	146.4
Central rows	32.80	100.0	19.90	100.0	22.92	100.0

As an average for three years the yields of outside rows of oats, spring wheat, and barley expressed in per cent based on the yields of the central rows is 199.8 and that for the middle rows 138.0 when the plots were surrounded by 18-inch clean-cultivated alleys. Border effect was relatively unimportant when extended to the third drill row. Knowledge that border effect is not uniform precludes the use of any percentage figures derived in one place to reduce yields secured in another location to a border-effect-free basis. Arny (1921) further showed that the rank of a variety may be changed due to border effect. In all cases, plot yields were higher than where these rows were eliminated before harvest. Hulbert and Remsburg (1927) found it necessary to discard two border rows from each side of small grain plots to remove the error in border effect in variety tests. Competition effects were noticeably increased when the adjacent plots were seeded at different rates. Hulbert, et al. (1931) obtained similar results. Robertson and Koonce (1934) studied border effect on Marquis wheat grown in plots irrigated at different stages in its relationship to yield when different numbers of border rows were included. The yield increased as the size of plot increased but the percentage increase was uniform for the three different treatments employed. Comparable yields were the same for plots of 10 rows, and for 10 plus 2, 4, or 6 border rows.

(b) Other Crops

In kafir and milo, Cole and Hallsted (1926) obtained marked increases in yield from outside rows. The excess yield was roughly proportional to the increased available soil area. Recently, Conrad (1937) has called attention to the fact that sorghum plants next to uncropped areas may use soil moisture six feet away laterally. A definite use of nitrates was made four feet away laterally for both sorgo and corn. The influence of border effect on total dry matter per plot was studied at the Central Experimental Farm (Ottawa) by McRostrie and Hamilton (1927). In all cases, border plants of Western rye grass gave an increased yield due to the influence of the two-foot pathway which surrounded the plots. The increase in yield differed with the strain under test, and varied from 6 to 54 per cent. The rank of the strains was materially changed due to the wide variation in border yields. When theoretical plots 1/72.6-acre in size were used for red clover and alfalfa forage yields, Hollowell and Heusinkveld (1933) found a serious experimental error in yield when border rows were included in the harvested plot. Their plots were composed of 8, 12, and 16-inch alleys. The inclusion of border rows increased the yield from 2.1 per cent to 20.0 per cent for red clover and from 1.8 to 14.0 per cent for alfalfa. Border effect was greater on the first than on the second alfalfa crop, but varied greatly from year to year under Ohio conditions. Rainfall appeared to be directly correlated with border effect. These investigators concluded that the discard of two border rows would effectively eliminate border competition on plots of this size. Similar results were obtained by Tysdal and Kiesselbach (1939) when they compared dissimilar adjacent alfalfa plots that differed as to spacing of rows or plants. A solid-drilled block with 7-inch row spacing was separated by a 7-inch alley space from a space-planted block with rows 24-inches apart. The adjacent border rows were compared with their respective types of interior rows. The solid-drilled rows gave an excess yield of 74 per cent because of reduced competition on one side, whereas the space-planted row was depressed 63 per cent in yield because increased competition. It is evident that great care must be exercised in taking yields from adjacent rows that are affected with respect to row-space or density of stand.

XIII. Control of Inter-plot Competition

Inter-plot competition can be controlled by several methods. Hayes and Garber (1927), Kiesselbach (1918) (1923) and others give these recommendations: (1) group varieties with similar growth habits, dates of maturity, etc., together; (2) use of multiple-row plots; and (3) discard outside border rows and ends at time of harvest. Alleys are sometimes used in closely-sown crops such as small grains and forage crops to facilitate harvest and to reduce mixtures. In small plots the borders should be removed, but in large field plots it is generally satisfactory to harvest the entire plot and to include the additional alley space in the plot area. Untreated interspaces of sufficient width to avoid serious soil translocation are recommended for permanent soil fertility, rotation, and tillage experiments. These alleys can either be cropped or left bare.

References

1. Arny, A. C. The Dry Matter Content of Field Cured and Green Forage. Jour. Am. Soc. Agron., 8:358-363. 1916.
2. __________ Border Effect and Ways of Avoiding It. Jour. Amer. Soc. Agron., 14:266-278. 1922.
3. __________, and Hayes, H. K. Experiments in Field Technic in Plot Tests. Jour. Agr. Res., 15:251-262. 1918.
4. __________ Further Experiments in Field Technic in Plot Tests. Jour. Agr. Res., 21:483-499. 1921.
5. Brewbaker, H. E., and Immer, F. R. Variations in Stand as Sources of Experimental Error in Field Tests with Corn. Jour. Amer. Soc. Agron., 23:469-481. 1931.
6. ______________, and Deming, G. W. Effect of Variations in Stand on Yield and Quality of Sugar Beets Grown under Irrigation. Jour. Agr. Res., 50:195-210. 1935.
7. Brown, B. A. Plot Competition with Potatoes. Jour. Amer. Soc. Agron., 14:257-258. 1922.
8. Bryan, A. A. Factors Affecting Experimental Error in Field Plot Tests with Corn. Ia. Agr. Exp. Sta. Bul. 163. 1933.
9. Christidis, B. G. Competition Between Cotton Varieties: A Reply. Jour. Amer. Soc. Agron., 29:703-705. 1937.
10. Clements, F. E., Weaver, J. E., and Hanson, H. C. Plant Competition: An Analysis of Community Functions. Carnegie Institution of Washington. 1929.
11. Cole, J. S., and Hallsted, A. L. The Effect of Outside Rows on the Yields of Plots of Kafir and Milo at Hays, Kansas. Jour. Agr. Res., 32:991-1002. 1926.

12. Coleman, D. A., and Boerner, E. G. The Brown-Duval Moisture Tester and How to Operate It. Dept. Bul. 1375, U. S. D. A. 1927.
13. Conrad, J. P. Distribution of Residual Soil Moisture and Nitrates in Relation to Border Effect of Corn and Sorgo. Jour. Amer. Soc. Agron., 29:367-378. 1937.
14. Cook, W. H., Hopkins, J. W., and Geddes, W. F. Rapid Determination of Moisture in Grain. Can. Jour. Res., 11:264-289, and 409-447. 1934.
15. Farrell, F. D. Basing Alfalfa Yields on Green Weights. Jour. Am. Soc. Agron., 6:42-45. 1914.
16. Garber, R. J., and Odland, T. E. Influence of Adjacent Rows of Soybeans on One Another. Jour. Amer. Soc. Agron., 18:605-607. 1925.
17. Grantham, A. E. The Effect of Rate of Seeding on Competition in Wheat Varieties. Jour. Amer. Soc. Agron., 6:124-128. 1914.
18. Hancock, N. I. Row Competition and its Relation to Cotton Varieties of Unlike Plant Growth. Jour. Amer. Soc. Agron., 28:948-957. 1936.
19. Hayes, H. K., and Arny, A. C. Experiments in Field Technic in Rod Row Tests. Jour. Agr. Res., 11:399-419. 1917.
20. ————, and Garber, R. J. Breeding Crop Plants, pp. 75-79. 1927.
21. Hollowell, E. A., and Heusinkveld, D. Border Effect Studies of Red Clover and Alfalfa. Jour. Amer. Soc. Agron., 25:779-789. 1933.
22. Hulbert, H. W., and Remsberg, J. D. Influence of Border Rows in Variety Tests of Small Grains. Jour. Amer. Soc. Agron., 19:585-590. 1927.
23. ————, et al. Border Effect in Variety Tests of Small Grains. Idaho Agr. Exp. Sta. Tech. Bul. No. 9. 1931.
24. Immer, F. R. Varietal Competition as a Factor in Yield Trials with Sugar Beets. Jour. Amer. Soc. Agron., 26:259-261. 1934.
25. Kiesselbach, T. A. Competition as a Source of Error in Comparative Corn Tests. Jour. Amer. Soc. Agron., 15:199-215. 1923.
26. ———— Corn Investigations. Nebr. Agr. Exp. Sta. Res. Bul. No. 20. 1922.
27. ————, and Weihing, R. M. Effect of Stand Irregularities upon the Acre Yield and Plant Variability of Corn. Jour. Agr. Res. 47:399-416. 1933.
28. ————, Studies Concerning the Elimination of Experimental Error in Comparative Crop Tests. Nebr. Agr. Exp. Sta. Res. Bul. 13. 1918.
29. Klages, K. H. Yields of Adjacent Rows of Sorghums in Variety and Spacing Tests. Jour. Amer. Soc. Agron., 29:582-599. 1928.
30. Livermore, J. R. A Critical Study of Some of the Factors Concerned in Measuring the Effect of Selection in the Potato. Jour. Amer. Soc. Agron., 19:857-896. 1927.
31. Love, H. H., and Craig, W. T. Investigations in Plot Technic with Small Grains. Cornell U. Memoir 214. 1938.
32. Mahoney, C. H., and Baten, W. D. The Use of the Analysis of Covariance and its Limitation in the Adjustment of Yields based upon Stand Irregularities. Jour. Agr. Res., 58:317-328. 1939.
33. McKee, R. Moisture as a Factor of Error in Determining Forage Yields. Jour. Amer. Soc. Agron., 6:113-117. 1914.
34. McRostrie, G. P., and Hamilton, R. I. The Accurate Determination in Dry Matter in Forage Crops. Jour. Amer. Soc. Agron., 19:243-251. 1927.
35. Nuckols, S. B. The Use of Competitive Yield Data from Sugar Beet Experiments. Jour. Amer. Soc. Agron., 28:924-934. 1936.
36. Robertson, D. W., and Koonce, D. Border Effect in Irrigated Plots of Marquis Wheat Receiving Water at Different Times. Jour. Agr. Res., 48:157-166. 1934.
37. Smith, H. Fairfield. The Variability of Plant Density in Fields of Wheat and its effect on Yield. Counc. Sci. and Ind. Res. Bul. 109 (Australia). 1937.
38. Stadler, L. J. Experiments in Field Plot Technic for the Preliminary Determination of Comparative Yields in the Small Grains. Mo. Agr. Exp. Sta. Res. Bul. No. 49. 1921.

39. Standards for the Conduct and Interpretation of Field and Lysimeter Experiments. Jour. Amer. Soc. Agron., 25:803-828. 1933.
40. Stewart, F. C. Missing Hills in Potato Fields: Their Effect upon Yields. New York State Agr. Exp. Sta. Bul. 459, pp. 45-69. 1919.
41. ____________, Further Studies on the Effect of Missing Hills in Potato Fields on the Variation in the Yields of Potato Plants from Halves of the same Seed Tuber. New York (Geneva) Agr. Exp. Sta. Bul. 489. 1921.
42. Stringfield, G. H. Intervarietal Competition among Small Grains. Jour. Amer. Soc. Agron., 19:971-983. 1927.
43. Tysdal, H. M., and Kiesselbach, T. A. Alfalfa Nursery Technic. Jour. Am. Soc. Agron., 31:83-98. 1939.
44. Vinall, H. N., and McKee, Roland. Moisture Content and Shrinkage of Forage and the Relation of these Factors to the Accuracy of Experimental Data. Dept. Bul. 353, U. S. D. A. 1916.
45. Werner, H. O., and Kiesselbach, T. A. The Effects of Vacant Hills and Competition upon the Yield of Potatoes in the Field. An. Proc. Potato Assn. America, 16:109-120. 1929.
46. Wiebe, G. A. The Error in Grain Yield Attending Misspaced Wheat Nursery Rows and the Extent of the Misspacing Effect. Jour. Amer. Soc. Agron., 29:713-716. 1937.
47. Wilkins, F. S., and Hyland, H. L. The Significance of Dry Matter Determinations in Yield Tests of Alfalfa and Red Clover. Ia. Agr. Exp. Sta. Res. Bul. 240. 1938.

Questions for Discussion

1. Why should pure seed be used in variety tests?
2. How may differences in acclimatization introduce errors in crop tests? How can they be avoided?
3. When or with what crops or under what conditions is plant individuality a factor to be considered in planning experiments?
4. When is moisture content of the crop a factor of importance? How may the error be eliminated or corrected?
5. Could you secure comparable forage yields by taking green weights? Why?
6. What are the advantages of the vacuum oven over an ordinary oven for securing moisture-free weights?
7. Compare rapid moisture determining devices for cereals.
8. What is meant by plant competition? Who have emphasized its importance?
9. Is competition universally present in experimental plots? Is it always objectionable? Explain.
10. What effect does severe competition have on plants?
11. How can you reconcile the fact that some workers claim plant competition is a fruitful source of error in experimental work, while others contend it is negligible?
12. Do stand irregularities in corn affect the yield so long as the same number of plants per unit area is involved? Explain.
13. Why may a variable stand in a wheat field yield as much as an evenly-spaced stand? Explain.
14. Why is intra-plot competition in small grains unimportant from the practical standpoint?
15. What is the general effect in corn hills surrounded by hills with different numbers of plants? Why?
16. What is meant by "normally competitive" in calculation of sugar beet yields?
17. What is the effect of adjacent blank hills on the weights of individual beets?
18. Under what conditions may yields from "competitive" beets result in errors in yield?

19. What recommendations would make as to correcting for uneven stands?
20. What is the general practice for the prevention of errors due to uneven stands in corn and sorghums?
21. How are stand errors generally corrected in corn plots?
22. What is meant by border competition?
23. How may errors be introduced in variety tests by use of single-row plots?
24. How could you possibly justify two-row plots in cotton variety tests with no borders removed? Single-row alfalfa plots?
25. How does competition introduce errors in rate and date tests?
26. Under what conditions may it be desirable to have blank alleys surrounding plots?
27. What influences do border rows have on plot yields?
28. Is it always necessary to remove borders for the determination of plot yields? Why?
29. What recommendations would you make for the control of inter-plot competition?

Problems

1. Explain how to arrange and conduct an experiment with 10 varieties of corn so as to control both intra and inter-plot competition.

2. The yield of field-cured hay on a 1/10-acre plot is 400 lbs. The shrinkage sample taken at that time weighed 3.8 lbs. After 3 weeks it weighed 3.4 lbs. Calculate the yield per acre of the plot on an air-dry basis.

3. The yields (marketable ears) and stands of 6 strains of sweet corn for 4 replications were as follows: Data from Mahoney and Baten.

Item	Yield and Stand for Strain Number: 1	2	3	4	5	6
Replication 1.						
Yield (x)	56	31	21	23	50	60
Stand (y)	77	68	61	83	70	84
Replication 2.						
Yield (x)	64	29	32	20	59	50
Stand (y)	80	76	72	88	89	92
Replication 3.						
Yield (x)	56	30	24	18	60	47
Stand (y)	74	83	82	78	78	78
Replication 4.						
Yield (x)	36	32	24	19	39	50
Stand (y)	57	61	73	78	52	88
Yield Totals	212	122	101	80	208	207
Stand Totals	288	288	288	327	289	342

Calculate the regression of yield on stand.

CHAPTER XV

DESIGN OF SIMPLE FIELD EXPERIMENTS

I. Criticisms of Agronomic Experiments

There are about 2300 agronomic projects in force in the different state experiment stations, besides those carried on by the U.S. Department of Agriculture, and those in related fields. In fact, two-thirds of all agricultural experimental projects in this country are agonomic. They have increased in number by 50 per cent since 1920. Frequently, this experimental work is criticised by farmers and others. The criticism may or may not be justified. Agriculture is sometimes looked upon as a "practical" field in which results are sought rather than knowledge concerning the phenomena of life. At other times, there is a genuine shortcoming in experimentation. Allen (1930) states that fully one-half of the agronomic experimental projects consist of tests and trials of different kinds. Very little ingenuity is involved in many of them. Variety and cultural experiments are popular while many genetic studies are merely field selection. Soil fertility experiments are often shallow. In many cases, old methods of experimentation are used while in others the experiments are carried too long.

A -- Basic Principles in Design

II. Outline of Experimental Tests

A review of literature on the subject should be the first step in the plans for an experiment. This should be followed by a detailed outline in order to crystallize the ideas of the investigator on the subject. Recently, Fisher (1937) has shown that design of an experiment is inseparable from the statistical analysis of the data.

Certain objectives must be kept in mind in all agricultural experiments. These may be enumerated as follows: (1) The tests should furnish a basis for recommendations to farmers; (2) They should furnish occular proof of the beneficial results attained; and lastly, (3) They should supply information on the fundamental causes of the phenomena which the results are expected to demonstrate.

Several factors need to be considered in the outline of an experiment. These are well described by Allen (1930): (1) It should be definite and limited in scope. (2) The problem should be subjected to competent persons for criticisms and suggestions. (3) Previous work on the subject should be familiar so that the investigator can start work where others left off. (4) Next, he should ascertain the data essential to the problem and devise means to secure and analyze them. (5) Then it remains to test their applicability or sufficiency to the problem. Sometimes it is found that progress is dependent upon the advance in related sciences.

III. Principle of the Extremes

Results should be secured over a wide range on either side of the optimum. Stapledon (1931) has made this statement: "I believe in all field experiments of a research nature we should go at each end far beyond what is deemed by practical men to be the economic limit." The situation may be illustrated in a rate of seeding test for wheat where the optimum rate is approximately 5 pecks per acre. The preliminary test should include rates at regular intervals from the very lowest to a maximum well beyond the point where the optimum is expected to fall, e.g.,

1 peck	5 pecks	10 pecks
minimum	optimum	maximum

The size of interval in tests is determined by the amount of land, facilities, character of the problem, or available finances. An increase in the size of the interval is justifiable as one goes from the optimum to either a minimum or to a maximum. In the final test, it may be advisable to throw out the extremes and conduct a precise test around the optimum rate.

IV. Simple vs. Complex Experiments

Experiments may be classified into several kinds based on the number of factors studied at the same time. The formal experiment is sometimes preceded by a preliminary test.

(a) Preliminary Tests

All preliminary experiments are necessarily empirical in nature. They give the investigator an opportunity to detect faulty technique, inadequate methods, etc. The final experiment can be planned to eliminate many of the shortcomings observed in the preliminary test. A survey is sometimes used for a preliminary test. A further use of the preliminary experiment is to reduce the error in subsequent tests. (See Wishart and Sanders, 1935).

(b) Simple Experiments

One thing is studied at a time in the simple experiment. All factors are kept constant or uniform, so far as possible, except the one under investigation. This is the classical method of experimentation, i.e., the essential conditions are varied only one at a time. R. A. Fisher has recently pointed out that this approach is inadequate for many research problems because the laws of nature may be controlled and influenced by several variables. In his book on "The Design of Experiments", Fisher (1937) makes this statement: "We are usually ignorant which, out of innumerable possible factors, may prove ultimately to be the most important, though we may have strong presuppositions that some few of them are particularly worthy of study. We have usually no knowledge that any one factor will exert its effects independently of all others that can be varied, or that its effects are particularly simply related to variations in these factors". The simple experiment is justified when the time, material, or equipment are too limited to allow for attention on more than one narrow aspect of the problem. As an illustration of this type, an experiment can be set up to determine the best variety of sugar beets to grow. Another could be designed to determine the best fertilizers to apply, while a third separate experiment could be relegated to the best cultural practices. The simple experiment is the one most commonly used by investigators. It is recommended to beginners because it is less involved.

(c) Combination and Complex Experiments

More than one variable is studied at a time in combination experiments. Examples of some of the more simple experiments of this type are: (1) rate and date of planting tests, (2) the relation between time of planting and date of maturity, (3) depth and rate of planting in relation to yield, (4) fertilizer tests, etc. Recently, the Rothamsted workers have advocated the complex experiment in which two or more treatments are studied in all possible combinations. Yates (1935) states complex experimentation is due primarily to R. A. Fisher who first suggested it in 1926. It is extensively practiced at Rothamsted and to a lesser extent elsewhere. Fisher (1937) claims two advantages of the complex experiment (factorial arrangement) over experiments that involve single factors, viz., greater efficiency and greater comprehensiveness. A further advantage is that a wider inductive basis for conclusions is available. As an example, a complex experiment could be set up to determine the responses of several fertilizers and methods of land preparation.

V. Replication

As previously pointed out, soil heterogeneity is the principal source of error in the field experiment. It can be overcome theoretically by replication which tends to diminish the experimental error as well as to provide for an estimate of the magnitude of such errors. Fisher (1931) gives a diagram to show these relationships:

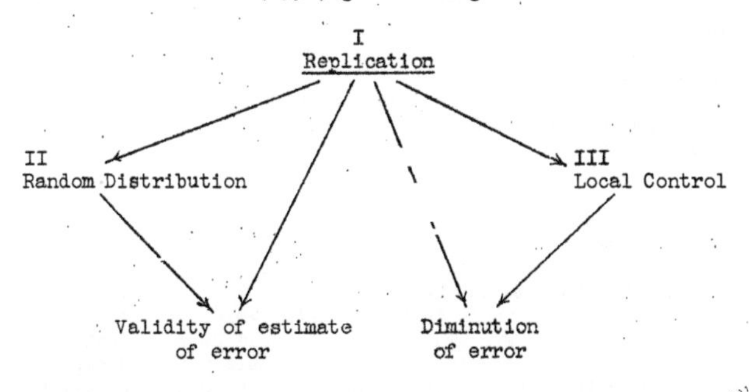

(a) Relation to Soil Heterogeneity

The decrease in the standard error of the mean of one variety or treatment is proportional to the square root of the number of replications. Some workers have argued that increased replication results in more heterogeneity due to the occupation of a larger land area with the result that a point will be reached beyond which further replication will give no further increase in accuracy. Fisher (1931) points out that the experimental error is due only to the irregularities within blocks and that this difficulty is not effective when different treatments are compared locally within relatively small pieces of land. The number of blocks or replicates makes no difference because the block effect may be removed by the experimental arrangement (e.g. randomized blocks and Latin squares). Large blocks presents a problem in itself. The situation of large blocks led Hayes (1923) to make the statement that, when a large number of strains are being tested, it is necessary to use a large number of replications to attain the same degree of accuracy as when a smaller number of strains are being compared. Special designs are advisable for tests of a large number of varieties or treatments.

(b) Duration of Tests

Replication in time is a necessary consideration in experimental tests. Comparative results from various treatments or varieties are frequently modified or even reversed in different seasons in response to climatic and soil variations and to the prevalence of plant diseases, insects, and other pests. The American Society of Agronomy (1933) recommends the continuation of a field experiment over a number of years so as to give a random sample of such seasonal effects. As an illustration, seasonal variability at the Hays (Kansas) substation is greater than that due to soil.

Crop	Variable Factor	Acre Yield (Bu.)	s'
Wheat	Season	18.4	12
Wheat	Soil		4

The standard deviation due to soil and season would be: $s' = \sqrt{12^2 + 4^2}$. The reduction in seasonal variation would require a replication of the test over a greater number of years. Ordinarily, a minimum of 5 years should be required in a field experiment where a seasonal influence is important.

VI. Plot Arrangements

Each variety or treatment may be arranged either (1) in the same order in each replicate, or (2) entirely at random in each replicate. The former is called a systematic distribution while the latter is designated as a random arrangement. Until rather recently, systematic distributions have been generally used in field experiments. Random arrangements have been advocated by Fisher (1931)(1937) and the Rothamsted workers who claim that randomization is necessary for a valid estimate of error. Regardless of the arrangement used, the various plots of a variety or treatment should be arranged so as to adequately sample the experimental area. This usually leads to certain restrictions on the arrangement.

(a) Random Arrangements

To justify random arrangements, Fisher (1931) states that uniformity trials have quite generally established the fact that soil fertility cannot be regarded as distributed at random but to some extent systematically. As an average, nearby plots are known to be more alike than those farther apart. Moreover, soil fertility distribution is seldom or never so systematic that it could be represented by a single mathematical formula. As to the estimate of error, Goulden (1931) explains that it depends upon differences in plots treated alike. Such an estimate will be valid only when pairs of plots treated alike are not nearer together or farther apart than pairs of plots treated differently. The total variance is made up of differences between plots in both directions. When the differences between plots treated differently are reduced by any sort of systematic arrangement one must automatically increase the differences between plots treated alike, and vice versa, e.g.

V (total variance) = A (plots treated alike) + B (plots treated differently)

An alteration in either A or B will result in a similar alteration in the opposite direction. Systematic arrangements which attempt to distribute the plots of any one variety or treatment as widely as possible over the experimental area tend to reduce B and increase A. Thus, the real differences between varieties or treatments are reduced and the experimental error increased. An example of a random arrangement for 8 "varieties" in 4 replicates is as follows:

Replicate I: 5-7-2-4-8-6-3-1
Replicate II: 1-3-5-6-2-8-7-4
Replicate III: 1-8-2-3-5-7-6-4
Replicate IV: 7-4-2-1-8-3-5-6

In practice, a set of random numbers such as those compiled by Tippett (1927) is useful to effect randomization of treatments or varieties. One may draw numbered chips at random or shuffle cards to obtain a random arrangement.

(b) Systematic Arrangements

A systematic arrangement is the repetition of the varieties in the same order in each replicate. Correlation between adjacent varieties is likely under such arrangements. However, systematic arrangements may be more practical in some experiments. Certain advantages have been given by the advocates of systematic arrangements: (1) Simplicity. It facilitates planting, harvesting, and note-taking operations. (2) It provides adequate sampling of the soil, i.e., allows for "intelligent placement" of the various varieties or treatments. (3) Varieties may be arranged in the order of maturity so as to facilitate machine harvest of field plots. (4) It may be desirable to alternate dissimilar varieties (bearded and beardless) so that mechanical mixtures can be detected in subsequent years. Systematic arrangement may be effective in such cases. Thru the use of plots which provide for the elimination of plant competition effects, systematic distribution loses one of its most serious sources of systematic error.

The plot scatter on the experimental area is a matter of simple repetition when the plots are all planted in a single series, viz.,

Replicate I	Replicate II	Replicate III
A B C D E F G H	A B C D E F G H	A B C D E F G H

As a rule, all plots cannot be placed exactly in one series, i.e., there are either too few or too many. It is advisable to commence each block with a different variety, especially when there is a soil gradient in the same direction as the series. This eliminates the possibility that one variety will fall on the best soil in each block. For compact blocks, the knight's move (one down and two over) is a common arrangement to secure an adequate scatter, viz.,

Replicate	Varieties
I	A B C D E F G H
II	G H A B C D E F
III	E F G H A B C D

(c) Influence of Arrangement on Error

Few data are available to show the relative accuracy of systematic and random arrangements. The "Student"---Fisher controversy in 1936 indicates that the problem has not been fully settled. In a comparison of diagonal with random arrangements, Tedin (1931) found that the degree of variability within 6 by 5 blocks was not influenced by either arrangement in the estimate of error. However, he advised random arrangements for the highest degree of scientific accuracy. In studies from uniformity trials with rice, Pan (1935) concluded that, with a systematic arrangement of varieties, the deviations from mathematical expectation were too great to be explained on the basis of random sampling. In a randomized arrangement, the number of differences in yield between all possible comparisons of hypothetical varieties that fell within a range of 0.5 σ, 1.0 σ, etc., were computed. Satisfactory agreement with mathematical expectation was obtained in two experiments, and poor agreement in one (P = less than 0.01). On the other hand, Odland and Garber (1928) obtained somewhat lower standard deviations from systematic arrangements than from the theoretical random arrangement. So far as small grains in nursery plots are concerned, Love and Craig (1938) found the relative yields to be about the same for systematic and random arrangements.

VII. Error Control

The differences between plots of a single treatment in a replicated experiment are due partly to experimental error and partly to the average differences between replicates. The variability between replicates is irrelevant to the experimental test when each variety or treatment occurs but once in a replicate. Therefore, the variance due to replicates or blocks is generally removed from the error. The precision of the experiment becomes greater when a large amount of the total variability can be removed in this way.

The shape of plots and blocks are also concerned in error control. Long narrow plots are preferable within the block so long as the blocks themselves approach a square in shape. The basic experimental designs are the randomized block and Latin square arrangements. (See Goulden, 1939).

VIII. Randomized Blocks

The randomized block test is the simplest type of experiment where satisfactory error control is obtained. This type of design is extremely flexible and can be used for as many as 30 treatments. The principal restriction in this test is that the same treatment should fall only once in each block, the treatments or varieties being arranged at random. The number of replicates or blocks depends somewhat upon the number of treatments included in a block and the degree of precision desired. It is preferable for the test area to be square in shape, altho this is not absolutely necessary.

(a) Field arrangement

A field arrangement for 10 varieties in 4 blocks could be as follows: [1]

I	5	10	7	2	4	8	9	6	3	1
II	9	1	8	2	3	10	5	7	6	4
III	6	1	2	9	8	3	10	5	7	4
IV	8	7	5	4	9	6	2	3	10	1

For more than 30 varieties, special designs should be used. (See later chapters).

(b) Computation of Sums of Squares

The yield data can be arranged conveniently for computation as in the table below. (Data from Goulden, 1929).

	Varieties										
Blocks	1	2	3	4	5	6	7	8	9	10	Totals
I	34.0	16.0	34.1	14.5	18.5	29.9	28.6	16.0	17.3	23.1	232.0
II	14.0	11.0	20.5	13.5	15.6	28.2	27.6	8.3	12.1	29.5	180.3
III	26.6	9.0	29.3	7.9	13.4	25.5	23.5	5.6	8.1	22.6	171.5
IV	18.5	11.9	21.0	15.2	8.9	28.8	16.5	9.5	10.3	17.7	158.3
Totals	93.1	47.9	104.9	51.1	56.4	112.4	96.2	39.4	47.8	92.9	742.1
Means	23.12	11.98	26.22	12.78	14.10	28.10	24.05	9.85	11.95	23.22	18.55

First, it is necessary to compute the sums of squares for total, varieties, blocks (or replicates), and error. The correction factor is $(Sx)^2/N$, or $(742.1)^2/40$ = 550,712.41/40 = 13,767.81.

$$\text{Total} = S(x^2) - \frac{(Sx)^2}{N} = 16,279.27 - 13,767.81 = 2511.46$$

$$\text{Varieties} = \frac{S(x_v^2)}{n} - \frac{(Sx)^2}{N} = \frac{62,114.01}{4} - 13,767.81 = 1760.69$$

In this case, it is necessary to square the total for each variety and divide by the number of values that make up each total to reduce the results to a single-plot basis.

---

[1] A set of random numbers such as table 6 in the appendix is useful to randomize the varieties. In fact, columns I, III, V, and VII were used.

$$\text{Blocks} = \frac{S(x_b^2)}{m} - \frac{(Sx)^2}{N} = \frac{140{,}803.23}{10} - 13{,}767.81 = 312.51$$

$$\text{Error} = \text{Total} - (\text{varieties} + \text{blocks})$$
$$= 2511.46 - (1760.69 + 312.51) = 438.26$$

The data are assembled in a convenient table as follows:

Variation due to	D.F.	Sums Squares	Mean Square	s	F-value
Blocks	3	312.51	104.17		6.42**
Varieties	9	1760.69	195.63		12.05**
Error	27	438.26	16.23	4.029	
Total	39	2511.46			

**Exceeds 1.0 per cent point, i.e., the value of "F" which has a probability of 0.01 of occurring due to chance.

$$F = \frac{\text{larger variance}}{\text{smaller variance}} = \frac{195.63}{16.23} = 12.05$$

By reference to the F-table, it is observed that the obtained F-value exceeds the 1.0 per cent point in both cases.

The other computations are as follows:

Standard error of a single determination $(s) = \sqrt{16.23} = 4.029$

Standard error of the mean for each variety $(\sigma_{\bar{x}}) = s/\sqrt{n} = 4.029 / \sqrt{4} = 2.0143$

Standard error of a difference $(\sigma_d) = \sigma_{\bar{x}}\sqrt{2} = (2.0143)(1.1414) = 2.8486$

Level of significance for 5 pct. point $= (\sigma_d)(t)$ (for 27 d.f.)

$$= (2.8486)(2.052) = 5.8453$$

In this case, 2.052 times the standard error of the difference gives odds of 19:1. This value can be obtained from the "t" table by Fisher (1934) where "t" is taken for the degrees of freedom for error at the 5 per cent point.

(c) Application to Mean Comparisons

Tests are sometimes found in which the value of z or F, for the comparison of variances due to varieties and error, just fails to reach the 5 per cent level of significance. This would indicate that the differences between variety means were of doubtful significance. In spite of this, certain differences between variety means can often be found which exceed twice the standard error. The use of twice the standard error (which gives approximately odds of 19:1 as the degrees of freedom approach 60) would indicate that certain differences might be significant. However, in such cases the testimony of the "z" or "F" test should be accepted as correct. Twice the standard error is not a sufficiently stringent test for the comparison of the greatest yield difference found in a large set of possible differences. Student (1927) and Tippett (1937) have both pointed out that, when the highest and lowest values are compared, the conventional use of twice the standard error to obtain odds approximately equivalent to the 5 per cent level of significance is no longer valid. For example, with 10 varieties in the test the difference between the highest and lowest varieties would need to reach 3.2 times the standard error to lie on the 5 per cent level of significance. On the other hand, when "F" is determined significant, the practice of using twice the standard error of the difference of two means as a criterion for significance may be too stringent when the means under consideration are contiguous in an arrangement of the variety (or treatment) means in order of magnitude.

IX. The Latin Square

The Latin square design is very efficient where a small number of varieties or treatments is being tested, but it becomes unwieldy for more than 10. Two restrictions are imposed on the treatments in this design, i.e., the same treatment can occur only once in the same row or column. The treatments are arranged at random within these restrictions. The limitation of the Latin square for a large number of varieties is due to the requirement of the same number of replications as treatments. It should be emphasized that the plots need not be square in shape. (See Fisher and Wishart, 1930). This design gives error control across the field in two directions, which always takes care of soil gradients. The most generally used Latin squares vary from 4 by 4 to 10 by 10. Some data from an irrigation study with sugar beets will be used as an illustration of the Latin square arrangement in the field as well as for the statistical analysis.

(a) Field Plot Arrangement

The field lay-out for the 5 irrigation treatments (A,B,C,D, and E) was as follows:

		Columns				
		1	2	3	4	5
Rows	1	E	D	A	B	C
	2	C	E	D	A	B
	3	A	C	B	E	D
	4	D	B	E	C	A
	5	B	A	C	D	E

(b) Analysis of the Data

The data for the irrigation study are compiled below, followed by the statistical analysis.

	Tons Beets Per Acre					
Row	1	2	3	4	5	Row Totals
1	18.52 (E)	19.46 (D)	20.66 (A)	22.68 (B)	18.65 (C)	99.97
2	20.68 (C)	14.29 (E)	18.82 (D)	20.02 (A)	20.58 (B)	94.39
3	26.04 (A)	17.49 (C)	21.06 (B)	18.91 (E)	20.03 (D)	103.53
4	22.51 (D)	22.98 (B)	17.15 (E)	17.14 (C)	20.62 (A)	100.40
5	24.44 (B)	20.25 (A)	18.92 (C)	19.73 (D)	14.07 (E)	97.41
Column Totals	112.19	94.47	96.61	98.48	93.95	495.70

Treatment	A	B	C	D	E
Totals	107.59	111.74	92.88	100.55	82.94
Means	22.35	21.52	20.11	18.58	16.59

Correction factor = $(Sx)^2/N$ = $(495.70)^2/25$ = 9828.7396

Total = $S(x^2) - \frac{(Sx)^2}{N}$ = 10,007.8598 - 9.828.7396 = 179.1202

Rows = $\frac{S(x_r^2)}{n} - \frac{(Sx)^2}{N}$ = 9,838.1604 - 9,828.7396 = 9.4208

Columns = $\frac{S(x_c^2)}{n} - \frac{(Sx)^2}{N}$ = 9,873.9164 - 9,828.7396 = 45.1768

Treatments = $\frac{S(x_t^2)}{n} - \frac{(Sx)^2}{N}$ = 9,935.4952 - 9,828.7396 = 106.7556

Error = Total - (Rows + columns + treatments)

= 179.1202 - (9.4208 + 45.1768 + 106.7556) = 17.7670

The data are assembled to complete the analysis:

Variation due to	D.F.	Sums Squares	Mean Square	Standard Error	F-Value Actual	F-Value 5% Point
Rows	4	9.4208	2.3552		1.59	3.26
Columns	4	45.1768	11.2942		7.63	3.26
Treatments	4	106.7556	26.6889		18.03	3.26
Error	12	17.7670	1.4806	1.2168		
Total	24	179.1202				

$$F = \frac{\text{larger variance}}{\text{smaller variance}} = \frac{26.6889}{1.4806} = 18.03$$

Since the computed F-value is greater than that for the 5 per cent point, significant differences exist between treatments.

The other constants may be computed as follows:

Standard error of the mean ($\sigma_{\bar{x}}$) = $\frac{s}{\sqrt{n}}$ = $\frac{1.2168}{\sqrt{5}}$ = 0.5440 tons.

Standard error of the difference (σ_d) = $\sigma_{\bar{x}}\sqrt{2}$ = $0.5440\sqrt{2}$ = 0.77

Level of significance (for 5 pct. point) = 2.179 σ_d = (2.179)(0.77) = 1.68 tons.

The data may be arranged as follows in summary form:

Treatment	Mean Yield (tons)
A	22.35
B	21.52
C	20.11
D	18.58
E	16.59
Standard Error of the Mean	0.544
Level of Significance (5% point)	1.68

B -- Relation of Type of Experiment to Design

X. Variety and Similar Tests

The variety test is probably the most common type of agronomic field experiment. Crop varieties are tested for yield in most crop improvement programs to determine which ones are superior under given soil and climatic conditions. Varieties are known to differ as to the best rate of planting. Less seed is required under dryland than under irrigated conditions due to the moisture factor. Carleton (1909) points out that winter wheats tiller more than spring wheats and, when winter-hardy, may be sown at a thinner rate. It is not always possible to overcome the objection of differential response of varieties to different rates of seeding in a variety test. It is usually safer to use a somewhat higher rate than that recommended to farmers because variations due to unexpected causes will then have less effect.

Rate and date tests are sometimes combined with variety trials, or they may be conducted separately. The combined sort permits a study of differential variety response to different rates or dates. A rather wide range of rates on either side of the optimum is suggested for rate of planting tests in order to determine the point of maximum yield. A test to determine the most satisfactory dates for planting crops is usually an exploratory stage in field experimentation to secure this information for certain environmental conditions. Such tests are usually planted at a regular interval of time between dates from extremely early in the planting season to extremely late. Due to occasional differential varietal response to time of planting, consideration should be given to the question of planting a variety series at several different dates.

Experiments of these types can be designed as Latin squares for small precise tests with 5 to 10 varieties, while randomized blocks are commonly used for 10 to 30 varieties. For greater numbers in a single experiment, incomplete block designs should be investigated.

XI. Crop Rotation Experiments

In crop rotations, or other experiments in which a study of residual effects is made, it is necessary to grow all crops used in the rotation each year in order to obtain reliable results. Carleton (1909) early called attention to the fact that this simple but essential matter had been entirely overlooked in many of the older experiments. For accuracy in a rotation series, every stage or crop must experience every condition. Each year there must be as many plots as there are crops or stages in the rotation. For example, in a 4-year rotation of (1) oats seeded to red clover, (2) red clover, (3) corn, and (4) barley, there must be four plots. The plots must be at least in duplicate in order to allow for the removal of soil variability. Adequate replication is the greatest need in crop rotation experiments. In another block in the same test there may be a plot of each crop in continuous culture, although this is not always necessary. The crop rotation test must be conducted over a period of years so that the crop yields will be definitely influenced by the different rotation treatments. Such a test might be laid out as follows:

	4-year Rotation				Continuous Culture		
	Oats	Red Clover	Corn	Barley	Corn	Oats	Barley
Replicate I	(a)	(b)	(c)	(d)	(e)	(f)	(g)
	Continuous Culture				4-year Rotation		
	Oats	Corn	Barley	Corn	Barley	Oats	Red Clov
Replicate II	(f)	(e)	(g)	(c)	(d)	(a)	(b)

To compare this 4-year rotation with a 3-year rotation, it would be necessary to wait 12 years. For a 7 and 5-year rotation, the results could be compared at the end of 35 years, etc.

XII. Cultural Experiments

Cultural experiments include such tests as fall vs. spring plowing, methods of seed-bed preparation, surface vs. furrow planting, etc. Field plots are generally necessary for experiments of this type because of the use of farm machinery. Many dryland experiments are concerned with cultural methods. The same procedures for variety tests are generally satisfactory in tests of this kind.

XIII. Fertilizer Experiments

The most reliable information on the fertilizer needs of soils may be obtained from the field experiment. Nutrient solutions and sand cultures are used in special studies. The early fertility experiments at Rothamsted were concerned primarily with the fertilizer value of certain mineral fertilizers as shown by increased crop yields. The present long-time fertilizer experiments are concerned more with comparisons of similar fertilizers, effects on crop plants, and efficiency of fertilizer practices. The earlier workers often tested one fertilizer at a time, but many present workers are inclined to favor more comprehensive tests, i.e., the inclusion of several fertilizers at more than one level. Most investigators use crop yield as the major criterion of fertilizer response.

(a) General Types of Fertilizer Tests

Fertilizer Tests may be conducted for several definite purposes. (1) Deficiency of Fertilizer Elements in a Field: Results of such tests are applicable only to the field tested or, at most, to soil of similar type with similar previous cultural treatment. It is strictly applicable for the test year, since the crop grown may modify conditions for the next season. (2) Efficiency of single Fertilizer Elements: For this type of test it is desirable to have the fertilizer elements tested in minimum. (See Giles, 1914). Several rates of a standard fertilizer can be compared with one or more rates of a fertilizer that carries the elements in a different form. Equal rates of each fertilizer can be compared also. (3) Comparative Methods of Application: This type includes tests on depth of placement, time of application, placed to side vs. with the seed, etc. (4) Optimum Fertilizer Balance: This type is concerned with fertilizer balance for various crops. It involves many complications when made in the field because it is difficult to control or even measure the fertilizer balance in a field. In such a study it is necessary to estimate by chemical tests the amount of the fertilizer elements furnished by the soil as well as the amount applied. Probably the most practical method to make such a study is to vary each element separately over a wide range by several rates of application. The regression of yield on amount of the element available (amount in plant plus amount applied) may then be calculated. A further complication would be to test all possible combinations of several fertilizers at different levels. The triangle system suggested by Schreiner and Skinner (1918) may be useful for the computation of all possible combinations of three fertilizer elements (say P_2O_5, NH_3, and K_2O) at several levels. This triangle system should not be used as a basis for the field lay-out as originally advocated. Such a test should be designed as a factorial experiment. (See Chapter 19). (5) Long-Time Effect of Fertilizers: Such tests with various forms of fertilizers are concerned with the physical and chemical properties of soils as well as soil productivity.

(b) Design of Fertilizer Experiments

Several basic principles should be considered in soil fertility experiments. A soil profile to a depth of 3 feet is highly desirable for each series of plots.

Before soil treatment experiments are begun, the American Society of Agronomy (1933) recommends that "representative samples of the soil and subsoil should be carefully taken for such analyses as may be desired for future reference". In the matter of plot design the Society cautions that "the lateral translocation of soil or fertilizer beyond the plot interspaces of soil experiments should be avoided . . . . . . Since the manner of fertilizer application may affect yields materially, due consideration should be given to this problem".

For fertilizer tests where 2 or more fertilizers are applied at 2 or more levels, the factorial design is suitable. The factorial experiment, explained by Fisher (1934), Yates (1935), Summerby (1937) and others, involves all combinations of the fertilizers and levels (or amounts) of application. The study of interactions is an important consideration in such an experiment. For example, suppose a fertilizer test is to be conducted with nitrogen, phosphorus, and potassium at two different rates each. The rates can be designated by subscripts so as to give the 8 possible treatment variants as follows:

$N_0P_0K_0$, $N_1P_0K_0$, $N_0P_1K_0$, $N_0P_0K_1$, $N_1P_1K_0$, $N_1P_0K_1$, $N_0P_1K_1$, and $N_1P_1K_1$.

Such an experiment can be planned for a randomized block test or for some form of the incomplete block test. Goulden (1931) gives some suggestions on the design of more complicated fertilizer experiments. Residual effects due to past fertility treatments is discussed by Forester (1937).

XIV. Pasture Experiments

In experimental pasture work, the investigator may desire to: (1) determine the amount of herbage produced on an area by different pasture-grass mixtures, (2) to find out the influence of fertilizers on pastures as to yield and survival of the palatable species, or (3) he may desire to measure the influence of different grazing methods on yield and survival. Replication of treatments is vital in any case.

One of the important technique problems is the comparative results from grazing and mechanical harvest of herbage. Stapledon (1931) states that the animal is the master factor in pasture studies. He tethered sheep on small plots and moved them twice a day in the Aberystwyth pasture researches. Certain advantages are claimed for his tethering method: (1) replicated plots are possible; (2) the experimental sheep are handled and examined twice per day; (3) grazing will be uniform, and (4) the animal capacity is increased per unit area. Schuster (1929) recommends at least 4 replications and 3 animals per plot in pasture investigations. The use of grazing methods permits the effects of trampling on the vegetation to be measured. Pasture plots may be harvested mechanically, i.e., clipped with a mower or with shears. Brown (1929) advocates the use of grass shears for small cages, the lawn mower for grass less than 6 inches high, and a mowing machine for taller herbage. Several studies have compared grazing and mechanical harvest of pasture plots. Brown (1929) found that the herbage of grazed and mowed plots varied markedly in time due to animal preferences. Animals void a large proportion of the fertilizer elements consumed in feeds, particularly nitrogen and phosphorus. Thus, mowed pastures may be low in fertilizer elements when compared with grazed pastures. A high correlation between mowed and grazed yields was found when the mowed areas were changed to the previously grazed areas every two or three years. Continuously clipped cages have yielded less than annually mowed cages. Robinson, et al (1937) found a progressive decrease in the yields of clipped permanent quadrats in relation to grazed areas.

Sampling methods are often involved in the design of pasture experiments. See Chapter 16 for further details on this phase, as well as the report of Vinall and others (1934).

C -- Incomplete Experimental Records[1]

XV. Missing Values in Experiments

In general, replicated field experiments are so arranged that the mean yield for all plots that receive a given treatment provides the best estimate of the effects of that treatment. Sometimes the yields of some plots are lost or prove unreliable with the result that the orthogonality of the original design disappears. Since the treatment, block, etc., effects are computed from the total yield of all plots in a given treatment, block, etc., it is necessary to interpolate the yield of the missing plot in order to use the ordinary analysis of variance.

Allan and Wishart (1930) were the first to provide formulae for the estimation of the yield for a single missing plot in either randomized block or Latin square tests. They arrived at their formula by the procedure of fitting constants by least squares. Yates (1933) used a simpler solution by minimizing the error variance obtained when unknowns are substituted for the missing yields. The two formulae give the same results, but the one by Yates also provided a method appropriate for the estimation of the yields of several missing values. His formula is used here.

XVI. Calculation of Single Missing Value

A single missing value can be calculated for either a randomized block or latin square test.

(a) Randomized Block Test

Some data are given on the effect of date of planting on the yields of sugar beets in which a plot value is missing. The yields are in tons per acre.

Date Planted	Block Number 1	2	3	4	5	Totals	
Early	22.3	21.8	19.7	21.2	20.0	105.0	
Medium	18.3	18.4	18.5	21.5	17.3	94.0	
Late	17.2	17.2	17.9	(18.8)	16.7	(87.8)	69.0
Very Late	14.9	12.6	13.1	14.4	12.4	67.4	
Totals	72.7	70.0	69.2	(75.9) 57.1	66.4	(354.2)	335.4

It is assumed that the yield of the late-planted plot in block 4 is missing. The sums for block 4 and for late planting are given below or to the right of the appropriate block or treatment to show that they are the sums of only the known plots. The values in brackets are filled in later.

The formula for the estimation of yield of this value in a randomized block test is as follows:

$$x = \frac{mM + m'M' - Tx}{(m - 1)(m' - 1)} \quad (1)$$

where x = yield of missing plot,
m = number of treatments
m'= number of blocks
M = sum of known yields of treatment with missing plot
M'= sum of known yields of block with missing plot
T_x = total yield of known plots.

[1] This portion is taken entirely from an outline prepared by Dr. F. R. Immer, U. of Minnesota.

In the sugar beet test used as an example,

$$x = \frac{4(69.0) + 5(57.1) - 335.4}{(4 - 1)(5 - 1)} = 18.8$$

The yield, x = 18.8, is inserted in the table after which the block, treatment, and general sum are corrected accordingly. These figures are in the brackets. The analysis of variance will be computed in the usual way, except that the degrees of freedom for error and total have been reduced one. The degrees of freedom must be reduced by one for each plot value interpolated. The analysis of variance is as follows:

Variation due to	Degrees Freedom	Sums Squares	Mean Square	Standard Error (s)	F-Value Observed	F-Value 5 pct. point
Blocks	4	13.043	3.2608		3.71	3.36
Treatments	3	149.618	49.8727		56.69	3.59
Error	11	9.677	0.8797	0.9379		
Total	18	172.338				

(b) Latin Square Test

The formula to be used for the interpolation of a single value in a latin square test is as follows:

$$x = \frac{m(M_r + M_c + M_t) - 2\,T_x}{(m-1)\ (m-2)} \quad - (2)$$

Where x = missing plot yield;

M_r, M_c, M_t = totals of known yields of the row, column, and treatment from which the plot is missing;

m = number of treatments (also equals number rows or columns);

T_x = total yield of all known plots.

XVII. More than One Missing Value

A method of approximation may be used for more than one missing plot yield. Three plots are missing in the randomized block trial given below:

Date Planted	1	2	3	4	5	Total	
Early	22.3	21.8	(21.2)	21.2	20.0	(106.5)	85.3
Medium	18.3	(18.6)	18.5	21.5	17.3	(94.2)	75.6
Late	17.2	17.2	17.9	(18.7)	16.7	(87.7)	69.0
Very Late	14.9	12.6	13.1	14.4	12.4	67.4	
Totals	72.7	(70.2) 51.6	(70.7) 49.5	(75.8) 57.1	66.4	(355.8)	297.3

The plot yields given in brackets have been assumed to be missing. As it is possible to interpolate the yield of only one plot at a time, one must assume yields for all missing plots except the one to be interpolated. First, suppose the medium planting in block 2 is interpolated. For the early plot in block 3 and the late plot in block 4, one must insert the mean yield of the known plots for those two dates, or 21.3 and 17.2 tons, respectively.

The formula for interpolation, in which have been substituted the values for the first approximation for the yield of the medium planting in block 2, is as follows:

$$x = \frac{mM + m'M' - T_x}{(m-1)(m'-1)} = \frac{4(75.6) + 5(51.6) - 335.8}{(4-1)(5-1)} = 18.7$$

The same procedure can be followed for the early planting in block 3, except that the guess of 21.3 used before should be removed. The interpolated value (18.7) is used for the medium planting in block 2, and the guessed value (17.2) for the late planting in block 4. The grand total is corrected accordingly. The value of x in this case is 21.3.

In like manner the yield of the late planting in block 4 is interpolated. This is found to be 18.7.

Since it was necessary to estimate the yields of two plots in order to start the interpolation process, the values obtained will be somewhat in error. Therefore, the values are re-interpolated, using the values obtained by the first interpolation for all but the plot yield being calculated. This is repeated until no further changes take place. The values obtained in this case were as follows:

Treatment	Block	Approximations 1st	2nd	3rd
Medium	2	18.7	18.6	18.6
Early	3	21.3	21.2	21.2
Late	4	18.7	18.7	18.7

The interpolated values did not change after the second approximation.

The interpolated yields are inserted in the above table (as shown in brackets) after which the correct treatment and block totals are determined. The analysis of variance is then computed as shown below:

Variation due to	Degrees Freedom	Sums Squares	Mean Square	Standard Error (s)	F-Value Obtained	5% Point
Blocks	4	11.923	2.9808		3.23	3.63
Treatments	3	160.306	53.4353		57.88	3.86
Error	9	8.309	0.9232	0.9608		
Totals	16	180.538				

Three degrees of freedom have been subtracted from error and from total because three plot yields were interpolated.

XVIII. Tests of Significance

The error calculated from analyses of variance, in which one or more plot values have been interpolated, is a valid estimate of experimental error when the degrees of freedom have been reduced by one for each value interpolated. However, the variance due to treatments is not entirely without bias, being always higher than it should be. The significance of the test is accentuated, but the correction for this condition is quite trivial for cases in which only a single value is missing. The bias is more pronounced where many plots are missing.

Tests of significance by means of the analysis of variance are generally all that are required. For a single missing plot, the treatment mean with the estimated value of the missing plot will have an error as follows for a randomized block test:

$$\sigma_{\bar{x}t} = \frac{1}{m'} \left[1 + \frac{m}{(m-1)(m'-1)} \right] s^2 \text{ - - - - - - - - - - - - - - - - - (3)}$$

Where m' = number of blocks, m = number of treatments, and s^2 = variance of a single plot calculated from error. The variance of the treatment mean would be s^2/m' where no plot was missing.

For a single missing plot in a latin square, the variance of the treatment mean with the missing value would be as follows:

$$\sigma_{\bar{x}t} = \frac{1}{m} \left[1 + \frac{m}{(m-1)(m-2)} \right] s^2 \text{ - - - - - - - - - - - - - - - - - (4)}$$

For more than one missing plot, these formulae are strictly applicable only to comparisons between means where one contains no missing plot. To find the variance of the difference between two means, both of which contain missing values, is rather difficult.

References

1. Allen, E. W. Initiating and Executing Agronomic Research. Jour. Am. Soc. Agron., 22:341-348. 1930.
2. Allen, F. E., and Wishart, J. A Method of Estimating the Yield of a Missing Plot in Field Experiments. Jour. Agr. Sci., 20:399-406. 1930.
3. Brown, B. A. Technic in Pasture Research. Jour. Am. Soc. Agron., 29:468-476. 1937.
4. Carleton, M. A. Limitation in Field Experiments. Soc. Prom. Agr. Sci., pp. 55-61. 1909.
5. Fisher, R. A. The Technique of Field Experiments. Rothamsted Conf., 13:11-13. 1931.
6. ____________ Statistical Methods for Research Workers. Oliver and Boyd. 5th Ed. pp. 199-231. 1934.
7. ____________ Design of Experiments. Oliver and Boyd. 2nd Ed. pp. 13-100. 1937.
8. ____________ The Half-Drill Strip System of Agricultural Experiments. Nature, 138:1101. 1936.
9. ____________, and Wishart, J. The Arrangement of Field Experiments and the Statistical Reduction of the Results. Imp. Bur. Soil Sci., Tech. Comm. No. 10. 1930.
10. Forester, H. C. Design of Agronomic Experiments for Plots Differentiated in Fertility by Past Treatments. Ia. Agr. Exp. Sta. Res. Bul. 226. 1937.
11. Giles, P. L. On the Plans of Fertilizer Experiments. Jour. Am. Soc. Agron., 6:36-42. 1914.
12. Goulden, C. H. Modern Methods of Field Experimentation. Sci. Agr., 11:681-701. 1931.
13. ____________ Statistical Methods in Agronomic Research. Can. Seed Growers Assn. 1929.
14. ____________ Methods of Statistical Analysis. John Wiley, pp. 45-51, and 142-148. 1939.
15. Hayes, H. K. Controlling Experimental Error in Nursery Trials. Jour. Am. Soc. Agron., 15:177-192. 1923.
16. Love, H. H. and Craig, W. T. Investigations in Plot Technic with Small Grains. Cornell Memoir 214. 1938.
17. Odland, T. E., and Garber, R. J. Size of Plat and Number of Replications in Field Experiments with Soybeans. Jour. Am. Soc. Agron., 20:93-108. 1928.
18. Pan, C. Uniformity Trials with Rice. Jour. Am. Soc. Agron., 27:279-285. 1935.

19. Paterson, D. D. Statistical Technique in Agricultural Research. McGraw-Hill. pp. 156-188. 1939.
20. Robinson, R. R., Pierre, W. H., and Ackerman, R. A. A Comparison of Grazing and Clipping for Determining the Response of Permanent Pastures to Fertilization. Jour. Am. Soc. Agron. 29:349-359. 1937.
21. Schreiner, O., and Skinner, J. J. The Triangle System of Fertilizer Experiments. Jour. Am. Soc. Agron., 10:225-246. 1918.
22. Schuster, G. L. Methods of Research in Pasture Investigations. Jour. Am. Soc. Agron., 21:666-673. 1929.
23. Standards for the Conduct and Interpretation of Field and Lysimeter Experiments. Jour. Am. Soc. Agron., 25:803-828. 1933.
24. Stapledon, R. G. The Technique of Grassland Experiments. Rothamsted Conf. 13, pp. 22-28. 1931.
25. Student. Errors of Routine Analysis. Biometrika, 5:351. 1927.
26. ________ The Half-Drill Strip System of Agricultural Experiments. Nature, 138: 971-972. 1936.
27. Summerby, R. The Use of the Analysis of Variance in Soil and Fertilizer Experiments with a Particular Reference to Interactions. Sci. Agr., 17:302-311. 1937.
28. Tedin, O. Influence of Systematic Arrangement upon the Estimate of Error in Field Experiments. Jour. Agr. Sci., 21:191-208. 1931.
29. Tippett, L. H. C. Tracts for Computers XV --Random Sampling Numbers. Cambridge U. Press. 1927.
30. Tippett, L. H. C. The Methods of Statistics. Williams and Norgate. 2nd Ed. pp. 125-139. 1937.
31. Vinall, H. N., et al. Report of the Joint Committee on Pasture Research. Am. Soc. Agron. (mimeographed) 1934.
32. Wishart, J., and Sanders, H. G. Principles and Practice of Field Experimentation. Emp. Cotton Growing Corp., pp. 60-85. 1935.
33. Yates, F. The Analysis of Replicated Experiments when the Field Results are Incomplete. Emp. Jour. Exp. Agr., 2:129-142. 1933.
34. ________ Complex Experiments. Suppl. Jour. Roy. Stat. Soc., 2:181-247. 1935.

Questions for Discussion

1. What criticisms have been made of agronomic experiments in general? Are they justified?
2. What justification is there for the antagonism sometimes found between scientific theory and practical facts?
3. What are the principal objectives in agricultural experiments?
4. What factors should be considered in the outline of an experiment?
5. What is the principle of the extremes? Illustrate.
6. In laying out field experiments in which one variable is continuous, what principle or rule should be followed with respect to the extremes?
7. Distinguish between preliminary and permanent experiments.
8. What is a simple experiment? Its limitations? Advantages?
9. What are combination or complex experiments? Are they desirable? Why?
10. What sources of variation or error, other than that due to soil or season, may occur in field experiments?
11. How does a random arrangement differ from a systematic arrangement?
12. Is soil heterogeneity systematic or random? Explain.
13. Upon what is the estimate of error based? How influenced by a systematic plot arrangement?
14. What are the advantages usually given for systematic arrangement? Random arrangement?

15. What is meant by the "knight's move"?
16. Discuss the relative efficiency of systematic and random plot arrangements.
17. What is a randomized block test? What restrictions are imposed? Its limitations?
18. What is the Latin square arrangement of plots? What is the primary objective in this arrangement?
19. What conditions should be observed in planning variety tests? What is a check variety?
20. What precautions are necessary in crop rotation tests?
21. What are the limitations in fertilizer tests?
22. In rotation and soil treatment tests what should be the treatment of the check?
23. What is the law of the minimum? Its application to fertilizer tests?
24. What is a factorial experiment? Give an example.
25. Discuss grazing vs. mechanical harvest of herbage.
26. Why is it necessary to calculate missing values for the analysis of variance to apply?
27. How are the degrees of freedom modified when a missing value is computed?

Problems

1. Different amounts of fertilizer were applied to sugar beets by the Colorado Experiment Station in 1930 (Data from D. W. Robertson) in a randomized block trial. The yields in pounds of sugar per plot for various amounts of treble superphosphate applied per acre were as follows:

Phosphate Treatment	Block I	II	III	Total
None	343	185	208	736
100 lbs.	358	415	483	1256
200 lbs.	393	435	463	1291
300 lbs.	427	468	487	1382
Totals	1321	1503	1641	4665

(a) Compute the analysis of variance for a randomized block experiment.
(b) Determine significance by use of the "F" test.
(c) Compare the average yields of the no treatment and 200 lb. treatment by means of the standard error.

2. A rate of planting test with sugar beets was conducted in 1931 by H. E. Brewbaker. The rates used were: 15, 20, 25, and 30 lbs. per acre. The experiment was designed as a 4 by 4 Latin square, the data for which follow:

	Tons beets per acre				Row totals
	(3) 16.73	(1) 15.38	(4) 16.35	(2) 15.27	63.73
	(4) 17.74	(2) 17.20	(3) 18.83	(1) 16.94	70.71
	(1) 17.52	(3) 18.15	(2) 17.97	(4) 18.31	71.95
	(2) 18.21	(4) 19.53	(1) 17.74	(3) 16.61	72.09
Column Totals	70.20	70.26	70.89	67.13	278.48

(a) Compute the analysis of variance.
(b) Obtain "F" for a comparison of error with rows, columns, and treatments.
(c) Test the significance of "F".
(d) Continue the analysis and compute the standard error of the mean, standard error of a difference, and level of significance in case it is justified.

3. Design a crop rotation experiment to show the effects of a legume in a rotation. The rotations are as follows: (a) barley (seeded to alfalfa), alfalfa, alfalfa, corn, and sugar beets; and (b) barley, corn, and sugar beets.

4. Four varieties of wheat were grown in 1938 in a randomized block trial in 5 blocks. The yield of one plot was lost. (a) Calculate the yield of the missing plot. (b) Complete the analysis of variance.

Variety	Block 1	2	3	4	5
Kanred	54.4	41.7	52.1	56.1	61.0
Cheyenne	40.7	----	46.5	59.9	53.7
Tenmarq	61.7	51.7	43.5	61.9	58.7
Hays No. 2	55.5	50.6	61.9	45.1	72.4

5. The same 4 varieties were grown in a randomized block test in 1937. The records on 2 plots were lost. Calculate the missing values and complete the analysis of variance.

Variety	Block 1	2	3	4	5	6
Kanred	54.6	53.7	68.0	55.2	58.5	62.1
Cheyenne	66.3	60.9	64.8	67.6	----	66.2
Tenmarq	58.5	57.5	44.1	65.6	52.9	51.6
Hays No. 2	57.3	----	60.5	62.2	58.8	54.3

CHAPTER XVI

QUADRAT AND OTHER SAMPLING METHODS

I. Sampling in Agronomic Work

There are times when it is impractical to use the whole plot or plant population to obtain a numerical determination of some characteristic of the experimental material. In such cases as tiller number, yield, percentage dry matter, nitrogen or sugar in the crop, it is more practical to sample only a proportion of the whole. To quote Wishart and Sanders (1935): "The object is to obtain as close an estimate as we can of the measure, which would be obtained accurately, within the limits of experimental error, had the produce of the whole plot been counted, weighed, or analyzed." The sample must be representative and taken in such a manner as to assure that end. It is also necessary to take into account the further source of error due to the sampling process. Yields determined by sampling procedure are not determined as accurately as when the entire plot is taken, but it is often advantageous to sacrifice some accuracy to save labor.

II. Theory of Sampling

The sampling distributions so far considered have been based on the assumption of independence. The simple theory of errors does not apply when the variation is heterogenous and the extent to which the sources of variation are represented is not left to chance. It has been shown in a randomized block trial that the variance due to error is an unbiased estimate of the error variance of the infinite population from which the data under consideration are a sample. The other items in the mean square column (blocks and varieties) are not unbiased estimates of the respective variances of the population. In fact, they contain the variance due to error as the degrees of freedom become indefinitely large. For example, the estimated variance due to varieties, in the theory of large samples, is made up of the true variance due to varieties plus the variance due to error. This becomes important in statistics of estimation as shown by Tippett (1937), Immer (1932, 1936), and others.

Suppose some data on protein in relation to different rate-of-planting treatments in corn for 1931 be used to illustrate the computations:

Variety	Method Planted	Rate Planted	Protein Per cent per Sample [1] Block I (1)	Block I (2)	Block II (1)	Block II (2)	Block III (1)	Block III (2)	Totals
Golden Glow	Hills	3 [2]	10.357	10.408	10.425	10.522	10.100	10.043	61.855
"	"	4	9.525	9.422	9.223	9.342	9.667	9.548	56.732
"	"	5	8.995	8.983	9.325	9.211	9.422	9.479	55.415
Pride North		3	10.363	10.351	9.713	9.553	9.605	9.627	59.212
"	"	4	9.171	9.245	9.399	9.376	9.057	9.052	55.300
"	"	5	9.166	9.120	9.171	9.211	8.527	8.504	53.699
Golden Glow	Drills	12 [3]	10.072	10.038	10.528	10.408	10.380	10.488	61.714
"	"	9	9.730	9.605	9.696	9.559	9.433	9.451	57.474
"	"	6	8.835	8.778	8.812	8.364	9.143	9.030	53.512
"	"	3	8.482	8.590	9.479	9.422	9.543	9.502	55.023
Pride North		12	9.872	9.929	10.009	9.884	9.827	9.724	59.245
"	"	9	9.325	9.462	8.858	8.892	9.114	9.149	54.800
"	"	6	8.789	8.761	9.525	9.565	8.852	8.892	54.384
"	"	3	8.641	8.767	9.263	9.428	8.453	8.510	53.[illegible]
Totals			131.323	131.459	133.236	133.237	131.128	131.049	791.432

[1] Protein = N (nitrogen) x 5.7 [2] Plants per hill [3] Inches between plants in row.

The analysis of variance is set forth for the experiment in which two protein determinations were made on the shelled corn per plot. For simplicity, treatments will be considered without regard to variety or method of planting.

The calculations for the sums of squares are as follows, the two samples per plot being added together for the plot determinations:

$S(x) = 791.432$ $\qquad (Sx)^2/N = 7,456.7216$

$S(x_s)^2 - (Sx)^2/N = 7,480.8643 - 7,456.7216 = 24.1427$

$S(x_p)^2 - (Sx)^2/N = 14,961.4426/2 - 7,456.7216 = 23.9997$

$S(x_b)^2 - (Sx)^2/N = 208,799.0186/28 - 7,456.7216 = 0.3862$

$S(x_t)^2 - (Sx)^2/N = 44,852.7957/6 - 7,456.7216 = 18.7444$

The summary for the analysis of variance is as follows:

Variation due to	Degrees Freedom	Sums Squares	Mean Square	Standard Error	F-Value
Blocks	2	0.3862	0.1931		1.03
Treatments	13	18.7444	1.4419		7.70**
Error	26	4.8691	0.1873	0.4328	
Total for Plots	41	23.9997			
Samples within Plots	42	0.1430	0.0034	0.0583	
Total samples	83	24.1427			

In the simple case where one sample is drawn from each plot with the treatment replicated for m plots, the variance of a treatment mean is V_p^2/m, where V_p^2, the mean variance between plots approaches σ^2, the true variance of an individual plot as m approaches infinity. However, when n samples are drawn from each plot, the variance of a treatment mean is V_p^2/mn, where V_p^2/n estimates σ_p^2, the true variance of an individual plot plus the true variance of an individual plot mean or σ_s^2/n. This follows because a plot mean is now subject to variation due to more than one sample. It is evident that σ_s^2 is the true variance of an individual sample taken from a plot. The relationship may be shown as follows:

$$\frac{V_p^2}{mn} \longrightarrow \frac{\sigma_p^2}{m} + \frac{\sigma_s^2}{mn} = \frac{1}{m} \left(\sigma_p^2 + \frac{\sigma_s^2}{n}\right) \qquad (1)$$

It should also be noted that:

$$V_s^2 \longrightarrow \sigma_s^2 \qquad (2)$$

It is clear that σ_p^2 can be estimated from the above formula because V_p^2 and V_s^2 are both obtainable from the analysis of variance.

In the present experiment,

$V_p^2 = 0.1873$, $V_s^2 = 0.0034$, $n = 2$, and $m = 3$.

$\frac{0.1873}{6} \longrightarrow \frac{1}{M} \left(\sigma_p^2 + \frac{\sigma_s^2}{n}\right)$, and $0.0034 \longrightarrow \sigma_s^2$.

Therefore,

$$\frac{0.1873}{6} \longrightarrow \frac{1}{3} \quad (\sigma_p^2 + \frac{0.0034}{2}) \text{ or } 0.0919 \longrightarrow \sigma_p^2 .$$

The standard errors for plot and sample means are then calculated.

$\sqrt{0.0919} = 0.3032 \longrightarrow \sigma_p$, and $\sqrt{0.0034} = 0.0583 \longrightarrow \sigma_s$.

The ratio, σ_p/σ_s is estimated as 0.3032/0.0583 = 5.20. This indicates that the variation between plots greatly exceeds that within plots or between samples, being 5.20 times as great.

III. Economy in Sampling

It is of considerable importance to analyze how the precision of an experiment, as measured inversely by $1/m\ (\sigma_p^2 + \sigma_s^2/n)$ is affected by varying m, the actual plot replications in the field, and n, the number of samples drawn from a plot. The most important inference to be drawn is that the precision is mainly controlled by m, the number of plot replications. Increasing the number of samples taken from the different plots can only appreciably affect the precision when σ_s^2 is not relatively small as compared with σ_p^2. In the present problem 0.0034, the estimated value of σ_s^2 is small compared with 0.0919, the estimated value of σ_p^2. Hence, it must be concluded that to make more than one analysis on a sample from a plot was unwarranted by the small gain that would result.

(a) Computation of Number of Samples or Replicates

The required variance of the mean of a treatment (K) would be:

$$K = \frac{1}{m} \left(\sigma_p^2 + \frac{\sigma_s^2}{n}\right) \quad \text{------------------------} \quad (3)$$

For the data in this problem,

$$K = \frac{1}{3} \left(0.0919 + \frac{0.0034}{2}\right) = 0.0312$$

The computation, for different values of m or n, will give the number of replications and number of samples per plot that will be necessary to reduce the variance of the mean to a given level, i.e., K = 0.0312. The K values for the estimation of the variance for treatments are as follows when the number of analyses per sample are varied for three replications:

Number of Samples (n)	K-values for m = 3
1	0.0318
2	0.0312
3	0.0310
4	0.0309
5	0.0307

These data indicate the negligible effect when 1, 2, 3, 4, or 5 protein analyses are made from shelled corn samples.

(b) Determination of Minimum Expense

Technical difficulties often prevent plot replication beyond a certain degree. In such cases it is frequently worthwhile to strengthen the precision of the experiment by drawing replicate samples from the different plots. The number that should be drawn depends on several factors. These factors are: (1) The variation between plots as measured by σ_p^2 in relation to σ_s^2, and (2) the cost of growing a plot or compared with the cost of obtaining and analyzing replicate samples per plot. The time factor instead of the cost factor, or the combination of the two, should be considered in many types of experiments.

It is proposed to investigate how these relative costs determine a balance between plot replicates (m) and sample replicates (n) in order that a stated precision for an experiment may be obtained at a minimum expense. Let C represent the cost per plot replicated, and c the cost per sample replicate in the conduct of an experiment. For a given treatment the total cost of plot replications will be mC, while the total cost of sample replications will be mnc. Hence, E, the total expense per treatment, is given by:

$$E = mC + mnc \qquad (4)$$

A certain criterion of precision to be obtained may be represented by $K = 1/m\ (\sigma_p^2 + \sigma_s^2/n)$, where K = the required variance of the mean for a treatment. In order to reduce (4) to an equation with one variable, it is found that $m = 1/K\ (\sigma_p^2 + \sigma_s^2/n)$, a value which is substituted.

Then,

$$E = 1/K \quad (\sigma_p^2 + \sigma_s^2/n)(C + nc)$$

To reduce the total cost to a minimum, differentiate E with respect to n, and set the equation to equal zero, viz.,

$$\frac{dE}{dn} = \frac{1}{K}\left[(\sigma_p^2 + \sigma_s^2/n)c + (C + nc)(-n^{-2}\sigma_s^{\ 2})\right] = 0$$

$$= c\sigma_p^2 + \frac{c\,\sigma_s^2}{n} - \frac{C\,\sigma_s^2}{n^2} - \frac{c\sigma_s^2}{n} = 0$$

$$n^2c\sigma_p^2 = C\sigma_s^2$$

$$n^2 = \frac{C\,\sigma_s^2}{c\,\sigma_p^2} \quad \text{or} \quad n = \sqrt{\frac{C\,\sigma_s^2}{c\,\sigma_p^2}}$$

Thus, the total cost will be a minimum when,

$$n^2 = \frac{C\,\sigma_s^2}{c\,\sigma_p^2} \qquad (5)$$

In this case, n, and hence m, are determined to afford a most economical design. It is worthwhile to note that n is determined to be independent of K, the precision desired.

In the present experiment, the values are substituted in the above equation (5), viz., $n = 2$, $\sigma_p^2 = 0.0919$, and $\sigma_s^2 = 0.0034$. The ratio of costs will be:

$$\frac{C}{c} = \frac{0.0919 \times 2^2}{0.0034} = 108.12$$

From the standpoint of expense, the analysis of a duplicate sample from each plot would have been justifiable to produce the most economical design only if the cost per additional plot had been 108 times the cost per analysis.

IV. Sampling Practices

The important practical consideration in sampling is that sufficient units should be taken to give a reasonably accurate representation of the whole. In sampling processes a small representative amount of the material is analyzed. For field plots, Wishart and Sanders (1935) advise that the samples should amount to 5 per cent of the plot at the very least. It must be recognized that the use of the entire plot is the most reliable where it is feasible, as sampling can afford only an estimate of the plot yield. Yates (1935) found 31 per cent loss of information as an average of several experiments where sampling technic was employed.

Some of the practices for different crop material are discussed below. To determine sampling errors it is necessary to draw at least two independent samples.

(a) Quadrat Methods

Some form of quadrat is generally used for sampling yield trials, or for the detailed study of vegetation. It was early pointed out by McCall (1917) that, while the harvest of the entire plot is most satisfactory in yield trials, it is attended with difficulties that make it practically impossible for plots away from the main station. A form of quadrat was suggested as a solution. The quadrat may be linear for a certain length of row, often being units of one foot, one yard, or one rod in length. The type of quadrat most frequently used in range and pasture experiments is a square area, usually a square meter or yard. The different kinds of area quadrats are described by Weaver and Clements (1929).

1. Small Grains: The rod-row unit is widely used by American investigators to harvest small grain plots by sampling methods. The English workers prefer one-foot or one-meter lengths of drill row.

The use of the rod-row, to secure a yield estimate for the entire plot, was studied by Arny and Garber (1919). They harvested 9, 5, and 4 rod-row samples from 1/10-acre plots of wheat and oats, and subsequently the entire plot. They concluded that increases over the mean yield of the checks of 15.76 per cent for triplicate 1/10-acre plots, 9.49 per cent for the nine rod rows, 12.73 per cent for the five rod rows, and 14.44 per cent for the four rod rows were (on the average) probably significant. Nine rods removed from 1/10-acre plots were concluded to give practically as accurate yield determinations as for the harvest of the entire plot. It was admitted that the amount of labor required to remove nine rod rows was about the same as for the harvest of the entire plot. In studies with wheat and barley, Clapham (1929) obtained a standard error of less than 6 per cent for the yield estimate where 30 one-meter-length drill rows were harvested from 1/30-acre plots. The rod row method was compared with meter lengths with six sets of five contiguous meter lengths. In barley, the standard error for 30 meter-lengths was 5.99 per cent, while the standard error calculated from sets (rod rows) was 7.20 per cent.

The use of the square yard, in addition to the rod row, has been studied by various workers. Kiesselbach (1917) determined the yield on 14 entire 1/30-acre plots compared to 20 areas 32x32 inches (quadrat areas). It was concluded that 20 systematically distributed areas may be safely substituted for the yield of the entire plot. Arny and Steinmetz (1919) concluded that four or five systematically distributed square-yard areas removed from 1/10-acre plots gave approximately the same error for yield as harvesting the entire plot.

2. Potatoes and Other Hill Crops: Yields of potatoes at Rothamsted were analyzed both by sampling and by harvesting the entire plot. There were 180 plants on each plot. Wishart and Clapham (1929) reasoned that the individual plant was the logical unit rather than a metrical unit. The actual number of plants necessary to determine

the plot yield depended upon the uniformity of the crop and the plot size, but rarely was less than 10. A one-in-10 sample was inadequate for plots 1/90-acre in size, it being necessary to take every third or fourth plant on a plot that size to give an error of four per cent. It was concluded that there was little to gain by the use of sampling methods on plots less than 1/20-acre in size. It was better to harvest the entire plot for yield. A pattern method has been used to sample sugar beet plots at harvest time, the individual beet being the unit. Wishart and Sanders (1935) make this explanation: "Suppose that there are 200 beets in a plot, and that it is proposed to take two sampling units of 10 beets each. Two numbers between 1 and 20 (inclusive) are drawn--say 4 and 16. The plot is then covered by walking along one row, back on the next, and so on, and the 4th, 24th, 44th.....beet pulled for one sampling unit, and the 16th, 36th, 56th.....beet for the other sampling unit, totals only being recorded in each case." Small samples of 10 roots for sugar analysis were criticized by Johnson (1929). Five samples of 10 roots from 1/40-acre plots gave a difference as high as 2.2 per cent sugar. He concluded the results were unreliable within one per cent of sugar each way. For 50-beet samples taken in groups of 10 he reduced the estimate of significance to 0.534 per cent and 0.593 per cent sugar in two experiments.

3. Pastures: The area quadrat has been widely used in pasture investigations, the usual size being a square meter. For the determination of the abundance and frequency of species in range pastures, Hanson (1930) concluded that a quadrat two-meters in size was desirable. Stewart and Hutchings (1936) have suggested the Point-Observation-Plot method for vegetation surveys. These plots are 100 square feet in area, being marked off by a circle 5.64 feet in radius. This method is claimed to be more rapid than the ordinary quadrat method. It is also suitable for statistical analysis. The vertical point method has been advocated recently by Tinney et al. (1937). Two horizontal pipes are mounted on legs 12 inches high, with a linear row of 10 holes spaced two inches apart through which needles 14 inches long are moved up and down. The point is pushed down until it touches a plant or a bare spot. The number of times a species is hit per 100 readings (needles) is expressed directly in per cent. The needle may hit a number of different species, e.g., bluegrass 52, timothy 30, redtop 12, red clover 3, and bare space 8. A modification is the inclined point method. Others who have published on quadrats for pastures are Brown (1937), and Robinson, et al (1937).

(b) Sampling from Bulk Material

So far sampling has dealt with plots during growth and harvest. Sampling of harvested produce, or bulk material, is another important form found in experimentation. In laboratory determinations, a sample of the material is mixed, and one or two sub-samples taken. A good method is to mix the heap of material thoroughly, to divide it into four quarters and to reject, for example, the N.E. and S.W. quarters mixing the other two again. The process is repeated until the bulk is reduced to the size required for a sample.

1. Protein Determinations: Duplicate samples were taken in 241 cars of wheat by Coleman, et al (1926) to determine the accuracy of sampling for protein. The cars were sampled twice in 5 different areas with a grain probe. The contents of both samples were composited separately and each reduced to 75 grams in size. Over 96 per cent of the tests varied less than 0.25 per cent in protein. In a study of sample size the error was found to be less than 0.10 per cent when the samples weighed 30 grams or more. The error was higher for smaller samples. Bartlett and Greenhill (1936) and Leonard and Clark (1936) found that one protein determination per replicate reduced the error more rapidly when the number of replicates was increased than by the use of duplicate laboratory analyses. The latter workers found the cost ratio of plot replicate to sample replicate for protein determinations in corn. The analysis of a duplicate sample from each plot would have been justified in producing the

most economical design only if the cost per plot had been 108 times the cost per analysis.

2. Shrinkage Samples in Forage: Two or three samples per plot were found by Wilkins and Hyland (1938) to accurately measure the water content of forage on individual plots of alfalfa and red clover. Samples 2 to 4 pounds in size were considered best.

3. Purity and Germination Tests in Seeds: Tests for germination and purity in seed analyses are affected by personal, sample, and random errors. Standard rules (1927) for seed testing specify minimum sample sizes as follows: (1) Two ounces of grass seeds; (2) Five ounces of red or crimson clover, alfalfa, rye grasses, brome grasses, millet, flax, rape, or seeds of similar size; (3) one pound of cereal, vetches, or seeds of similar size. Collins (1929) has set forth the procedure for statistical analyses of purity and germination tests.

References

1. Anonymous. Rules for Seed Testing, Dept. Cir. 406, U.S.D.A. 1937.
2. Arny, A. C., and Garber, R. J. Field Technic in Determining Yields of Plots of Grain by the Rod Row Method. Jour. Am. Soc. Agron., 11:33-47. 1919.
3. Arny, A. C., and Steinmetz, F. H. Field Technic in Determining Yields of Experimental Plots by the Square Yard Method. Jour. Am. Soc. Agron., 11:81-106. 1919.
4. Bartlett, M. S., and Greenhill, A. W. The Relative Importance of Plot Variation and of Field and Laboratory Sampling Errors in Small Plot Pasture Productivity Experiments. Jour. Agr. Sci., 258-262. 1936.
5. Brown, B. A. Technic in Pasture Research. Jour. Am. Soc. Agron., 29:468-476. 1937.
6. Clapham, A. R. The Estimation of Yield of Cereals by Sampling Methods. Jour. Agr. Sci., 19:214-235. 1929.
7. Coleman, D. A., et al. Testing Wheat for Protein with a Recommended Method for Making the Test, Dept. Bul. 1460. U.S.D.A. 1926.
8. Collins, G. N. The Application of Statistical Methods to Seed Testing, Cir. 79. U.S.D.A. 1929.
9. Fisher, R. A. Statistical Methods for Research Workers, Oliver and Boyd. (4th edition). 1933.
10. Hanson, H. C. Size of List Quadrat for Use in Determining Effects of Different Systems of Grazing Upon Agropyron Smithii Mixed Prairie, Jour. Agr. Res., 41:549-560.
11. Immer, F. R. Sampling Technic. Mimeographed outline, U. of Minnesota. 1936.
12. Immer, F. R. A Study of Sampling Technic with Sugar Beets. Jour. Agr. Res., 44:633-647. 1932.
13. Immer, F. R., and LeClerg, E. L. Errors of Routine Analysis for Percentage of Sucrose and Apparent Purity Coefficient with Sugar Beets taken from Field Experiments. Jour. Agr. Res., 52:505-515. 1936.
14. Johnson, S. T. A Note on the Sampling of the Sugar Beet, Jour. Agr. Sc., 19:311-315. 1929.
15. Kiesselbach, T. A. Studies Concerning the Elimination of Experimental Error in Comparative Crop Tests, Nebraska Research Bul. 13. 1917.
16. Leonard, W. H., and Clark, A. Protein Content of Corn as Influenced by Laboratory Analyses and Field Replication. Colo. Exp. Sta. Tech. Bul. 19. 1936.
17. McCall, A. G. A New Method of Harvesting Small Grain and Grass Plots. Jour. Am. Soc. Agron., 9:138-140. 1917.
18. Robinson, R. R., Pierre, W. H., and Ackerman, R. A. A Comparison of Grazing and Clipping for Determining the Response of Permanent Pastures to Fertilization. Jour. Am. Soc. Agron., 29:349-359. 1937.

19. Stewart, G., and Hutchings, S. S. The Point-Observation-Plot (Square-Foot-Density) Method of Vegetation Survey. Jour. Am. Soc. Agron., 28:714-722. 1936.
20. Tinney, F. W., Aamodt, O. S., and Ahlgren, H. L. Preliminary Report of a Study on Methods used in Botanical Analyses of Pasture Swards. Jour. Am. Soc. Agron., 29:835-840. 1937.
21. Tippett, L. H. C. The Methods of Statistics. Williams and Norgate, pp. 125-130, and 204-226. 1937.
22. Weaver, J. E., and Clements, F. E. Plant Ecology, pp. 10-42. 1929.
23. Wilkins, F. S., and Hyland, H. L. The Significance and Technique of Dry Matter Determinations in Yield Tests of Alfalfa and Red Clover. Ia. Agr. Exp. Sta. Res. Bul. 240. 1938.
24. Wishart, J., and Clapham, A. R. A Study of Sampling Technic: The Effect of Artificial Fertilizers on the Yield of Potatoes, Jour. Agr. Sci., 19:600-618. 1929.
25. Wishart, J., and Sanders, H. G. Principles and Practice of Yield Trials. Empire Cotton Growing Corp., pp. 42-45, and 85-95. 1935.
26. Yates, F., Some Examples of Biased Sampling. Ann. Eugenics, 6:202-213. 1935.
27. Yates, F., and Zacopanay, I. The Estimation of the Efficiency of Sampling with Special Reference to Sampling for yield in Cereal Experiments. Jour. Agr. Sci., 25:545-577. 1935.

Questions for Discussion

1. Under what conditions may it be desirable to take samples rather than use all the available material?
2. Are yields determined from samples as accurate as when the entire plot is harvested for yield? Why?
3. In a randomized block trial what makes up the variance due to varieties?
4. When can an increase in the number of samples effect an appreciable increase in precision?
5. What were the conclusions in the shelled corn data on the number of samples for protein analysis?
6. What factors influence the number of samples that should be drawn so far as the precision of the experiment is concerned?
7. Explain how to determine the most economical design from the standpoint of cost.
8. Under what conditions are quadrat methods used for small grain harvest? How many quadrats are usually advised for 1/10-acre plots?
9. Compare the rod, meter-length, and area quadrats for small grains.
10. What is the logical sampling unit for potatoes or sugar beets? How many samples should be taken?
11. Describe how to use a pattern method of sampling for sugar beets and similar crops.
12. Describe the different quadrat methods used in pasture studies.
13. Give sampling precautions and technic to use in bulk material.
14. What errors may be introduced in seed testing? Give the sample sizes generally used for analyses.

Problems

Four varieties of crested wheat grass were grown in a randomized block trial in plots 1/80-acre in size. Four replicated plots of each variety were grown, the yield data being obtained from six quadrats per plot. The yields in pounds per square yard sample are given below. (Data from Dr. T. M. Stevenson, U. of Saskatchewan).

Yields of Crested Wheat Grass

Square Yard	Block	Mecca:S-1	C.W.G.:S-10	C.W.G.:S-11	C.W.G.:Unsel.
(No.)	(No.)	(lbs.)	(lbs.)	(lbs.)	(lbs.)
1	I	0.52	0.68	0.48	0.58
2		0.49	0.62	0.55	0.58
3		0.59	0.70	0.46	0.61
4		0.36	0.70	0.58	0.63
5		0.28	0.62	0.51	0.65
6		0.49	0.66	0.38	0.71
1	II	0.61	0.77	0.44	0.68
2		0.49	0.91	0.48	0.48
3		0.52	0.89	0.49	0.75
4		0.56	0.95	0.61	0.71
5		0.57	0.77	0.58	0.65
6		0.49	0.77	0.41	0.68
1	III	0.52	0.85	0.27	0.42
2		0.42	0.77	0.61	0.51
3		0.66	0.46	0.44	0.58
4		0.57	0.81	0.51	0.54
5		0.59	0.58	0.61	0.66
6		0.56	0.33	0.41	0.58
1	IV	0.42	0.70	0.55	0.58
2		0.31	0.37	0.72	0.39
3		0.47	0.33	0.68	0.44
4		0.50	0.66	0.65	0.66
5		0.35	0.64	0.48	0.65
6		0.26	0.33	0.56	0.41

1. Calculate the analysis of variance for the crested wheat grass yields for a subdivision of the total variation into blocks, varieties, error, and square yards within plots.

2. Compare the variance due to samples and that due to replications. Make a statement on the number of samples and number of plot replicates that you would recommend in a subsequent experiment.

3. Determine the most economical design from the data at hand, i.e., the ratio of replicate cost to sample cost.

CHAPTER XVII

COMPLEX EXPERIMENTS [1]

I. Use of Complex Experiments

The present trend in field experiments is toward somewhat complicated designs which permit the study of several factors in one large-scale comprehensive experiment. There are several advantages of the complex experiment. (1) One that includes several treatments in all possible combinations permits a broad basis for generalization due to the fact that the interactions, as well as the main effects, can be studied. It is obvious that field experimental results may be influenced materially by environmental conditions with the result that a combination of factors may provide a more satisfactory answer to the problems under study. (2) The degrees of freedom for error variance are higher than would be the case for single experiments designed to study each factor separately. This leads to greater precision in the results. (See Paterson, 1939).

The value of a complex experiment depends upon a careful analysis of the problem and the various treatment combinations to be tested. The amount of complexity introduced depends also upon the facilities and funds available. It is a safe precaution to key-out the degrees of freedom for the various factors to be tested before field work begins to be sure that the proposed plan is satisfactory. After the data are collected the investigator should make certain that the data are sufficiently homogenous to combine in a single test. (See paragraph VII).

II. Application to a Barley Variety Trial

In agronomic tests of cereal crop varieties it is often desirable to conduct the trials at various points in the area under consideration and to carry them on for a period of years. Some data collected by Immer and others (1934) on the yield of barley varieties tested in randomized blocks in 4 locations in Minnesota for a 2-year period will be used to illustrate the method of computation. The data are based on 6 square-yard samples harvested from each plot of approximately 1/40-acre each. Each test consisted of 3 randomized blocks. The same 5 varieties were tested at University Farm, Waseca, Crookston, and Grand Rapids for the years 1932 and 1935. The yields in bushels per acre for each plot of each variety are given below in Table 1.

[1] From Dr. F. R. Immer, University of Minnesota, with minor modifications.

Table 1. Yields of five varieties of barley, replicated 3 times in each of 4 locations in 1932 and 1935.

	Block Number								Tot. for
	I	II	III	Tot.	I	II	III	Tot.	both years
	Univ. Farm - 1932				Univ. Farm - 1935				
Manchuria	19.7	31.4	29.6	80.7	45.5	50.3	60.0	155.8	236.5
Glabron	28.6	38.3	43.5	110.4	47.5	41.4	49.4	138.0	248.4
Velvet	20.3	27.5	32.6	8 .4	54.2	52.3	64.5	171.0	251.4
Wis. #38	27.9	40.0	46.1	110.0	62.2	53.1	74.7	190.0	304.0
Peatland	22.3	30.8	31.1	84.2	47.4	57.8	50.5	155.7	239.9
Total	118.8	168.0	182.9	469.7	256.8	254.6	299.1	810.5	1280.2
	Waseca - 1932				Waseca - 1935				
Manchuria	40.8	29.4	30.2	100.4	53.9	58.8	47.7	160.4	260.8
Glabron	44.4	34.9	33.9	113.2	63.7	61.1	52.2	177.0	290.2
Velvet	44.6	41.4	26.2	112.2	53.9	59.1	56.4	169.4	281.6
Wis. #38	39.8	39.2	29.1	108.1	74.2	75.6	67.0	216.8	324.9
Peatland	71.5	47.6	55.4	174.5	51.1	47.3	45.0	143.4	317.9
Total	241.1	192.5	174.8	608.4	296.8	301.9	268.3	867.0	1475.4
	Crookston - 1932				Crookston - 1935				
Manchuria	34.7	29.1	35.1	98.9	42.1	47.1	30.8	120.0	218.9
Glabron	28.8	28.7	21.0	78.5	38.8	29.4	30.5	98.7	172.2
Velvet	29.8	38.4	28.0	96.2	42.1	40.0	39.8	121.9	218.1
Wis. #38	27.7	27.6	20.4	75.7	44.3	43.5	47.7	135.5	211.2
Peatland	43.0	32.7	32.0	107.7	53.9	51.8	50.3	156.0	263.7
Total	164.0	156.5	136.5	457.0	221.2	211.8	199.1	632.1	1089.1
	Grand Rapids - 1932				Grand Rapids - 1935				
Manchuria	20.2	30.2	16.0	66.4	26.6	26.5	32.7	85.8	152.2
Glabron	13.2	20.5	9.6	43.3	21.4	18.7	24.1	64.2	107.5
Velvet	24.5	41.6	30.6	96.7	20.7	26.8	30.4	77.9	174.6
Wis. #38	19.0	18.4	24.6	62.0	20.7	23.6	30.9	85.2	137.2
Peatland	27.6	30.0	22.7	80.3	32.6	40.0	34.2	106.8	187.1
Total	104.5	140.7	103.5	348.7	122.0	135.6	152.3	409.9	758.6
Total 4 Stations	628.4	657.7	597.7	1883.8	896.8	903.9	918.8	2719.5	4603.3

III. Analysis of Test into Components

The analysis of a complex experiment of this type is merely an extension of the analysis of variance as applied to the randomized block test. The various factors together with their degrees of freedom may be represented as follows: It is noted that all block x variety[1] interactions are included in error.

[1]The symbol (x) in this connection denotes interaction.

Variation due to:		Degrees of Freedom	
Blocks		2	
Stations		3	
Years		1	
Varieties		4	
Interactions of			
Varieties x Stations		12	
" x Years		4	
" x Stations x Years		12	
Stations x Years		3	
Blocks x Stations		6	
" x Years		2	
" x Stations x Years		6	
" x Varieties	)	8)	
" x Varieties x Stations	) Error	24)	64
" x Varieties x Years	)	8)	
" x Varieties x Stations x Years	)	24)	
Total		119	

There will be a total of 119 degrees of freedom for the combined test since there are 120 plots. The degrees of freedom for the main effects will be one less than the number of blocks, stations, years, and varieties, respectively. The degrees of freedom for interaction is obtained by the multiplication of the degrees of freedom for the variables involved. For example, varieties x stations will be (4)(3) = 12. For the second order interaction, varieties x stations x years, the degrees of freedom will be (4)(3)(1) = 12. After the degrees of freedom are keyed-out, the remainder of the computation must be made in accordance with this plan.

IV. Computation of the Sums of Squares

The correction factor, $(Sx)^2/N$, is computed for the total yield of the 120 plots, i.e. $(4603.3)^2/120 = 176,586.4241$. This factor will be used for the entire test.

For the total sum of squares, the 120 individual plot yields are squared and summed. This value (Sx^2) is equal to 200, 879.35. The correction factor is then subtracted, viz., 200,879.35 - 176,586.4241 = 24,292.9259 ----------------------------XIII.[*]

The data for the remainder of the computations are grouped from table 1 into tables, each with two variables. The sums of squares are computed the same as for a simple randomized block test. It is to be noted that totals are used for the variable concerned. For this reason, the sums of squares obtained must be divided by the number of basic plots included in the respective totals (to reduce the variables to a single plot basis) in order for the common correction factor to apply.

The combined data for a comparison of varieties and stations are given in table 2.

[*]The roman numeral refers to the line in Table 8.

Table 2. Total Yields grouped for Varieties and Stations.

Barley Variety	Station U.Farm	Waseca	Crookston	Grand Rapids	Total
Manchuria	236.5	260.8	218.9	152.2	868.4
Glabron	248.4	290.2	177.2	107.5	823.3
Velvet	251.4	281.6	218.1	174.6	925.7
Wis. #38	304.0	324.9	211.2	137.2	977.3
Peatland	239.9	317.9	263.7	187.1	1008.6
Total	1280.2	1475.4	1089.1	758.6	4603.3

These data are taken directly from the right-hand column of table 1. The computation of the sums of squares is carried out as follows:

$$\text{Total} = \frac{S(x^2_{vs})}{6} - \frac{(Sx)^2}{N} = \frac{1,129,020.73}{6} - 176,586.4241$$

$$= 188,170\cdot1216 - 176,586\cdot4241 = 11,583.6975$$

$$\text{Varieties} = \frac{S(x^2_v)}{24} - \frac{(Sx)^2}{N} = \frac{4,261,251.19}{24} - 176,586\cdot4241$$

$$= 177,552\cdot1329 - 176,586\cdot4241 = 965\cdot7088$$

$$\text{Stations} = \frac{S(x^2s)}{30} - \frac{(Sx)^2}{N} = \frac{5,577,329\cdot97}{30} - 176,586\cdot4241$$

$$= 185,910.9990 - 176,586\cdot4241 = 9,324\cdot5749.$$

Varieties x Stations = Total - (Varieties + Stations)

$$= 11.583.6975 - 10,290.2837 = 1,293.4138\cdot$$

The values for varieties, stations, and varieties x station total are included in table 7, where the steps for computation are indicated. The interaction values are included in table 8.

The sums of squares for the other factors are computed in a similar manner, the data being given in Tables 3, 4, 5, and 6, with the results included in table 7.

In table 3 are given the data for comparisons of varieties and years, the yields at the four stations being totaled.

Table 3. Total yields grouped for varieties and years.

Variety	Year 1932	1935	Total
Manchuria	346.4	522.0	868.4
Glabron	345.4	477.9	823.3
Velvet	385.5	540.2	925.7
Wis. #38	359.8	617.5	977.3
Peatland	446.7	561.9	1008.6
Total	1883.8	2719.5	4603.3

In table 4 are assembled the data for comparisons of blocks and stations, the block totals for the two years of each station being added.

Table 4. Total yields of blocks and stations

Block	Stations				Total
	U. Farm	Waseca	Crookston	Grand Rapids	
I	375.6	537.9	385.2	226.5	1525.2
II	422.6	494.4	368.3	276.3	1561.6
III	482.0	443.1	335.6	255.8	1516.5
Total	1280.2	1475.4	1089.1	758.6	4603.3

In table 5 are the totals for comparison of blocks and years. This table is assembled from the totals at the bottom of table 1.

Table 5. Total yields of blocks and years

Block	Year		Total
	1932	1935	
I	628.4	896.8	1525.2
II	657.7	903.9	1561.6
III	597.7	918.8	1516.5
Total	1883.8	2719.5	4603.3

One other table is necessary, that of stations and years.

Table 6. Total yields of stations and years

Year	Station				Total
	U. Farm	Waseca	Crookston	Grand Rapids	
1932	469.7	608.4	457.0	348.7	1883.8
1935	810.5	867.0	632.1	409.9	2719.5
Total	1280.2	1475.4	1089.1	758.6	4603.3

The calculation of the sums of squares for the complete analysis can be performed with the least difficulty and confusion when the steps are carried thru in a routine manner. Many of the calculations are given in table 7. The remainder follow easily and logically.

Table 7. Calculation of sums of squares.

Variate	Total of Squares	Calc. from table	No. of Varia-bles squared	Single plots in each tot. squared		$\frac{(Sx)^2}{N}$	Sum of Squares	Key to table 8
	1	2	3	4	5 = 1÷4	6	5-7	
$S(x^2)$	200,879.35	1	120	1	200,879.3500	176,586.4241	24,292.9259	III
$S(x^2_s)$	5,577,329.97	2	4	30	185,910.9990	"	9,324.5749	II
$S(x^2_y)$	10,944,382.69	3	2	60	182,406.3782	"	5,819.9541	III
$S(x^2_v)$	4,261,251.19	2	5	24	177,552.1329	"	965.7088	IV
$S(x^2_b)$	7,064,601.85	4	3	40	176,615.0462	"	28.6221	I
$S(x^2_{sy})$	2,897,377.01	6	8	15	193,158.4673	"	16,572.0432	--
$S(x^2_{vs})$	1,129,020.73	3	20	6	188,170.1216	"	11,583.6975	--
$S(x^2_{vy})$	2,206,627.61	3	10	12	183,885.6342	"	7.299.2101	--
$S(x^2_{bs})$	1,871,824.37	4	12	10	187,182.4370	"	10,596.0129	--
$S(*^2_{by})$	3,650,180.03	5	6	20	182,509.0015	"	5,922.5774	--
$S(x^2_{vsy})$	593,855.03	1	40	3	197,951.6767	"	21,365.2526	--
$S(x^2_{bsy})$	974,322.93	1	24	5	194,864.5860	"	18,278.1619	--

A notation found to be very convenient in practice is to let $S(x^2)$ be the sum of the squares of the individual plots. The station totals are designated x_s, the variety totals x_v, etc. The totals for varieties at the separate stations are designated x_{vs}, varieties in different years by x_{vy}, etc. These are given in table 7. In column 1 of this table are the sums of the squares of the variates concerned and in column 2 is given the table from which they have been computed. Thus, $S(x^2_s)$ is calculated from the 4 station totals of table 3. $S(x^2_v)$ is calculated from the variety totals of table 2. The value of $S(x^2_{vs})$ is the sum of the squares of the 20 yields for each variety at each station separately in table 2

Column 3 of table 7 gives the number of figures squared under column 1. Column 4 gives the number of single plots contained in each figure squared. Column 5 is simply columns 1 divided by 4. This is necessary to reduce the sums of squares to a single plot basis throughout. The key numbers refer to that sum of squares in the complete analysis of variance given as table 8.

The sums of squares for total. blocks, stations, years and varieties are transferred directly from table 7 to table 8. The interaction sums of squares can be obtained from table 7 by subtraction.

The sum of squares for interaction of varieties x stations will be found by subtraction of the sum of squares for varieties and stations separately from the sums of squares opposite $S(x^2_{vs})$ in table 7. e.g.

11,583.6975	(19 D.F.)
- 965.7088	(4 D.F. for varieties) -----------IV
-9,324.5749	(3 D.F. for stations) ------------II
1,293.4138	(12 D.F. for varieties x stations)- V

Since there were 20 figures used to obtain $S(x^2_{vs})$ there would be 19 degrees of freedom. The interaction degrees of freedom are obtained by subtraction, e.g., 19 - (4 + 3) = 12. All other first order interactions are obtained in the same manner.

The second order interaction of varieties x stations x years is obtained by subtraction from the sum of squares opposite $S(x^2_{vsy})$ in table 7 the sums of squares for varieties, stations and years separately and their first order interactions in all possible combinations. Thus:

21,365.2526	(39 D.F.)
-965.7088	(4 D.F. for varieties)--------------------------IV
-9,324.5749	(3 D.F. for stations)---------------------------II
-5,819.9541	(1 D.F. for years)------------------------------III
-1,293.4138	(12 D.F. for varieties x stations)---------------V
-513.5472	(4 D.F. for varieties x years)------------------VI
-1,427.5142	(3 D.F. for stations x years)-----------------VIII
2,020.5396	(12 D.F. for stations x years)-----------------VII

The sums of squares for the main effects and the first order interaction for the computation of the second order interaction are taken from table 8 opposite the appropriate key number. The interaction of blocks x stations x years is obtained in a similar manner.

The complete analysis is now carried out in table 8, the error sum of squares being obtained as a remainder.

Table 8. Complete analysis of variance.

Key No.	Variation due to:	D.F.	Sums of Squares	Mean Square	F-Value[1]
I	Blocks	2	28.6221	14.3110	
II	Stations	3	9,324.5749	3,108.1916	162.85**
III	Years	1	5,819.9541	5,819.9541	304.92**
IV	Varieties	4	965.7088	241.4272	12.65**
	Interaction of:				
V	Varieties x stations	12	1,293.4138	107.7845	5.65**
VI	Varieties x years	4	513.5472	128.3868	6.73**
VII	Varieties x stations x years	12	2,020.5396	168.3783	8.82**
VIII	Stations x years	3	1,427.5142	475.8381	24.93**
IX	Blocks x stations	6	1,242.8159	207.1360	10.85**
X	Blocks x years	2	74.0012	37.0006	1.94
XI	Blocks x stations x years	6	360.6795	60.1132	3.15**
XII Error	(Blocks x varieties 8) (Blocks x varieties x stations 24) (Blocks x varieties x years 8) (Blocks x varieties x sta.x yrs.24)	64	1,221.5546	19.0868	
XIII	Total	119	24,292.9259		

**Exceeds the 1 per cent point in Snedecor's table of "F".

[1]For comparison with error.

In practice, it is unnecessary to key-out the complete analysis as given in table 8. The variation due to blocks, blocks x stations, blocks x years, and blocks x stations x years should be grouped as one quantity, being designated as "Blocks within stations and years" or "Blocks within tests" (for 16 D.F.). The reason for this is readily apparent. The blocks are numbered I, II, and III arbitrarily. Block I at University Farm has no relation to Block I at Waseca or any other station. Thus, it is an error to regard blocks as a factor that occurs at several definite levels (3 in this case). The correct procedure is therefore to compute the block sums of squares for each experiment and combine them to present in the final analysis. The analysis of variance may then be presented as in table 9.

Table 9. Analysis of Variance (in summary form)

Variation due to:	D.F.	Sums Squares	Mean Square	F-Value
Blocks within Tests	16	1,706.1187	106.6324	5.59**
Stations	3	9,324.5749	3,108.1916	162.85**
Years	1	5,819.9541	5,819.9541	304.92**
Varieties	4	965.7088	241.4272	12.65**
Interactions:				
Varieties x stations	12	1,293.4138	107.7845	5.65**
Varieties x years	4	513.5472	128.3868	6.73**
Varieties x stations x years	12	2,020.5396	168.3783	8.82**
Stations x years	3	1,427.5142	475.8381	24.93**
Error	64	1,221.5546	19.0868	
Total	119	24,292.9259		

The analysis may be summarized still further in case the investigator is not interested in the variation due to stations, years, and stations x years. He may group these factors into variation due to "tests within stations and years" or simply "tests" (for 7 D.F.). The variety factor and its interactions would be given in detail because it has definite biological significance.

V. Sums of Squares in Simple vs. Complex Experiments

It will be useful at this stage to relate the complete analysis in Table 8 with the simple randomized block tests computed for each of the 8 tests separately. The sums of squares for total, blocks, varieties, and error are given in table 10 for the 8 tests.

Table 10. Sums of squares calculated from the tests separately

Test	Year	Sum of squares for total	Sum of squares for blocks	Sum of squares for varieties	Sum of squares for error
U. Farm -	1932	867.2973	450.9073	375.6106	41.5894
U. Farm -	1935	1031.7133	251.6253	506.3600	273.7280
Waseca -	1932	1907.1360	471.3960	1196.3293	239.4107
Waseca -	1935	1203.5600	131.1480	993.8400	78.5720
Crookston-	1932	487.8733	80.8333	252.6266	154.4134
Crookston-	1935	807.8360	49.2040	595.8227	162.8093
G. Rapids-	1932	905.5573	179.6853	536.1640	189.7080
G. Rapids-	1935	509.9093	92.1293	336.4560	81.3240
Total		7,720.8825	1,706.1185	4,793.2092	1,221.5548

It is noted that the sums of squares for error for the 8 separate tests adds to 1,221.5548. This agrees with the sum of squares for error of table 8 (1,221.5546) the discrepancy being due to dropping of decimals. There were 8 D.F. for error in each separate test or 8 by 8 = 64 in all tests. These same 64 D.F. were used for error in the complete analysis in table 8. The error used in table 8 is, therefore, simply the average error of the separate tests.

The sums of squares added for blocks, in the 8 tests, gives 1,706.1185 (see table 10). Addition of the sums of squares for blocks, blocks x stations, blocks x years, and blocks x stations x years from table 8 gives a total of 1,706.1187, which agrees also. Further comparisons are given in table 11.

Table 11. Comparison of degrees of freedom and sums of squares of the 8 separate tests with the complete analysis of table 8.

Variation due to:	Calculated from 8 separate tests		Calculated from the complete analysis		
	D.F.	Sum of sq.	D.F.	Sum of Sq.	Key to table 8
Blocks	(8)(2) = 16	1,706.1185	16	1,706.1187	I, IX, X, XI
Varieties	(8)(4) = 32	4,793.2092	32	4,793.2094	IV, V, VI, VII
Error	(8)(8) = 64	1,221.5548	64	1,221.5546	XII
Total	(8)(14) =112	7,720.8825	112	7,720.8827	I,IV,V,VI,VII,IX, X,XI,XII

From the above table the analogy between the separate analyses of variance for each test and the complete analysis is clear. The 112 D.F. for total in table 11 is the total sums of squares <u>within tests</u>. When the 7 D.F. between tests (i.e., stations = 3, years = 1 and stations x years = 3) are added, the full 119 D.F. is obtained. The same is true for the sums of squares.

VI. Interpretation of the Data

The manner in which the data can be interpreted will now be illustrated. From table 8 it is seen that the variance (mean square) due to varieties compared with variance due to error exceeds the 1 per cent point. Therefore, some of the varietal differences are significant. The mean yields for varieties, computed from Table 1, are given in table 12.

Table 12. Mean yields of 3 plots of each variety, average yields for both years and average yields of varieties for all tests.

Year	Variety				
	Manchuria	Glabron	Velvet	Wis. #38	Peatland
		University Farm			
1932	26.9	36.8	26.8	38.0	28.1
1935	51.9	46.0	57.0	63.3	51.9
Mean yield	39.4	41.4	41.9	50.7	40.0
		Waseca			
1932	33.5	37.7	37.4	36.0	58.2
1935	53.5	59.0	56.5	72.3	47.8
Mean Yield	43.5	48.4	46.9	54.2	53.0
		Crookston			
1932	33.0	26.2	32.1	25.2	35.9
1935	40.0	32.9	40.6	45.2	52.0
Mean Yield	36.5	29.5	36.4	35.2	44.0
		Grand Rapids			
1932	22.1	14.4	32.2	20.7	26.8
1935	28.6	21.4	26.0	25.1	35.6
Mean Yield	25.4	17.9	29.1	22.9	31.2
Mean for all stations	36.2	34.3	38.6	40.7	42.0

The variance due to error was 19.0868 (table 8). The standard error of a single plot would be $\sqrt{19.0868} = 4.369$ bu. Since 24 plots are involved in the variety averages for all stations, the standard error of the mean of 24 plots is $\frac{4.369}{\sqrt{24}} = 0.892$ bu.

The standard difference between two such means would be $0.892\sqrt{2} = 1.26$ bu. With 64 D.F. for error one may accept twice the standard error of the difference as a level for odds of approximately 19:1 against the chance occurrence of a difference of $(2)(1.26) = 2.52$ bu.

Since the mean yield of Peatland for all stations and years was 42.0 bu., any variety that differs from it by more than 2.5 bu. may be judged as probably significantly lower in yield, on the basis of these tests alone. On this basis Manchuria, Glabron and Velvet are significantly lower in yield than Peatland.

The interaction of varieties x stations was also significant (table 8). A first order interaction is essentially a difference between two differences. The mean yield of Peatland at University Farm, for an average of both years, was 10.7 bushels less than the yield of Wisconsin #38 (50.7 - 40.0 = 10.7). The mean yield of Peatland at Grand Rapids exceeded the yield of Wisconsin #38 by 8.3 bu. (31.2 - 22.9 = 8.3). The question then is whether these two differences are significantly different. This difference between two differences will be given by Wisconsin #38 minus Peatland at University Farm less Wisconsin #38 minus Peatland at Grand Rapids, or (50.7 - 40.0) - (22.9 - 31.2) = 19.0 bu. The standard error of this "cross difference" will be $\sqrt{\frac{19.0868 \times 2 \times 2}{6}} = 3.567$.

It may also be computed as follows: The standard error of the mean ($\sigma_{\bar{x}}$) is equal to $\sqrt{\frac{19.0868}{6}}$ = 1.784 bu., since 6 plots are contained in each mean. The standard error of the difference between two differences then is $1.784\sqrt{2}\cdot\sqrt{2} = 3.567$ bu. Twice this is 7.13 bu. and any "cross difference" that exceeds this value is expected to occur less than once in 20 trials by random sampling alone. The cross differences for Peatland and Wisconsin #38 at University Farm and Grand Rapids greatly exceed 7.13 bu., being 19.0 bu. It is clear, therefore, that these two varieties responded in a differential manner at University Farm and Grand Rapids as an average of 1932 and 1935. Other significant cross differences could be found in the same way.

Significant interactions of varieties x years could be determined by application of the general procedure outlined above. Since only two years are involved these interactions of varieties x years can have very little practical significance.

While the second order interaction of varieties x stations x years was also significant, it is of secondary interest. This significant second order interaction merely means that certain differential responses of varieties x stations were not constant in different years. To illustrate the types of comparisons which must be made to show this, take the means of Glabron and Velvet at University Farm in 1932 and 1935 separately and the same yields at Grand Rapids. Then: $[(36.8 - 26.8) - (46.0 - 57.0)] - [(14.4 - 32.2) - (21.4 - 26.0)] = 34.2$ with an error of $\sqrt{\frac{19.0868 \times 2 \times 2 \times 2}{3}}$ or 7.13 bu.

Since the difference of 34.2 exceeds (2)(7.13) = 14.26 bu. it is obviously significant.

For a complete understanding of a complex analysis, of which that given in table 8 is an example, one further comparison can be made. Suppose that V, S and Y are designated to represent variance due to varieties, stations and years and V x S, V x Y and V x S x Y the interaction variances. Then one may determine whether the variance due to:

$$\left.\begin{matrix} V > V \times S > V \times S \times Y \\ V > V \times Y > V \times S \times Y \end{matrix}\right\} > \text{Error}$$

by means of the "F" test. When the variance due to varieties significantly exceeds the interaction, varieties x stations, there is evidence that varietal performance generally was consistent enough to demonstrate that some varieties were the best in all stations, as an average of the years in which tests were made. When the variety variance significantly exceeds that of varieties x years, one may conclude that, as an average of all stations, some varieties were consistently better in yield in the different years.

Further, when the interaction of varieties x stations significantly exceeds varieties x stations x years, it is plain that the differential response of the varieties at the separate stations were sufficiently similar in the different years to warrant the conclusion that these differential responses may be permanent features of these localities.

Unless the variance for varieties significantly exceeds that of varieties x stations or varieties x years, no general recommendations can be made for the entire state or for future years. To make such recommendations the stations (of which tests were made) are considered as random samples of all places in the state and the years in which tests were conducted must be considered as a random sample of all future years. It is only when the number of stations and years can be considered an adequate sample of all possible places and years that worthwhile predictions can be made for all places in the state and for future years. (See Summerby, 1937).

VII. The Homogeneity Test

The question may be raised as to whether the data afforded by the several experiments are sufficiently alike to assume that they may have resulted from a single population. In case this is true, the data from the experiments may be consolidated and analyzed as one complex experiment. Homogeneity tests have been suggested by Snedecor (1937) and by Stevens (1936).

The formula may be explained as follows:

Let n = the number of experiments.
n - 1 = the number of degrees of freedom between experiments.
M = the number of degrees of freedom for error within an experiment.
e = the sum of squares for error in a single experiment.
v = the observed variance
V = the theoretical variance
L = the Lexis ratio

The observed variance of the sums of squares due to error for all experiments is:

$$v = \frac{S(e - \bar{e})^2}{n - 1} = \frac{S(e^2) - (Se)^2/n}{n - 1}$$

The theoretical variance, where the total, S(e), is assumed to be the population of error sums of squares, is as follows:

$$V = 2M \left[\frac{S(e)}{M\ (n-1)}\right]^2 \quad \text{or} \quad \frac{2}{M}\left[\frac{S(e)}{n}\right]^2$$

The Lexis ratio (L) is the ratio of the observed to the theoretical standard error, so that its square is $L^2 = v/V$. When the ratio is greater than one, the series of sums of squares due to error is called supernormal. When "L" is less than one, it is called subnormal.

A certain degree of supernormality or subnormality can be attributed to chance. The limits for significance can be determined by the χ^2 test. viz.,

$$\chi^2 = (n-1)L^2 \text{ or } (n-1)\ \frac{(v)}{(V)}$$

When χ^2 corresponds to a probability of less than 0.05, the series is too supernormal to admit that they resulted from a single population. A χ^2 that corresponds to a probability greater than 0.95 indicates that the series is too subnormal for consolidation of the data.

The homogeneity test can be applied to the data on the barley yield trials as compiled in the separate tests in table 10:

Test	Year	Sums squares due to error
U. Farm	1932	41.5894
U. Farm	1935	273.7280
Waseca	1932	239.4107
Waseca	1935	78.5720
Crookston	1932	154.4134
Crookston	1935	162.8093
Grand Rapids	1932	189.7080
Grand Rapids	1935	81.3240

n = 8, (n-1) = 7, M = 8. S(e) 1221.5548

The sums of squares for the eight "error sums of squares" is as follows:

$$S(e^2) = S\left[(41.5894)^2 + (273.7280)^2 + \ldots\ldots(81.3240)^2\right]$$

$$= 233,100.8233$$

$$(Se)^2/n = 186,524.5773$$

$$S(e - \bar{e})^2 = S(e^2) - (Se)^2/n = 46,576.2460$$

$$v = \frac{S(e - \bar{e})}{n-1} = \frac{46,576.2460}{7} = 6,653.7494$$

$$V = 2M\left[\frac{S(e)}{nM}\right]^2 = 16\left[\frac{1221.5548}{(8)(8)}\right]^2 = 16\ (19.0868)^2$$

$$= (16)(364.3059) = 5828.8944$$

$$\chi^2 = (n-1)\left(\frac{v}{V}\right) = \frac{(7)(6653.7494)}{5828.8944} = 7.9906$$

When the χ^2 table is entered for 7 degrees of freedom, $P = 0.3335$.

Therefore, the data are sufficiently homogenous to permit the calculation of one generalized standard error for all tests.

VIII. Transformation of Percentage Data

Some types of discrete data cannot be combined to provide a valid estimate of a generalized standard error. This applies particularly to some forms of percentage data wherein each variate represents a certain number of observations of a given type or condition out of a total number of trials or cases (N). The variance of a single variate of this type is pqN. It is clearly dependent upon p, the estimated ratio of existence of the type or condition in question, as well as upon N. Bliss (1937), Salmon (1938), Cochran (1938), Clark and Leonard (1939), and others have recognized that each variate in discrete data of this kind does not have the same opportunity to contribute equally to a general experimental error.

(a) The Angular Transformation

R. A. Fisher has supplied a mathematical transformation for such data which will equalize the estimated variance of each variate so that it is functionally dependent only on N, the total number of trials. In this transformation, each estimate of p is replaced by $\sin^2 \theta$ whence,

$$\theta = \text{Sin}^{-1}\sqrt{p} \quad \text{or } 1/2\ \text{Cos}^{-1}(1-2p)$$

This transformation must be applied to discrete data of this type so that the analysis of variance may be valid. However, it is of little practical importance when the percentage values are between 30 and 70. Bliss (1937) has compiled a convenient table for the transformation of percentage values to angles, the latter being measured in degrees (See Table 5, appendix).

(b) Classification of Percentage Data

The type of discrete data, rather than its expression in percentages, determines whether or not the transformation should be employed. The types of percentage data are classified by Clark and Leonard (1939) as follows: (1) Continuous data from an experimental study may be expressed as percentages when each variate is divided by an arbitrary constant value, whereby each variate becomes a percentage of some standard or average. Clearly such a procedure merely transforms the unit of measurement. Percentages of this type should be treated statistically exactly as though the data were in their raw form. For example, yield data might be expressed in percentage of the check instead of actual yield in pounds. (2) Continuous data are often expressed in percentages to show concentrations. This type of percentage is very common. Some examples are: seed purity given by weight of pure seed/ total weight of seed, leafiness given by leaf weight/ total plant weight, protein content given by weight of protein/ total weight, sugar content given by weight of sugar/weight of root, etc. Such concentrations should not, as a rule, be subjected to any transformation to equalize the variance. (3) The third type of percentage is where the original data are discrete, being based upon a determinate number of trials or cases (N). The transformation, $p = \sin^2\theta$ should be applied to this type where it is desired to construct a generalized standard error. Illustrations of this type are as follows: Germination percentages given by number of seeds germinated/ total seeds, disease percentages given by number of plants diseased/ total plants, etc.

References

1. Bliss, C. I. The Analysis of Field Experimental Data Expressed as Percentages. Plant Protection Bul.12, U.S.S.R. (Leningrad). 1937.
2. Clark, Andrew, and Leonard, Warren H. The Analysis of Variance with Special Reference to Data Expressed as Percentages. Jour. Am. Soc. Agron., 31:55-66. 1939.
3. Cochran, W. G. Some Difficulties in the Statistical Analysis of Replicated Experiments. Emp. Jour. Exp. Agr., 6:157-175. 1938.
4. Fisher, R. A. Design of Experiments. Oliver and Boyd. 2nd Ed. pp. 75-76. 1937.
5. Goulden, C. H. Methods of Statistical Analysis. John Wiley. p. 139. 1939.
6. Immer, F. R., Hayes, H. K. and Powers, L. R. Statistical Determination of Barley Varietal Adaptation. Jour. Am. Soc. Agron., 26:403-419. 1934.
7. Paterson, D. D. Statistical Technique in Agricultural Research. McGraw-Hill. pp. 58-63, 66-67, and 205-208. 1939.
8. Salmon, S. C. Generalized Standard Errors for Evaluating Bunt Experiments for Wheat. Jour. Am. Soc. Agron., 30:647-663. 1938.
9. Snedecor, G. W. Statistical Methods. Collegiate Press, Inc. pp. 196-197. 1937.
10. Stevens, W. L. Heterogeneity of a Set of Variances. Jour. Gen., 33:398-399. 1936.
11. Summerby, R. The Use of the Analysis of Variance in Soil and Fertilizer Experiments with Particular Reference to Interactions. Sci. Agr., 17:302-311. 1937.

Questions for Discussion

1. What are the advantages of a complex experiment over separate single tests?
2. As a matter of design, would it be necessary to have all varieties in all locations in each year? Why?
3. Why does the total sum of squares for the simple tests fall short of the total sum of squares for the complex experiment? What would make them check?
4. How would you interpret a significant interaction such as, for example, varieties x stations?
5. Explain why a first order interaction is essentially a difference between two differences.
6. What is a homogeneity test? Why should it be made?
7. Under what conditions should percentage data be transformed to degrees of an angle to admit valid use of a pooled estimate of error?

Problems

1. The yields in bushels per acre for five spring wheat varieties tested in 3 randomized blocks for 3 years are given below. (Data from F. R. Immer).

Variety	Block Number I	II	III	Tot.	Block Number I	II	III	Tot.	Block Number I	II	III	Tot.	Grand Tot.
	U. Farm - 1931				U. Farm - 1932				U. Farm - 1935				
Thatcher	17.0	20.0	19.7	56.7	33.6	37.7	31.2	102.5	32.4	34.3	37.3	104.0	263.2
Ceres	16.1	18.9	20.3	55.3	29.7	30.0	35.9	95.6	20.2	27.5	25.9	73.6	224.5
Reward	21.1	23.1	21.8	66.0	24.1	26.9	29.8	80.8	29.2	27.8	30.2	87.2	234.0
Marquis	15.4	20.9	18.4	54.7	26.3	31.8	29.8	87.9	12.8	12.3	14.8	39.9	182.5
Hope	20.3	21.0	14.2	55.5	28.1	25.4	31.5	85.0	21.7	24.5	23.4	69.6	210.1
Total	89.9	103.9	94.4	288.2	141.8	151.8	158.2	451.8	116.3	126.4	131.6	374.3	1114.3
	Waseca - 1931				Waseca - 1932				Waseca - 1935				
Thatcher	26.8	33.6	26.4	86.8	22.8	20.7	23.5	67.0	28.5	30.1	30.1	88.7	242.5
Ceres	29.2	32.4	26.0	87.6	24.3	26.2	26.7	77.2	15.6	14.5	14.3	44.4	209.2
Reward	23.3	22.8	18.5	64.6	27.2	24.9	24.6	76.7	18.0	22.4	25.2	65.6	206.9
Marquis	26.2	28.8	23.3	78.3	27.8	26.3	24.0	78.1	13.4	6.4	4.9	24.7	181.1
Hope	22.6	21.0	24.2	67.8	24.0	23.7	23.3	71.0	28.0	29.0	25.3	82.3	221.1
Total	128.1	138.6	118.4	385.1	126.1	121.8	122.1	370.0	103.5	102.4	99.8	305.7	1060.8
	Crookston - 1931				Crookston - 1932				Crookston - 1935				
Thatcher	39.0	37.6	30.4	107.0	23.1	15.2	20.8	59.1	27.0	24.2	17.5	68.7	234.8
Ceres	34.6	37.4	33.7	105.7	31.1	19.5	20.9	71.5	15.4	11.0	11.5	37.9	215.1
Reward	32.5	31.3	29.8	93.6	23.1	22.8	19.8	65.7	16.8	16.4	14.6	47.8	207.1
Marquis	31.4	26.4	30.5	88.3	20.1	19.2	15.5	54.8	5.4	3.9	8.4	17.7	160.8
Hope	27.8	30.6	29.4	87.8	19.5	25.2	20.8	65.5	18.0	18.3	15.0	51.3	204.6
Total	165.3	163.3	153.8	482.4	116.9	101.9	97.8	316.6	82.6	73.8	67.0	223.4	1022.4
Total	383.3	405.8	366.6	1155.7	384.8	375.5	378.1	1138.4	302.4	302.6	298.4	903.4	3197.5

(a) Calculate the analysis of variance for the complete study.

(b) Test the significance of the different mean squares compared with the error variance, using the F test.

(c) Compare Thatcher with Ceres, as an average of all tests, using the standard error of the difference.

(d) What would be the standard error for testing the significance of the interaction between Thatcher and Ceres in 1932 and 1935, as an average of all stations? Make the proper test of significance.

(e) Do the same, as under (d), for comparing Thatcher and Ceres at University Farm and Waseca, as an average of all years. Is this interaction signigicant?

(f) Calculate the sums of squares for blocks, varieties, error and total for each of the 9 separate tests and add the different components for all 9 tests. Compare these sums of squares with appropriate combinations in the complete analysis of variance table.

2. Test the data in problem 1 for homogeneity.

3. The relative infection in different varieties for 5 bunt collections were as follows (Data from Salmon. 1938):

Bunt Collection	Variety							
	Oro		Ridit		Albit		Turkey	
	(1)[1]	(2)	(1)	(2)	(1)	(2)	(1)	(2)
	(Pct.)	(Pct.)	(Pct.)	(Pct.)	(Pct.)	(Pct.)	(Pct.)	(Pct.)
1	0.0	0.9	6.3	3.9	0.0	0.0	8.3	6.5
2	2.5	3.6	8.7	2.2	92.4	90.6	89.0	84.3
3	1.5	0.0	6.0	0.7	93.7	90.1	3.3	6.0
4	1.5	6.3	4.1	3.1	14.0	4.5	81.7	87.2
5	0.6	1.7	3.9	3.6	4.2	3.2	7.5	2.4

(a) Transform these percentage data to degrees of an angle and compute the analysis of variance for varieties, replicates, and collections.

(b) Compute the data without the transformation and compare the results with (a).

[1] These numbers refer to replicates.

CHAPTER XVIII

THE SPLIT-PLOT EXPERIMENT[1]

I. Use of Split Plot Experiments

Sometimes it is an advantage to use relatively large plots for one series of treatments and sub-divide these whole plots into a number of sub-plots to superimpose a second series of treatments. This type of design, called the split-plot experiment, was first proposed by Yates (1933, 1935). It is particularly useful in spacing tests with crop plants, some fertilizer trials, and in cultural studies. Le Clerg (1937) used this type of experimental design to ascertain the effect of 5 fertilizer mixtures (main treatments) on the seedling stand in plots sown with treated and untreated seed (sub-treatments). Goulden (1939) gives a more complicated split-plot design in which he studied the incidence of root-rot on wheat varieties, kinds of dust for seed treatment, method of dust application, and efficacy of soil inoculation with the root rot organism.

The split-plot design provides a more critical comparison of the sub-plot treatments than it does for the whole-plot units. This is due to the larger number of replications of the small units which, in turn, provide a larger number of degrees of freedom for error. Paterson (1939) advises that the less important treatment effects be allocated to the whole plots and the more important treatment effects to the sub-plots in order to obtain the maximum precision where it is most desired.

The split-plot design leads to two or more errors. To simplify the computation, all treatment values should be expressed in sub-plot units.

II. Data used for Computation

Two designs are outlined below together with the method of computation. These data are from a corn uniformity trial conducted at Waseca (Minnesota) in 1933 by C. W. Doxtator. They are for yield in pounds for the central two rows of four-row plots 12 hills long. For purposes of calculation, it was supposed that these data were obtained from 10 hybrids which are designated 1,2,3.....10. It was supposed further that these varietal plots had been split into three parts to test the yield of those crosses obtained from F_1, F_2, and F_3 generation seed. These are designated a,b, and c, respectively. The yields in the tables that follow are in the same order as in the field. The hypothetical hybrids and generations were superimposed on the data by random arrangement.

III. Sub-treatments Randomized within Main Plot (Plan A)

The field arrangement of the plots is given below. The 10 hybrids are assumed to have been planted in rows of 36 hills, using F_1 seed for 12 hills, F_2 seed for 12 hills and F_3 seed for the remaining 12 hills. The order of the hybrids in the field is random and the three generations of seed for each hybrid are planted in a random order within each long row.

Hybrid Number									
3	8	2	1	6	7	10	9	4	5
a	c	a	a	c	b	c	b	c	b
c	a	b	c	b	a	b	c	a	a
b	b	c	b	a	c	a	a	b	c

[1]This chapter is a modification of one prepared by Dr. F. R. Immer, University of Minnesota, for his Applied Statistics course.

The yields of each plot are given below in table 1. Data from two blocks are used to illustrate the calculations.

Table 1. Yield of corn per 12 hill plot and sums of yield of 36 hill plots.

Block I										
Hybrid Number										Total
3	8	2	1	6	7	10	9	4	5	
a 48	c 46	a 46	a 42	c 43	b 47	c 48	b 46	c 46	b 49	
c 46	a 45	b 44	c 46	b 45	a 49	b 45	c 48	a 48	a 49	
b 43	b 42	c 42	b 44	a 44	c 47	a 45	a 47	b 47	c 48	
T. 137	133	132	132	132	143	138	141	141	146	1375

Block II										
Hybrid Number										Total
4	3	9	5	1	7	2	8	6	10	
c 46	a 45	a 46	b 45	b 43	c 48	c 44	a 44	b 47	c 43	
a 48	b 44	c 46	a 45	c 50	b 51	a 48	c 46	a 48	b 43	
b 42	c 42	b 44	c 43	a 44	a 48	b 47	b 46	c 44	a 42	
T.136	131	136	133	137	147	139	136	139	128	1362

(a) Calculation of Sums of Squares

The analysis of variance is given in table 2.

Table 2. Analysis of Variance

Variation due to:	D.F.	Sums of Squares	Mean Square
Blocks	1	2.8166	2.8166
Hybrids	9	77.6833	8.6315
Error (a)	9	81.0167	9.0019
Plots of hybrids	(19)	161.5166	
Generations	2	7.2333	3.6167
Hybrids x generations	18	88.7667	4.9315
Error (b)	20	40.6667	2.0333
Total	59	298.1833	

The total sum of squares is calculated from the squares of the 60 individual plot yields as $S(x^2) - (Sx)^2/N$ which, numerically is 125,151.0000 - 124,852.8167 = 298.1833 (59 D.F.).

The sum of squares for blocks is $\frac{1375^2 + 1362^2}{30} - 124,852.8167 = 2.8166$ (1 D.F.)

Sum of squares for total plots of hybrids is calculated from the marginal totals for hybrids in the above table. Thus,

$$\frac{137^2 + 133^2 + ---------128^2}{3} - 124,852.8167 = 161.5166 \text{ (19 D.F.)}$$

To obtain the sums of squares for hybrids, generations and the interaction between them it is necessary to set up another table with the yields of the two replicates of each treatment combined.

Generation:	Hybrid Number										Sum
	1	2	3	4	5	6	7	8	9	10	
a	86	94	93	96	94	92	97	89	93	87	921
b	87	91	87	89	94	92	98	88	90	88	904
c	96	86	88	92	91	87	95	92	94	91	912
Sum	269	271	268	277	279	271	290	269	277	266	2737

Sum of squares for hybrids will be:

$$\frac{269^2 + 271^2 + -------266^2}{6} - 124,852.8167 = 77.6833 \text{ (9 D.F.)}$$

Sum of squares for generations is obtained from

$$\frac{921^2 + 904^2 + 912^2}{20} - 124,852.8167 = 7.2333 \text{ (2 D.F.)}$$

The sum of squares of the 30 yields in the above table will be equal to $\frac{86^2 + 94^2 + ---------91^2}{2}$ - 124,852.8167 or 173.6833 (29 D.F.)

Sum of squares for interaction of hybrids x generations will be:

173.6833	(29 D.F.)	
-7.2333	(2 D.F.)	for generations
-77.6833	(9 D.F.)	for hybrids
88.7667	(18 D.F.)	= sum of squares for interaction

(b) Errors to Test Significance

These sums of squares are brought together in table 2. The sum of squares for error (a) are obtained by subtracting the sums of squares for blocks (1 D.F.) and hybrids (9 D.F.) from "plots of hybrids" (19 D.F.). The sum of squares for error (b) is obtained by subtracting "plots of hybrids", generations and hybrids x generations from the total.

Error (a) is an ordinary randomized block error and may be used to test the significance of differences between hybrids.

Error (b) is obtained from the sum of the interactions between generations and blocks within hybrids. Thus, a table could be arranged for the data from hybrid No. 3 (see table 1) as follows:

	Generation		
	a	b	c
Block I	48	43	46
Block II	45	44	42

The interaction of blocks x generations, for 2 D.F., could be used as error for this simple comparison. However, a table similar to the above could be set up for each hybrid. There would be, then, 10 x 2 = 20 degrees of freedom for error. This is what is used as error (b). The mean square for error (b) is, then, the average error of blocks x generations within hybrids. It will be the legitimate error to use for comparing differences between generations and testing the interaction of hybrids x generations. In practice this sum of squares is obtained by subtraction.

IV. Sub-treatments in Strips Across Blocks (Plan B)

The same yield figures are used in this plan as in Plan A. The location of the hybrids is also the same. Instead of randomizing the generations within the plots for each hybrid as in Plan A, the generations are now considered planted in long strips crosswise of the entire block. However, randomization of the generations in the different blocks is used. The field plan is given below.

Table 3. Yields of corn plots and the field arrangement of these plots.

Block I										
Hybrid Number										Total
3	8	2	1	6	7	10	9	4	5	
a 48	a 46	a 46	a 42	a 43	a 47	a 48	a 46	a 46	a 49	461
c 46	c 45	c 44	c 46	c 45	c 49	c 45	c 48	c 48	c 49	465
b 43	b 42	b 42	b 44	b 44	b 47	b 45	b 47	b 47	b 48	449
Tot. 137	133	132	132	132	143	138	141	141	146	1375

Block II										
Hybrid Number										Total
4	3	9	5	1	7	2	8	6	10	
b 46	b 45	b 46	b 45	b 43	b 48	b 44	b 44	b 47	b 43	451
c 48	c 44	c 46	c 45	c 50	c 51	c 48	c 46	c 48	c 43	469
a 42	a 42	a 44	a 43	a 44	a 48	a 47	a 46	a 44	a 42	442
Tot. 136	131	136	133	137	147	139	136	139	128	1362

The same plots are used here as in table 1. The hybrids occur in the same order as in the previous table, the only difference being the arrangement of the generations. In table 3 the generations occur in strips crosswise of the blocks.

Table 4. Analysis of variance from the data of table 3

Key number for D.F. and sums of squares	Variation due to	D.F.	Sum of Squares	Mean Square
1	Blocks	1	2.8166	2.8166
2	Hybrids	9	77.6833	8.6315
3 = 4-1-2	Error (a)	9	81.0167	9.0019
4	Plots of hybrids	19	161.5166	
1	Blocks	1	2.8166	2.8166
5	Generations	2	35.4333	17.7167
6 = 7-1-5	Error (b)	2	16.2334	8.1167
7	Plots of "generations"	5	54.4833	
4	Plots of hybrids	19	161.5166	
8 = 5 + 6	Deviation of generation plots from blocks)	4	51.6667	
9	Hybrids x generations	18	61.5667	3.4204
10 = 11-9-8-4	Error (c)	18	23.4333	1.3019
11	Total	59	298.1833	

The total sum of squares (298.1833), sum of squares for blocks, hybrids, error (a) and total plots of crosses will be the same as under Plan A. The position of the "generation" plots has been changed, however, and the other sums of squares must be recalculated.

As far as the test of the three generations a, b and c is concerned there are but six plots as given in the marginal total of table 3. The sum of squares for those six "plots of generations" will be

$$\frac{461^2 + 465^2 + 449^2 + 451^2 + 469^2 + 442^2}{10} - 124{,}852.8167 = 54.4833 (5\ D.F.)$$

To obtain the sum of squares for generations and for interaction, a table is made up by combining the two yields of each treatment.

Generation	Hybrid Number 1	2	3	4	5	6	7	8	9	10	Tot.
a	86	93	90	88	92	87	95	92	90	90	903
b	87	86	88	93	93	91	95	86	93	88	900
c	96	92	90	96	94	93	100	91	94	88	934
Tot.	269	271	268	277	279	271	290	269	277	266	2737

The sum of squares for generations will be $\frac{903^2 + 900^2 + 934^2}{20} - 124{,}852.8167 =$ 35.4333 (2 D.F.)

The yield figures in the above table are squared, i.e., $86^2 + 93^2 \ldots\ldots + 88^2$. The sum is divided by 2, the correction factor then being subtracted. This gives 174.6833 as the sum of squares for those 29 degrees of freedom. The sum of squares for the interaction, hybrids x generations will be: 174.6833 - 77.6833 - 35.4333 = 61.5667 (18 D.F.).

In table 4 it is noted that the comparison of hybrids is the same as under Plan A. Error (b) will be obtained by the subtraction of blocks and generations from the "plots of generations." It is seen from table 3 that the analysis of variance to test the significance of generations involves only 6 large plots. The total yields of these could be set down from the marginal totals of table 3 as:

	Generations			
	a	b	c	Total
Block I	461	449	465	1375
Block II	442	451	469	1362
Total	903	900	934	2737

An analysis of variance of this 2-by-3 table would give the second section in the complete analysis of table 4. Error (b) is appropriate for the comparison of differences between generations.

Error (c) is obtained by subtraction from the total the items listed in table 4 opposite error (c). Error (c) is the second order interaction of blocks x hybrids x generations, the degrees of freedom being 1 by 9 by 2 = 18. It was obtained in table 4 by subtraction but has the above meaning. Error (c) is appropriate for testing the significance of the interaction of hybrids x generations.

Since these were uniformity trial data no attempt will be made to determine significance of the different mean squares. In a practical experiment these tests are carried out in the ordinary way, the appropriate errors given in the tables being used.

Yates (1933) has discussed the above two designs rather fully.

V. Comparison of Two Designs

Suppose the 10 hybrids and 3 generations of the seed of each (F_1, F_2 and F_3) had been considered as simply 30 treatments and completely randomized within the blocks without reference to split plot arrangements. The analysis of the data would have taken the form:

Variation due to:	D.F.
Blocks	1
Hybrids	9
Generations	2
Hybrids x generations	18
Error	29
Total	59

The degrees of freedom for error given above (29) is equal to the sum of the degrees of freedom for errors (a) and (b) under Plan A and the sum of degrees of freedom for errors (a), (b) and (c) under Plan B.

Plan B is the same as Plan A insofar as precision of tests of the hybrids is concerned. It differs from Plan A in that precision for the comparison of generations is sacrificed in order to obtain greater precision for the interaction.

The design of an experiment will depend entirely on what element of the treatments the highest degree of precision is desired. When the primary emphasis is to be placed on the interactions, at the expense of higher errors for the main effects, Plan B is to be preferred. When the main effects are of major interest either the complete randomized block or Plan A are to be preferred.

In practice the relative differences in magnitude of the different errors under Plans A and B will depend on the dimensions of the blocks. In this case the blocks were 40 rows wide, or 140 feet, and the 36 hill rows of hybrids were 126 feet long. Consequently the "generation" plots tended to be closer together than the most distant hybrids, in the same block.

Plan A is particularly applicable to studies of space relationship between plants in relation to yield. In a recent study of the effect of spacing on yield of soybeans, conducted by the Division of Agronomy and Plant Genetics, U. of Minnesota, Plan A was found to be admirably suited to the test. The soybeans were planted in 4 row plots, the rows being 16, 20, 24, 28, 32, and 40 inches apart. Then, the soybeans were planted at 4 different rates within each spacing, being 1/2, 1, 2, and 3 inches apart within rows. The only easy way to lay out such a test was to plant the plots of different width rows crosswise of the regular 132-foot series. The 4 different rates of seeding were then randomized within these long rows, the ultimate plots being 33 feet long.

Plan A could be laid out as follows also, using the same notation as employed in table 1.

Hybrid Number									
3			8			2			etc.
a	c	b	c	a	b	b	a	c	etc.

Here the hybrids are planted in groups of 3 plots with the three generations in a random arrangement within each hybrid plot but they occur side by side instead of end to end. By this plan it is obvious that the comparisons between generations (a, b, and c) will have a lower error than comparisons between hybrids (1,2,3 ----). The data from this arrangement would be analyzed exactly in the same manner as given under Plan A.

VI. Randomized-Block vs. Split-Plot Experiments

The relative efficiency of randomized-block and split-plot experiments was studied on uniformity trial data with sugar beets by Le Clerg (1937) both in the field and in the greenhouse. He compared the magnitude of the variance of the sub-plots within main plots in the split-plot design with the variance of sub-plots within blocks in the randomized block arrangement. The variance for sub-plots within main plots in the split-plot design was markedly less than that for sub-plots within blocks in the randomized arrangement. The split-plot design was 71 per cent more efficient in one set of uniformity data and 53 per cent more efficient in another. For comparisons of main plots within blocks there was a decrease in efficiency by the use of the split-plot arrangement. Similar results were obtained for greenhouse trials, altho less marked.

References

1. Goulden, C. H. Methods of Statistical Analysis. John Wiley. pp. 151-159. 1939.
2. LeClerg, E. L. Relative Efficiency of Randomized-Block and Split-Plot Designs of Experiments Concerned with Damping-off Data for Sugar Beets. Phytopath., 27:942 - 945. 1937.

3. Paterson, D. D. Statistical Technique in Agricultural Research. McGraw-Hill. pp. 209-214. 1939.
4. Yates, F. The Principles of Orthogonality and Confounding in Replicated Experiments. Jour. Agr. Sci., 23:108-145. 1933.
5. ———— Complex-Experiments. Jour. Roy. Stat. Soc. Suppl., 2:181-223. 1935.

Questions for Discussion

1. What is a split-plot design? Where used to advantage? List at least 3 situations.
2. Explain the differences in field lay-outs that lead to two and three errors.
3. Under what conditions would you use Plan "A"? Plan "B"?
4. Compare the relative efficiency of split-plot and randomized-block designs super-imposed on uniformity trial data.

Problems

The following data are from a randomized block experiment with "split plots" designed to test the differences in yield of soybeans planted at different spacings between and within rows. Four row plots were used, one row being harvested for hay and one for seed.

(A) Yield of soybeans in bushels per acre

Block No.	Width of rows	1/2"	1"	2"	3"	Total	Block Total
I	16"	25.1	21.3	22.3	22.1	90.8	
	20"	21.8	22.7	22.2	22.8	89.5	
	24"	21.9	21.8	21.2	20.6	85.5	
	28"	21.2	20.4	20.4	17.9	79.9	
	32"	20.7	20.0	18.3	20.0	79.0	
	40"	19.5	18.3	17.5	16.3	71.6	496.3
II	16"	25.2	19.9	22.1	22.7	89.9	
	20"	21.9	21.3	22.1	22.9	88.2	
	24"	19.7	19.8	20.1	19.8	79.4	
	28"	20.8	21.2	18.8	20.6	81.4	
	32"	18.5	20.7	17.5	16.4	73.1	
	40"	18.5	18.2	19.8	15.9	72.4	484.4
III	16"	15.7	21.6	22.9	20.3	80.5	
	20"	22.0	20.4	22.4	20.7	85.5	
	24"	25.5	20.7	20.7	20.5	87.4	
	28"	21.5	19.9	20.5	20.9	82.8	
	32"	22.0	19.3	18.1	17.8	77.2	
	40"	20.5	16.4	17.5	18.5	72.9	486.3
IV	16"	23.8	29.0	12.3	23.5	88.6	
	20"	27.0	21.2	20.5	20.7	89.4	
	24"	23.5	20.0	22.3	19.8	85.6	
	28"	22.5	21.5	22.7	18.9	85.6	
	32"	23.9	18.4	20.7	18.7	81.7	
	40"	19.9	17.8	16.9	18.5	73.1	504.0
	Total	522.6	491.8	479.8	476.8	1971.0	1971.0

(B) Yield of dry hay in tons per acre

Block No.	Width of rows	1/2"	1"	2"	3"	Total	Block Total
I	16"	2.91	2.59	2.41	2.74	10.65	
	20"	2.96	2.35	2.31	2.10	9.72	
	24"	2.34	2.30	2.21	2.23	9.08	
	28"	2.59	2.47	2.16	2.10	9.32	
	32"	2.21	2.12	2.05	1.90	8.28	
	40"	2.24	1.90	1.82	1.79	7.75	54.80
II	16"	2.85	2.42	2.45	2.31	10.03	
	20"	2.42	2.48	2.31	2.27	9.48	
	24"	2.40	2.19	2.29	2.08	8.96	
	28"	2.48	2.22	2.30	2.08	9.08	
	32"	2.32	2.10	2.23	2.06	8.71	
	40"	2.34	2.07	1.76	1.78	7.95	54.21
III	16"	2.81	2.61	2.65	2.25	10.32	
	20"	2.66	2.52	2.78	2.52	10.48	
	24"	2.57	2.41	2.28	2.13	9.39	
	28"	2.03	2.22	2.39	2.01	8.65	
	32"	2.68	2.21	1.97	1.96	8.82	
	40"	2.13	2.09	1.84	1.96	8.02	55.68
IV	16"	2.83	3.10	2.12	2.38	10.43	
	20"	3.27	2.71	2.33	2.42	10.73	
	24"	2.71	2.31	2.22	1.97	9.21	
	28"	2.52	2.53	2.24	2.09	9.38	
	32"	2.37	2.29	2.20	1.85	8.71	
	40"	2.10	2.13	1.92	2.06	8.21	56.67
	Total	60.74	56.34	53.24	51.04	221.36	221.36

The actual field arrangement of plots in this experiment, in block number III was as follows: The plot arrangement in the other blocks was randomized in a similar manner.

Width of rows	Spacing within rows:			
16"	1/2"	1"	3"	2"
32"	1"	3"	2"	1/2"
28"	3"	1/2"	1"	2"
40"	2"	1"	1/2"	3"
24"	1"	3"	2"	1/2"
20"	3"	1"	1/2"	2"

1. Analyze the data on yields of soybeans in bu. per acre.

   (a) Calculate the complete analysis of variance. Test the significance of the different mean squares, compared with the appropriate error variances, by means of the F test.

   (b) Determine the significance of the difference between 20' and 32" rows by means of the standard error.

(c) Determine the significance of the difference between 1/2" and 2" spacings by means of the standard error.

2. Analyze the data on yields of soybeans for tons of dry hay per acre in a similar manner.

3. Key out the degrees of freedom for a split plot experiment (two errors) for 3 spacings, 4 blocks, and 5 widths of rows.

4. A split-plot experiment was designed to determine the effect of seed treatment on the stand and ultimate yield of dryland corn planted at 3 different dates. a, b, c. Each plot consisted of 2 sub-plots, the seed being treated (T) with an organic mercury compound in one-half, and untreated (U) in the other half. There were 3 main plots in each block. All treatments were randomized. The field design of Block I was as follows:

T ' U	T ' U	U ' T
b	c	a

The yield data for the 6 blocks of the experiment were as follows in bushels per acre:

Date Planted	Seed Treatment	Block 1	2	3	4	5	6	Total
a	U	2.3	4.6	5.4	2.8	5.8	3.3	24.2
	T	2.8	4.7	4.2	3.6	5.0	4.6	24.9
b	U	4.3	4.3	3.3	6.1	4.5	4.0	26.5
	T	5.1	6.1	3.1	4.3	5.3	5.9	29.8
	U	2.7	1.4	2.3	3.8	2.9	3.9	17.0
	T	2.0	1.8	1.8	4.7	3.4	1.5	15.2

(a) Calculate the complete analysis of variance.

(b) Determine the significance between treated and untreated seed, and also between planting dates.

CHAPTER XIX

CONFOUNDING IN FACTORIAL EXPERIMENTS

I. Factorial Experiments

The randomized-block and Latin-square designs are widely used in field experiments, both being very efficient for simple studies. However, there are situations in experimentation where a large number of varieties or treatments are to be compared at two or more levels. The factorial experiment is useful in such situations. Suppose that three fertilizers, Nitrogen (N), Phosphorus (P), and Potash (K) are to be tested at two or more levels. The classical method of approach would be to vary the two levels for each element only one at a time, i.e., the investigator would set up separate experiments to test each element alone at its respective level. The single factor could then be studied under controlled conditions at each of the two levels. To test these factors simultaneously in the same experiment, would permit one to study the effects of different amounts of one fertilizer on the others in all combinations. Thus, a wider base of inductive reasoning is provided. The experimental argument is also strengthened by the larger total number of plots in the test. (See Fisher, 1935).

Goulden (1937) describes a factorial experiment as one made to study simultaneously various treatment factors. Thus, an experiment designed to study at the same time rate and depth of seeding of a cereal crop would be a factorial experiment in which two factors, rate and depth, are represented at two or more levels. The study of interactions is an important consideration in such an experiment. The introduction of factors is limited by space and cost of experimentation.

Suppose a fertilizer test is to be conducted with N, P, and K at two different rates each. The rates can be designated by subscripts so as to give the eight possible treatment variants as follows:[1]

$N_0P_0K_0$, $N_1P_0K_0$, $N_0P_1K_0$, $N_0P_0K_1$, $N_1P_1K_0$, $N_1P_0K_1$, $N_0P_1K_1$, and $N_1P_1K_1$

The degrees of freedom, i.e., the number of comparisons free to vary, may be keyed out as follows:

Variation due to	Degrees freedom	Remarks
Nitrogen (N)	1	Main effects
Phosphorus (P)	1	
Potassium (K)	1	
N x P[2]	1	First order interactions
N x K	1	
P x K	1	
N x P x K	1	Second order interactions
Total	7	

[1] Note: The subscripts 0 and 1 represent the two fertilizer levels.
[2] The symbol (x) denotes interaction and not a variable as heretofore.

II. Data for Computation of Factorial Experiment

The computations for the Analysis of Variance for such a factorial experiment will be illustrated with uniformity trial data. Four complete replications will be used. The uniformity data on crested wheatgrass were furnished by Dr. R. M. Weihing. The plots are combined as 8-row plots, 16 feet long, with rows 6 inches apart. Thus, each plot is 4 by 16 feet in size. The yields are given in grams of air-dry field cured hay. The uniformity trial data follow.

Table 1. Uniformity Data for Crested Wheatgrass

Plot No.	Blocks I	II	III	IV
	(gm.)	(gm.)	(gm.)	(gm.)
1	5135	3175	4405	3750
2	4725	3980	4575	3920
3	4600	4420	3910	4175
4	4955	4580	4065	3280
5	3210	3970	3510	3190
6	3670	4255	4305	3575
7	3785	3665	3995	3530
8	3965	4315	4030	2900

III. Computation as Simple Randomized Block Experiment

The eight treatments will first be superimposed on the crested wheatgrass yield data for a randomized block test.[1]

Table 2. Yields of Crested Wheatgrass in Randomized Blocks

Replication	Treatment O	N	P	K	NP	NK	PK	NPK	Totals
I	3210	4955	5135	4600	3965	3785	3670	4725	34045
II	3970	3175	4420	3980	4255	3665	4315	4580	32360
III	4305	4405	4575	3910	4030	3510	3995	4065	32795
IV	3530	4175	3920	3280	2900	3575	3190	3750	28320
Totals	15015	16710	18050	15770	15150	14535	15170	17120	127520

The sums of squares for blocks, treatments, total, and error are computed in the ordinary manner. The Analysis of Variance can be summarized as follows:

Variation due to	D.F.	Sum Squares	Mean Square	"F" Value Observed	5 Pct. Point
Blocks	3	2,303,556	767,852	3.91	3.07
Treatments	7	2,628,237	375,462	1.91	2.57
Error	21	4,125,107	196,434		
Total	31	9,056,900			

The block effect removed is just enough to be statistically significant. Treatment effects are within the limits of error since the data are from a uniformity trial.

[1]Note: The same randomization for treatments is used here as in the confounded experiment to be mentioned later.

The crested wheat yield data will now be considered from the standpoint of confounding. This process is expected to accomplish several things: (1) A greater amount of the variability due to soil heterogeneity should be removed because more and smaller blocks will be used; (2) A chance to examine the second order interaction, N x P x K, will be forfeited; and (3) The reduction of experimental error in this manner should sharpen all treatment and interaction comparisons.

IV. Confounding in a 2 by 2 by 2 Experiment[1]

A few terms must first be made clear before the analyses are made.

(a) Explanation of Terms

Every effort is made to maintain orthogonality in an experiment. Yates (1933) defines orthogonality as follows: "Orthogonality is that property of the design which ensures that the different classes of effects shall be capable of direct and separate estimation without any entanglements." Thus, orthogonality is ensured in a randomized block experiment by the very nature of the design, i.e., each block contains the same kind and number of treatments. Non-orthogonality is introduced when some of the plots in one or more of the blocks are lost. Special methods may be required to separate treatment and block effects.

Non-orthogonality is sometimes deliberately introduced in factorial experiments that involve a fairly large number of combinations. This process is called confounding. The purpose is to increase the accuracy of the more important comparisons at the expense of the comparisons of lesser importance.

(b) Confounding the Second-order Interaction

The second order interaction (N x P x K) in this experiment may be considered the least important. Certainly, it would be difficult to interpret in terms of fertilizer practice, even though it were significant. Suppose it is desired to confound the one degree of freedom for this interaction with blocks. To accomplish this, it is necessary to determine the distribution of treatments in the blocks in a manner so as to confound this one treatment and, at the same time, leave the others intact.

Algebraically, the treatment effects can be represented as follows:

$$
\begin{aligned}
\text{Nitrogen (N)} &= (N_1 - N_o)(P_1 + P_o)(K_1 + K_o) \\
\text{Phosphorus (P)} &= (P_1 - P_o)(N_1 + N_o)(K_1 + K_o) \\
\text{Potassium (K)} &= (K_1 - K_o)(N_1 + N_o)(P_1 + P_o) \\
N \times P &= (N_1 - N_o)(P_1 - P_o)(K_1 + K_o) \\
N \times K &= (N_1 - N_o)(K_1 - K_o)(P_1 + P_o) \\
P \times K &= (P_1 - P_o)(K_1 - K_o)(N_1 + N_o) \\
N \times P \times K &= (N_1 - N_o)(P_1 - P_o)(K_1 - K_o)
\end{aligned}
$$

The last expression can be expanded as follows:

$$N \times P \times K = (N_1 - N_o)(P_1 - P_o)(K_1 - K_o) =$$
$$(N_1P_oK_o + N_oP_1K_o + N_oP_oK_1 + N_1P_1K_1) - (N_oP_oK_o + N_1P_1K_o + N_oP_1K_1 + N_1P_oK_1) =$$
$$(N + P + K + NPK) - (0 + NP + PK + NK)$$

Then, the blocks could be divided as follows so as to confound the second order interaction with block effect:

N P K NPK	0 NP PK NK
Sub-block A	Sub-block B

[1] For more complicated designs see Yates (1933), Fisher (1935) and Goulden (1937).

The contrasts between the two sub-blocks of each replicate will be contrasts of the second order interaction ($N_1P_1K_1$ and $N_oP_oK_o$). This interaction will have been confounded with blocks.

The sum of squares for the second order interaction will be given by: (See Goulden, 1937).

$$1/2\ k\ \left[(N + P + K + NPK) - (0 + NP + PK + NK) \right]^2$$

where k = number of plots represented in each total. The above sum of squares will contain not only the second order interaction effect but also the block effect.

In this case, blocks of four plots each have been used for error control instead of blocks of eight (as would be the case in a simple randomized block experiment), and only the second order interaction has been lost. The key-out for four complete replications will be as follows:

Variation due to	Degrees of freedom
Blocks	7
N	1
P	1
K	1
N x P	1
N x K	1
P x K	1
Error	18
Total	31

The treatments will be randomized in each sub-block. The field arrangement and plot yields follow:

Table 3. Field Plan with Plot Locations and Yields

	Sub-block A		Sub-block B	
Replication	Treatment	Yield	Treatment	Yield
I	$N_oP_1K_o$	5135	$N_oP_oK_o$	3210
	$N_1P_1K_1$	4725	$N_oP_1K_1$	3670
	$N_oP_oK_1$	4600	$N_1P_oK_1$	3785
	$N_1P_oK_o$	4955	$N_1P_1K_o$	3965
	Total	19415	Total	14630
II	$N_1P_oK_o$	3175	$N_oP_oK_o$	3970
	$N_oP_oK_1$	3980	$N_1P_1K_o$	4255
	$N_oP_1K_o$	4420	$N_1P_oK_1$	3665
	$N_1P_1K_1$	4580	$N_oP_1K_1$	4315
	Total	16155	Total	16205
III	$N_1P_oK_o$	4405	$N_1P_oK_1$	3510
	$N_oP_1K_o$	4575	$N_oP_oK_o$	4305
	$N_oP_oK_1$	3910	$N_oP_1K_1$	3995
	$N_1P_1K_1$	4065	$N_1P_1K_o$	4030
	Total	16955	Total	15840
IV	$N_1P_1K_1$	3750	$N_oP_1K_1$	3190
	$N_oP_1K_o$	3920	$N_1P_oK_1$	3575
	$N_1P_oK_o$	4175	$N_oP_oK_o$	3530
	$N_oP_oK_1$	3280	$N_1P_1K_o$	2900
	Total	15125	Total	13195

The yield data are summarize for main effects in Table 4 as follows:

Table 4. Total Yields for Four Replications per Treatment

	K_o	K_1	Sum		K_o	K_1	Sum
N_o P_o	15,015	15,770	30,785	N_1 P_o	16,710	14,535	31,245
P_1	18,050	15,170	33,220	P_1	15,150	17,120	32,270
Total	33,065	30,940	64,005	Total	31,860	31,655	63,515

The yields for the various interactions are totalled below in Table 5:

Table 5. Total Yields for Interactions

Comparison		Totals		
(a) N and K		K_o	K_1	Sum
	$(P_o + P_1)N_o$	33,065	30,940	64,005
	N_1	31,860	31,655	63,515
	Total	64,925	62,595	127,520
		K_o	K_1	Sum
(b) P and K	$(N_o + N_1)P_o$	31,725	30,305	62,030
	P_1	33,200	32,290	65,490
	Total	64,925	62,595	127,520
		P_o	P_1	Sum
(c) N and P	$(K_o + K_1)N_o$	30,785	33,220	64,005
	N_1	31,245	32,270	63,515
	Total	62,030	65,490	127,520

The sums of squares for the experiment are given in table 6. The sum of squares for blocks, N, P, K and total can be entered from table 6. The sum of squares for N x P is obtained by the subtraction of 7503 + 374,112 (N + P) from 443,743 which is $S(x^2_{NP})$. The result would be 62,128. The sums of squares for N x K and for P x K are obtained in a similar manner.

Table 6. Calculation of Sum of Squares

Symbol	Table	Total Sum Squares	Divide by	$(Sx)^2/N$	Corrected Sum Squares	D.F.
$S(x^2_t)$	3	517,224,100	1	508,167,200	9,056,900	31
$S(x^2_b)$	3	513,954,112	4	508,167,200	5,786,912	7
$S(x^2_N)$	4	8,130,795,250	16	508,167,200	7,503	1
$S(x^2_P)$	4	8,136,661,000	16	508,167,200	374,112	1
$S(x^2_K)$	4	8,133,389,650	16	508,167,200	169,653	1
$S(x^2_{NP})$	5	4,068,887,550	8	508,167,200	443,743	1
$S(x^2_{NK})$	5	4,067,676,450	8	508,167,200	292,356	1
$S(x^2_{PK})$	5	4,069,752,750	8	508,167,200	551,893	1

Table 7. Analysis of Variance

Variation due to	D.F.	Sum Squares	Mean Square	"F" Value Obtained	"F" Value 5 pct. point
Blocks	7	5,786,912	826,702	5.87	2.66
N	1	7,503	7,503	18.76	243.91
P	1	374,112	374,112	2.66	4.41
K	1	169,653	169,653	1.21	4.41
N x P	1	62,128	62,128	2.27	243.91
N x K	1	115,200	115,200	1.22	243.91
P x K	1	8,128	8,128	17.32	243.91
Error	18	2,533,264	140,737		
Total	31	9,056,900			

It is noted that the mean square for error has been decreased materially in the confounded experiment as compared to that in the simple randomized block experiment. In the former, the mean square is 140,737 while in the latter it is 196,434. It is also to be noted that more of the variability due to soil heterogeneity has been removed from the experimental error and drawn off in block effect which now appears as highly significant.

The real value of confounding as a means to bring out more closely significant treatment effects and interactions is not evidenced in this illustration because uniformity data have been employed. The confounding design is purely artificial.

V. Partial Confounding in a 2 by 2 by 2 Experiment

The above procedure resulted in the complete sacrifice of the second order interaction, but it may be argued that the experimenter has taken too much for granted. He may overcome this difficulty by partial confounding, i.e., confounding different interactions in different replications. Goulden (1937) states that the results are used from the blocks in which the particular effects are not confounded in order to recover a portion of the information desired. The fertilizer test used as an example can be partially confounded and at the same time recover a portion of the information on all the comparisons. Four replications will be required for this purpose. In each replication, one degree of freedom can be confounded with blocks for one of the interactions. There are four interactions, viz., N x P, N x K, P x K, and N x P x K.

The algebraic relations stated previously can be used to determine the treatments to place in each sub-block to gain the desired effect.

Interaction		Algebraic Relationship		Sub-Blocks A		Sub-Blocks B
N x P	=	$(N_1 - N_o)(P_1 - P_o)(K_1 - K_o)$	=	(N+P+NK+PK)	-	(O+NP+K+NPK)
N x K	=	$(N_1 - N_o)(K_1 - K_o)(P_1 + P_o)$	=	(N+K+NP+PK)	-	(O+P+NK+NPK)
P x K	=	$(P_1 - P_o)(K_1 - K_o)(N_1 + N_o)$	=	(P+K+NP+NK)	-	(O+N+PK+NPK)
N x P x K	=	$(N_1 - N_o)(P_1 - P_o)(K_1 - K_o)$	=	(N+P+K+NPK)	-	(O+NP+PK+NK)

The treatments within each sub-block will be randomized. Table 8 gives the field design together with the plot yields in grams for the fertilizer trial superimposed on crested wheatgrass uniformity trial data.

II (NPK confounded)	N	3175	NK	3970
	P	3980	NP	4255
	K	4420	O	3665
	NPK	4580	PK	4315
	Total	16155	Total	16205
III (P x K confounded)	NP	4405	NPK	3510
	P	4575	N	4305
	NK	3910	PK	3995
	K	4065	O	4030
	Total	16955	Total	15840
IV (N x K confounded)	N	3750	P	3190
	PK	3920	NPK	3575
	NP	4175	NK	3530
	K	3280	O	2900
	Total	15125	Total	13195

Grand Total = 127,520

The treatment totals required for the computation of the sums of squares are arranged in Table 8 for the totals of the four blocks, and for the omission of each replication.

Table 9. Treatment Totals Required for Calculation of Sums of Squares

Treat-ment	All Replications	Minus Replication I	Minus Replication II	Minus Replication III	Minus Replication IV
O	13805	10395	10140	9775	10905
N	16185	11230	13010	11880	12435
P	16880	11745	12900	12305	13690
K	15435	11765	11015	11370	12155
NP	16620	12835	12365	12215	12445
NK	16010	11410	12040	12100	12480
PK	16955	12230	12640	12960	13035
NPK	15630	11665	11050	12120	12055
(1)	(2)	(3)	(4)	(5)	(6)

The sums of squares can be computed as follows for the treatment effects (for 1 d.f.):

$$N = 1/2\ k\ [(N + NP + NK + NPK) - (O + P + K + PK)]^2$$
$$P = 1/2\ k\ [(P + NP + PK + NPK) - (O + N + K + NK)]^2$$
$$K = 1/2\ k\ [(K + NK + PK + NPK) - (O + N + P + NP)]^2$$
$$N \times P = 1/2\ k\ [(N + P + NK + PK) - (O + NP + K + NPK)]^2$$

$$N \times K = 1/2\ k\ [(N + K + NP + PK) - (0 + P + NK + NPK)]^2$$
$$P \times K = 1/2\ k\ [(P + K + NP + NK) - (0 + N + PK + NPK)]^2$$
$$N \times P \times K = 1/2\ k\ [(N + P + K + NPK) - (0 + NP + PK + NK)]^2$$

For example, the interaction N x P is calculated from the replications in which it is not confounded, i.e., from Column 3, Table 9. Note that k = 12.

$$N \times P = 1/24\ [(11230 + 11745 + 11410 + 12230) - (10595 + 11765 + 12835 + 11665)]^2$$
$$= 1/24\ [46615 - 46860]^2$$
$$= 1/24\ [245]^2 = 60025/24 = 2501.04$$

Similarly,

$$N \times P \times K = 1/24\ [(13010 + 12900 + 11015 + 11050) - (10140 + 12365 + 12040 + 12640)]^2$$
$$N \times P \times K = 1/24\ [(47975 - 47185)]^2 = 1/24\ [790]^2$$
$$= 624,100/24 = 26,004$$

The main effects are calculated from all the replications, i.e., k = 16. The calculation for N is as follows:

$$N = 1/32\ [(16185 + 16620 + 16010 + 15630) - (13805 + 16880 + 15435 + 16955)]^2$$
$$= 1/32\ [64445 - 63075]^2 = 1/32\ [1370]^2$$
$$= 1,876,900/32 = 58,653$$

The total sum of squares is calculated from all plot yields in all replications of the experiment, i.e., 32 plots. The block sum of squares is computed from the 8 block totals. The ordinary method of computation is used.

The analysis of variance can be set up as follows:

Table 10. Complete Analysis for Partially Confounded 2x2x2 Experiment

Variation due to	D.F.	Sum Squares	Mean Square	"F" Value Obtained	"F" Value 5 Pct. Point
Blocks	7	5,786,912	826,702	5.87	2.70
N	1	58,653	58,653	2.40	243.91
P	1	675,703	675,703	4.79	4.45
K	1	9,112	9,112	15.46	243.91
N x P	1	2,501	2,501	56.34	243.91
N x K	1	36,817	36,817	3.83	243.91
P x K	1	65,626	65,626	2.15	243.91
N x P x K	1	26,004	26,004	5.42	243.91
Error	17	2,395,572	140,916		
Total	31	9,056,900			

In this experiment, information is obtained on the main effects and on all interactions, including the second order interaction. However, there is a loss of one-fourth the information on each of the interactions, due to the fact that the replication in which an interaction was confounded was omitted in the calculation of its sum of squares. The error is of approximately the same magnitude as that for the experiment in which N x P x K was completely confounded.

References

1. Fisher, R. A. Design of Experiments, pp. 96-137. 1935.
2. Goulden, C. H. Methods of Statistical Analysis. Burgess Publ. Co., pp. 107-120. 1937.
3. Wiebe, G. A. Variation and Correlation in Grain Yield Among 1500 Wheat Nursery Plots. Jour. Agr. Res., 50:331-357. (Source of Data). 1935.
4. Yates, F. The Principles of Orthogonality and Confounding in Replicated Experiments. Jour. Agr. Sci., 23:108-145. 1933.
5. Yates, F. Complex Experiments. Suppl. Jour. Roy Stat. Soc., 2:181-247. 1935.

Questions for Discussion

1. What is a factorial experiment? Give an example.
2. Under what conditions may a factorial experiment be used?
3. What is meant by the term "orthogonality"?
   Give an example of an orthogonal experiment.
4. Explain the use of the term "confounding". What is done in confounding? Why?
5. Suppose a second-order interaction, N x P x K is to be confounded. How can this be done by design?
6. What is partial confounding? How does it differ from confounding?

Problems

Some uniformity data presented by Wiebe (1935) on wheat yields in grams per row are presented below as they occurred in the field:

Plot No.	Blocks I	II	III	IV
1	670	690	785	645
2	685	790	770	665
3	660	825	960	750
4	705	805	860	635
5	610	720	705	615
6	640	735	805	665
7	690	855	905	700
8	715	765	945	820

1. Calculate these data as a randomized block experiment using the 8 fertilizer treatments given in the text example.

2. Design an experiment so as to confound the second order interaction, N x P x K. Carry through the complete analysis. Compare the results with those obtained in problem 1.

3. Design an experiment to superimpose on these data so as to partially confound the second order interaction (N x P x K). Carry through the complete analysis. Compare the results with those obtained in problems 1 and 2.

CHAPTER XX

SYMMETRICAL INCOMPLETE BLOCK EXPERIMENTS

I. Incomplete Block Tests

It has been shown (Chapter 19) that greater accuracy is obtained in factorial experiments when certain degrees of freedom for the higher-order interactions are confounded with blocks, especially when the number of combinations is large. In variety trials it is sometimes desirable to test a large number of varieties in a single experiment. To compare them in an ordinary randomized block test leads to less accuracy due to the large size of the blocks. Methods have been developed by Yates (1936, 1937) to overcome this difficulty. The procedure is analogous to confounding in factorial experiments in that the replications are divided up into smaller blocks which are used as error control units. These small blocks contain only part of the total number of varieties, hence the name "incomplete blocks".

Incomplete block experiments have been shown to give increased efficiency by Yates (1936), and Goulden (1937). Weiss and Cox (1939) found the lattice square arrangement to result in a gain of 150 per cent on extremely heterogenous soil, but a loss of precision of 31.5 per cent on a very uniform soil.

One type of incomplete block experiment will be illustrated, i.e., the symmetrical incomplete block where all possible groups of sets are used. The computation procedure will follow closely that described by Weiss and Cox (1939). For other types of incomplete blocks, Goulden (1937,1939) should be consulted. These include the two dimensional quasi-factorial with two groups of sets, and the three dimensional quasi-factorial with three groups of sets. An excellent discussion of the lattice square design (quasi-Latin squares) is given by Weiss and Cox (1939) who applied it soybean variety test.

The computations will be illustrated with some uniformity trial data obtained from Dr. R. M. Weihing on forage yields of crested wheatgrass expressed in kilograms. The plots consist of 3-rows, 15 feet long, the individual rows being 6 inches apart.

II. Design of Symmetrical Incomplete Block Tests

In order to determine the details of an acceptable design with regard to the number of varieties and blocks to use, it is necessary to satisfy the condition that each variety occur with every other variety in the same number of blocks. Suppose that m varieties are replicated n times over a portion of the available blocks each of which is to contain n' plots. For example, suppose that one considers the n plots in which one certain variety occurs. The total number of plots contained in these n blocks is obviously (n)(n'), of which n corresponds to the one variety under consideration. Therefore, there are (n)(n'-n) = (n'-1)(n) plots available for the other m - 1 varieties in those blocks. To meet the above condition, these (n'-1)(n) plots must be distributed equally among the m-1 varieties that remain. For this reason, (n'-1)(n) must either equal m-1 or be a multiple of m-1. Thus, it becomes apparent that m-1 = (n'-1)(n) is a number that must be factorable, preferable into two numbers of nearly equal size. This can be effected in two different ways.

(1) First, one may use $m = k^2$, where k is an integer, from which $m-1 = k^2-1 = (k-1)(k+1)$. From this, it would appear that the choice of design could be either k-1 = n' - 1 (i.e., n' = k whence n will be k + 1), or k + 1 = n' - 1 (i.e., n' = k + 2) in which case n will be k - 1. However, when m' is equal to the total number of blocks, one must have mn = m'n'. Thus, it is clear that mn must be divisible by n'. The first choice gives $\frac{mn}{n'}$ to be $\frac{k^2(k+1)}{k}$ in which the divisibility is assured with

$m' = k(k + 1)$. The second choice gives $\frac{mn}{n'} = \frac{k^2(k-1)}{k + 2}$, which generally would not be an integer. Thus, only the first choice is acceptable.

(2) Second, one may choose $m = k^2 - k + 1$, from which $m - 1 = k^2 - k = k(k-1)$. From this relationship, it appears that one has a choice of design by the use of either $k - 1 = n' - 1$ (i.e., $n' = k$) from which n will also be k, or $k = n'-1$ (i.e., $n' = k + 1$) in which case n will be $k - 1$. In the analysis of this situation, $\frac{mn}{n'} = \frac{(k^2-k+1)k}{k}$. The divisibility is assured with the result that $m' = k^2-k+1$. For the second choice, $\frac{mn}{n'} = \frac{k^2-k+1}{k + 1}$, a value that is not generally divisible. Thus only the first choice is acceptable.

Therefore, it is obvious that designs of this nature can be constructed for $m = k^2$ where m varieties = 9,16,25,36,49,64, etc. The k^2-k+1 type can be designed for values of m = 7, 13, 21, 31, 43, 57, 73, etc. The structure of the arrangements is rather fully discussed by Yates (1936), Fisher and Yates (1938), and by Goulden (1937, 1939).

The first type, $m = k^2 = n'^2$, will be used to illustrate the process for a completely orthogonalized 5 by 5 square. This will give a series of symmetrical incomplete block arrangements.

1111	(1)	2222	(2)	3333	(3)	4444	(4)	5555	(5)
2345	(6)	3451	(7)	4512	(8)	5123	(9)	1234	(10)
3524	(11)	4135	(12)	5241	(13)	1352	(14)	2413	(15)
4253	(16)	5314	(17)	1425	(18)	2531	(19)	3142	(20)
5432	(21)	1543	(22)	2154	(23)	3215	(24)	4321	(25)

The explanation of the arrangement is taken directly from Weiss and Cox (1939). The numbers in parentheses designate the varieties which are to be compared in the experiment. "These variety numbers are arranged in 6 orthogonal groups as follows:

Group I (rows)

1	2	3	4	5
6	7	8	9	10
11	12	13	14	15
16	17	18	19	20
21	22	23	24	25

Group II (Columns)

1	6	11	16	21
2	7	12	17	22
3	8	13	18	23
4	9	14	19	24
5	10	15	20	25

Group III (first Number)

1	10	14	18	22
2	6	15	19	23
3	7	11	20	24
4	8	12	16	25
5	9	13	17	21

Group IV (second number)

1	9	12	20	23
2	10	13	16	24
3	6	14	17	25
4	7	15	18	21
5	8	11	19	22

Group V (third number)

1	8	15	17	24
2	9	11	18	25
3	10	12	19	21
4	6	13	20	22
5	7	14	16	23

Group VI (fourth number)

1	7	13	19	25
2	8	14	20	21
3	9	15	16	22
4	10	11	17	23
5	6	12	18	24

In group I the variety numbers are copied from the rows of the square, each row of the group specifying a block in the field. In like manner, the variety numbers in the blocks of group II are taken from the columns of the square. In group III the varieties in a block are specified by the numbers written first in the cells of the square. Thus, the varieties in the first block are those corresponding to number 1 wherever it occurs first in the cell; as examples, variety 1 is from row 1 column 1 of the completely orthogonalized square, variety 10 from row 2 column 5, variety

14 from row 3 column 4, etc. For group IV, the second numbers in the cells of the square are used to pick out the varieties. Thus, for the third block the number 3 is located in row 1 column 3 (variety 3), in row 2 column 1 (variety 6), etc.

"This set of six orthogonal groups constitutes a balanced incomplete block arrangement: in the 30 blocks of 5 plots, each of the 25 varieties occurs 6 times, once and once only with every other variety. The combination solution in the unreduced form would require a prohibitive number of blocks[1]".

The field arrangement for this type of symmetrical incomplete block design will be illustrated with the crested wheatgrass data. There are 25 varieties arranged in 6 replicates with 5 varieties in each block. The 5 varieties are randomized within each block. The block and replicate arrangement in the field may be as follows:[2]

I	II	III	IV	V	VI
5b	6b	13b	20b	24b	27b
4b	7b	11b	18b	25b	29b
1b	10b	12b	16b	21b	26b
2b	8b	14b	19b	22b	28b
3b	9b	15b	17b	23b	30b

III. Statistical Analysis of Incomplete Block Data

The symbols used in the discussion follow:

m = number of varieties (25)
n' = number of plots per block (5)
n = number of replicates of each variety (6)
m' = number of blocks (30)
N = mn = m'n' = total number of plots (150)
$\lambda = \frac{n(n'-1)}{m-1}$ = number of times any 2 varieties occur together in a block (1)
$E = \frac{1-1/n'}{1-1/m}$ = Efficiency Factor of Design, $\frac{(5)}{(6)}$
Sx = Sum of all N experimental values (217.79)
$S'x_V$ = Sum of n experimental values for any one variety.
$S'x_B$ = Sum of k experimental values for any one block.
s^2 = Error variance of a single experimental value.

[1] $b = \frac{C}{m\ n'} = \frac{C}{25\ 5} = \frac{25!}{5!\ 20!} = 53{,}130.$

[2] The blocks (5b, 4b, etc.) were arranged consecutively for the analysis of the data used in this problem, but they should be randomized (at least within replicates) in an actual field experiment. The Roman numerals refer to replicates.

(a) Computation of Block Totals

The yield data for the incomplete block experiment may be assembled as shown in Table 1 for the computation of the block totals. The numbers in parentheses refer to "varieties". The forage yields of crested wheatgrass are expressed as kilograms per plot.

Table 1. Plot Yields of the Symmetrical Incomplete Blocks Assembled for 25 Crested Wheatgrass "Varieties" in 6 Replicates.

Replicate	Set or Block	Plots in Block 1	2	3	4	5	Block Totals
I	1	(5) 1.25	(2) 1.52	(4) 1.30	(3) 1.83	(1) 1.64	7.54
	2	(10) 1.38	(7) 1.48	(8) 1.41	(9) 1.52	(6) 1.46	7.25
	3	(11) 1.42	(14) 1.40	(13) 1.35	(15) 1.32	(12) 1.19	6.68
	4	(20) 1.20	(18) 1.32	(17) 1.59	(16) 1.21	(19) 1.22	6.54
	5	(21) 1.48	(25) 1.14	(23) 1.16	(22) 1.54	(24) 1.67	6.99
II	6	(1) 1.27	(16) 1.54	(21) 1.66	(6) 1.81	(11) 1.96	8.24
	7	(12) 1.92	(17) 1.36	(2) 1.32	(22) 1.38	(7) 1.65	7.63
	8	(23) 1.61	(3) 1.16	(13) 1.47	(8) 1.24	(18) 1.32	6.82
	9	(19) 1.04	(9) 1.42	(14) 1.16	(24) 0.92	(4) 0.99	5.53
	10	(15) 1.34	(25) 1.38	(10) 1.02	(5) 1.22	(20) 1.72	6.68
III	11	(18) 1.86	(1) 1.94	(14) 1.84	(22) 1.96	(10) 1.92	9.52
	12	(15) 1.64	(2) 1.64	(23) 1.66	(19) 1.78	(6) 1.83	8.55
	13	(20) 1.84	(7) 1.72	(11) 1.20	(3) 1.29	(24) 1.29	6.89
	14	(4) 1.33	(12) 1.54	(16) 1.16	(8) 1.48	(25) 1.54	7.05
	15	(13) 1.50	(9) 1.18	(21) 1.24	(5) 1.35	(17) 1.53	6.80
IV	16	(1) 1.00	(20) 1.30	(23) 1.24	(12) 1.64	(9) 1.62	6.80
	17	(24) 1.62	(16) 1.48	(13) 1.66	(10) 1.62	(2) 1.83	8.21
	18	(3) 1.60	(6) 1.44	(17) 1.60	(14) 1.56	(25) 1.84	8.04
	19	(4) 1.30	(7) 1.31	(21) 1.45	(15) 1.51	(18) 1.72	7.29
	20	(22) 1.56	(5) 1.44	(8) 1.34	(11) 1.58	(19) 1.25	7.17
V	21	(15) 1.48	(8) 1.72	(17) 1.68	(24) 1.94	(1) 1.84	8.66
	22	(9) 1.48	(2) 1.48	(11) 1.29	(25) 1.36	(18) 1.67	7.28
	23	(19) 1.48	(12) 1.40	(21) 1.26	(3) 1.44	(10) 1.86	7.44
	24	(22) 1.40	(20) 1.38	(4) 1.42	(13) 1.74	(6) 1.55	7.49
	25	(16) 1.50	(7) 1.35	(5) 1.64	(23) 1.68	(14) 1.32	7.49
VI	26	(1) 1.22	(25) 1.22	(7) 1.70	(13) 1.56	(19) 1.60	7.30
	27	(2) 1.62	(8) 1.43	(20) 1.50	(21) 1.44	(14) 1.56	7.55
	28	(22) 1.32	(9) 0.93	(3) 1.45	(15) 1.36	(16) 1.41	6.48
	29	(10) 1.12	(17) 1.50	(11) 1.18	(4) 1.19	(23) 1.10	6.09
	30	(6) 1.08	(18) 1.00	(5) 1.08	(12) 1.32	(24) 1.31	5.79
Grand Total							217.79

(b) Computation of Variety Means

In symmetrical incomplete block designs, a preliminary step is required to obtain the sum of squares for varieties. Due to the fact that variety differences are partially confounded with block effects, it is necessary to compute each variety sum by a formula that involves both the yields of the plots planted to the variety and the yields of the blocks in which the variety occurs.

The first step is to accumulate the variety sums which are recorded in table 2, column 2. The yields for each variety are collected from table 1. For example, the total yield of variety 1 is:

$$S'_{V} = 1.64 + 1.27 + 1.94 + 1.00 + 1.84 + 1.22 = 8.91$$

For each variety total there is also a sum of block total ($S'_{V}S'_{B}x$) which is recorded in table 2. Since variety 1 appears in blocks 1, 6, 11, 16, 21, and 26, $S'_{V}S'_{B}x$ = 7.54 + 8.24 + 9.52 + 6.80 + 8.66 + 7.30 = 48.06

Table 2. Computation of Variety Means for the Crested Wheatgrass Experiment with 25 "varieties" in 6 Replications.

Variety	Replicate I	II	III	IV	V	VI	Variety Totals S'_{V}	Block tots. for each $S'_{V}S'_{B}x$	$n'S_{V}x - S'_{V}S'_{B}x$	Q/25	Variety Means
	Yields in Kg.						$S'_{V}x$	$S'_{V}S'_{B}x$	Q	d	Sx/N + d
1	1.64	1.27	1.94	1.00	1.84	1.22	8.91	48.06	-3.51	-0.14	1.31
2	1.52	1.32	1.64	1.83	1.48	1.62	9.41	46.76	+0.29	+0.01	1.46
3	1.83	1.16	1.29	1.60	1.44	1.46	8.78	43.21	+0.69	+0.03	1.48
4	1.30	0.99	1.33	1.30	1.42	1.19	7.53	40.99	-3.34	-0.13	1.32
5	1.25	1.22	1.35	1.44	1.64	1.08	7.98	41.47	-1.57	-0.06	1.39
6	1.46	1.81	1.83	1.44	1.55	1.08	9.17	45.36	+0.49	+0.02	1.47
7	1.48	1.65	1.27	1.31	1.35	1.70	8.76	43.85	-0.05	0.00	1.45
8	1.41	1.24	1.48	1.34	1.72	1.43	8.62	44.50	-1.40	-0.06	1.39
9	1.52	1.42	1.18	1.62	1.48	0.93	8.15	40.14	+0.61	+0.02	1.47
10	1.38	1.02	1.92	1.62	1.86	1.12	8.92	45.19	-0.59	-0.02	1.43
11	1.42	1.96	1.20	1.58	1.29	1.18	8.63	42.35	+0.80	+0.03	1.48
12	1.19	1.92	1.54	1.64	1.40	1.32	9.01	41.39	+3.66	+0.15	1.60
13	1.35	1.47	1.50	1.66	1.74	1.56	9.28	43.30	+3.10	+0.12	1.57
14	1.40	1.16	1.84	1.56	1.32	1.56	8.84	44.81	-0.61	-0.02	1.43
15	1.32	1.34	1.64	1.51	1.48	1.36	8.65	44.34	-1.09	-0.04	1.41
16	1.21	1.54	1.16	1.48	1.50	1.41	8.30	44.01	-2.51	-0.10	1.35
17	1.59	1.36	1.53	1.60	1.68	1.50	9.36	43.76	+2.54	+0.10	1.55
18	1.32	1.32	1.86	1.72	1.67	1.00	8.89	43.24	+1.21	+0.05	1.50
19	1.22	1.04	1.78	1.25	1.48	1.60	8.37	42.53	-0.68	-0.03	1.42
20	1.20	1.72	1.84	1.30	1.38	1.50	8.94	41.95	+2.75	+0.11	1.56
21	1.48	1.66	1.24	1.45	1.26	1.44	8.53	44.31	-1.66	-0.07	1.38
22	1.54	1.38	1.96	1.36	1.40	1.32	9.16	45.28	+0.52	+0.02	1.47
23	1.16	1.63	1.66	1.24	1.68	1.10	8.47	42.74	-0.39	-0.02	1.43
24	1.67	0.92	1.29	1.62	1.94	1.31	8.75	42.07	+1.68	+0.07	1.52
25	1.14	1.38	1.54	1.84	1.36	1.22	8.48	43.34	-0.94	-0.04	1.41
Totals							217.79↓	1088.95	0.00	0.00	

↓The sum of the $S'_{V}x$ column (217.79) is equal to Sx, while the sum of the $S'_{V}S'_{B}x$ column is equal to n'Sx. Therefore, the computations can be verified: (5)(217.79) = 1088.95.

For the computation of Q, the block sums are subtracted from 5 times the variety totals, i.e., $Q = n'S'x_{V} - S'S'x_{VB}$

For example, for variety 1,

$Q = 5(8.91) - 48.06 = -3.51$

The Q value is then divided by the number of varieties in the test (25) to give the values for d in table 2. Thus, d is the deviation of a variety mean from the mean yield of all the varieties in the experiment.

The best estimate of the variety means is $Sx/N + d$.

As an illustration, the mean of variety 1 is,

$Sx/N + d = 217.79/150 + (-0.14) = 1.31$

In the variety means, consideration has been given to the effect of partial confounding of variety differences with block effects. They are the best estimates of the yield performance.

(c) Derivation of Sums of Squares

The sums of squares may now be computed. The correction factor is the square of the total divided by the number of plots, viz.,

$$\frac{(Sx)^2}{N} = \frac{(217.79)^2}{150} = \frac{47,432.4841}{150} = 316.22$$

The total sum of squares is obtained in the usual manner, i.e., by the addition of the squares of each individual plot yield with the correction factor subtracted:

$$(1.25)^2 + (1.52)^2 + \ldots + (1.31)^2 - 316.22 = 8.23$$

The sums of squares between means of blocks is obtained by the addition of the squares of the block totals, these being divided by the number of plots which make up each block total. The correction term is subtracted from this value.

$$\frac{(7.54)^2 + (7.25)^2 + \ldots\ldots\ldots\ldots(5.79)^2}{5} - 316.22 = 4.17$$

The sum of squares between means of varieties is obtained from each Q value squared, added, and divided by N:

$$\frac{(-3.51)^2 + (0.29)^2 + \ldots\ldots + (-0.94)^2}{150} = 0.56$$

The analysis of variance is presented in table 3.

Table 3. Analysis of Variance of Symmetrical Incomplete Block Design

Source of Variation	D.F.	Sum of Squares	Mean Square	F-Value
Blocks	29	4.17	0.1438	3.94**
Varieties	24	0.56	0.0233	1.57
Error	96	3.50	0.0365	
Total	149	8.23		

The standard error of the plot yields is

$$s = \sqrt{0.0365} = 0.19 \text{ kilograms}$$

The standard error of the difference between two of the corrected means will be:

$$\sigma_d = \sqrt{\frac{2s^2}{n} \frac{n'+1}{n'}} = \sqrt{\frac{(2)(0.0365)}{6} \frac{6}{5}} = 0.12$$

IV. Efficiency Factor

The symmetrical incomplete block design is less efficient than the complete randomized block arrangement for equal numbers of replications when the soil is homogenous. This is because there has been no reduction in error variance due to the reduction of block size. The efficiency of the incomplete block design as compared to randomized complete blocks is expressed by the fraction, $\frac{1 - 1/n'}{1 - 1/m}$, when the replication numbers in each arrangement are equal. In soils that are heterogenous the reduction in block size usually more than compensates for the loss of information due to the arrangement. Goulden (1937) concluded that an increase of precision of 20 to 50 per cent was obtained over the complete randomized block arrangement.

In addition to the doubtful value of the symmetrical incomplete block design on very uniform soils, Weiss and Cox (1939) advise that the design not be employed to compare varieties which have an extremely large range in yields. However, poor varieties are usually eliminated in preliminary trials. The symmetrical incomplete block arrangement would provide a means to accurately determine relatively small differences between select varieties.

Goulden (1939) gives a list of the n' and n values for different numbers of varieties for which symmetrical incomplete blocks may be used:

No. Varieties	No. Plots in one block (n')	No. Replications for Each Variety (n)
13	4	4
16	4	5
21	5	5
25	5	6
31	6	6
49	7	8
57	8	8
64	8	9
73	9	9

References

1. Fisher, R. A. The Design of Experiments. Oliver and Boyd, 2nd Ed. pp. 100-171. 1937.
2. Fisher, R. A., and Yates, F. Statistical Tables for Biological, Medical, and Agricultural Research. Oliver and Boyd. 1938.
3. Goulden, C. H. Efficiency in Field Trials of Pseudo-Factorial and Incomplete Randomized Block Method. Can. Jour. Res. C, 15:231-241. 1937.
4. ——————— Modern Methods for Testing a Large Number of Varieties. Can. Dept. Agr. Tech. Bul. 9. 1937.

5. Goulden, C. H. Methods of Statistical Analysis. John Wiley, pp. 172-202. 1939.
6. Weiss, M. G., and Cox, G. M. Balanced Incomplete Block and Lattice Square Designs for Testing Yield Differences among Large Numbers of Soybean Varieties. Ia. Agr. Exp. Sta. Res. Bul. 257. 1939.
7. Yates, F. Complex Experiments. Suppl. Jour. Roy. Stat. Soc., 2:181-247. 1935.
8. ________ Incomplete Randomized Blocks. Ann. Eugenics, 7:121-140. 1936.
9. ________ A Further Note on the Arrangement of Variety Trials: Quasi-Latin Squares. Ann. Eugenics, 7:319-332. 1937.
10. ________ The Design and Analysis of Factorial Experiments. Imp. Bur. Soil Sci. Tech. Comm. 35. 1937.

Questions for Discussion

1. Why is an ordinary randomized block design inaccurate for comparisons of a large number of varieties?
2. What principles are involved in incomplete block tests? What is a symmetrical (or balanced) incomplete block?
3. Explain how to write out the sets for a completely orthogonalized 5 by 5 square.
4. What variations in field lay-out are permissable with a symmetrical incomplete block test?
5. How does the computation for variety sums of squares differ from that for an ordinary randomized block?
6. What is the efficiency factor? Compare the efficiency of a symmetrical incomplete block test with that for a randomized block trial.
7. What are the limitations in the use of the symmetrical incomplete block design?
8. How would you arrange a variety test so as to be able to fit 47 varieties into a symmetrical incomplete block test?

Problems

1. It is desired to conduct a symmetrical incomplete block test for 16 varieties of wheat. The form to be used will be $m = n'^2$. A 4 by 4 orthogonalized square is given below. Write out the sets for the different blocks for each replicate.

111	234	342	423
222	143	431	314
333	412	124	241
444	321	213	132

2. Some uniformity trial data on wheat nursery plots were as follows in grams per 15 foot row (Data from Dr. G. A. Wiebe):

695	860	960	725	615
735	910	975	775	680
645	745	815	700	605
630	810	730	635	535
680	745	840	730	645
620	730	775	680	610
620	745	660	565	520
560	675	690	635	525
625	700	725	645	645
700	765	725	615	640
685	785	655	600	570
625	550	590	590	605
745	790	675	600	625
680	670	630	640	645
655	730	615	650	640
625	700	675	720	695

Use the incomplete block sets written for problem 1 and apply the above yields to them. Calculate the data for a symmetrical incomplete block design.

CHAPTER XXI

MECHANICAL PROCEDURE IN FIELD EXPERIMENTATION

I. General Considerations

The experimental farm should be kept neat, clean, and in order at all times. Weeds should be hoed from plots and alleys and all trash destroyed. Alleys and roadways should be hoed or cultivated unless seeded to grass. Straight plot rows add to the general attractiveness and in some cases to accuracy.

(a) Crop Rotation Scheme

Zavitz (1912) states that it is essential to have a rotation plan for the entire experimental farm in order to maintain soil fertility. In addition, accurate maps should be kept for the different fields so that a continuous record exists as to the crops grown on each field for all past years. A rotation scheme prevents mixtures in small grain nurseries as well as on other plots since volunteer grain may contaminate seed plots where the same crop was on the land the previous year. On the Colorado Station farm it has been found advisable to fallow some of the fields to equalize the soil moisture due to the effect of irrigation and for weed control. However, many experiment stations prefer that a bulk crop always precede nursery plots. At the Nebraska station fallow has failed to equalize soil conditions.

(b) Preparation of Land for Experimental Crops

All plots for field trials should receive similar treatment except where the treatment itself is under study. Cultural operations should be at right angles to the direction of the plot rows so far as practicable. Thorne (1909) states that fertilizers should be applied by machinery rather than by hand methods because of the more uniform distribution. A two-way plow is useful in seedbed preparation as a means for the elimination of dead-furrows and back-furrows in the middle of the experimental area. Seeding machinery used in experimental work must be accurate and, for that reason, should be calibrated wherever possible. Many machines are unfit for such work. A drill that fails to drop seed uniformly may cause a serious error in field plot yields. Moreover, it is very desirable to have a drill that can be cleaned out readily. Plot rows should be made straight because crooked rows cause irregularity in plot shape.

A -- Methods for Planting Experimental Crops

II. Seed Preparation

The best seed obtainable should be used in variety trials, i.e., pure as to variety, free from weed seeds and foreign material, high germination, and uniform in size.

(a) Seed Source

Seed from entirely different sources may entirely upset the small differences commonly found in yield trials. All seed used in such trials should have been grown, harvested, and stored under uniform conditions for at least two years, according to Engledow and Yule (1926). This is usually impossible. Under those conditions, Parker (1931) advises "all that can be done is to see that the seed of the several varieties is approximately of equal germination and is equally sound and healthy in other ways." Adapted seed is highly desirable for self-fertilized crops and often even more so in cross fertilized crops like corn.

(b) Other Considerations

Unless disease reaction is under study, seeds of cereals should be treated for control of fungus diseases such as smut. New Improved Ceresan, a dust treatment, may be used at the rate of one-half ounce per bushel for the covered smuts. All seed should be of the same age when possible. It should be weighed out for the particular test on the same scales, especially when planted by weight per unit area. The procedure on many stations is to measure out the seed for both nursery and field plots. For rod-row or nursery trials the seed is placed in coin envelopes and numbered to correspond to the plots. When a drill is used, a little more than enough seed is desirable because the drill itself measures the seed planted.

III. Rate of Seeding

Considerable error may be introduced in some crops through variation in rate of seeding.

(a) Small Grains

In small grains the investigator must either plant equal weights or equal numbers of seeds per unit area. Up until 1910, the "centgener" method was extensively used in small-grain nurseries for the determination of yields. The kernels were space-planted 10 inches apart each way in blocks and contained 100 seeds. Aside from the theoretical objections in genetics, this was an absurd practice from the standpoint of field yields because the seeds were planted approximately 14 times as far apart as ordinarily occur in a drill-planted field. In addition, a great amount of detailed hand labor was required. The method has been discarded in this country in favor of the rod-row. (1) Rod-Row Trials: The general procedure in rod rows is to measure the seed per row. Kiesselbach (1928) summarizes the situation very well. Fortunately, he states, there may be considerable variation in the rate of seeding without material effect on the yield per acre. For instance, Turkey wheat planted at 3,4,5,6, and 8 pecks per acre at Nebraska yielded 22.2, 23.6, 23.7, 24.4, and 24.5 bushels per acre, respectively, for 9 years. Seed of average size, or screened seed, should be used for machine planters. Measurement of the seed gives results more comparable with field conditions than where individual seeds are space planted as in the centgener method. Seed for hand-planting should be weighed. (2) The "Checkerboard" Trial: The English workers use the "checkerboard" to some extent in their variety trials. It is essentially a modified centgener plan in which the seeds are spaced 2 x 6 inches apart. They admit it differs from field conditions and, for this reason, use larger "observation" plots to supplement the checkerboard trials. The checkerboard is precise but requires too much time and labor where many varieties are under test.

(b) Other Crops

Corn is generally planted by farmers in rows 3.0 to 3.5 feet apart. The usual rate is three plants per hill for checked corn or with single plants 14 inches apart when drilled in the row. Under dryland conditions, the plants are usually drilled 20 to 30 inches between plants in the row. This is the practice in experimental work except that the seed is often planted at double the required rate, later thinning the plants to the desired stand. Without this precaution, Kiesselbach (1928) points out, competition between adjacent rows that differ materially in stand may lead to faulty results. In sugar beets the seed is generally planted very thick. They are later thinned to the desired interval between plants, usually 12 inches. Sugar beets are ordinarily planted in rows 20 inches apart.

IV. Methods to Plant Field Plots

The ordinary grain drill is often used to plant field plots of small grain and forage crops.

(a) Calibration of Grain Drills

The necessity for drill calibration was shown by the work of Bonnett and Burkart (1923). The drill may be jacked up, the seed rate set as desired, and the wheels turned 50 revolutions at the rate they would turn over in the field. The amount of grain collected for each drill should be weighed. A mark should be made on the wheel to facilitate the count. It is only a matter of arithmetic to calculate the rate that the seed will be planted.

(b) Use of the Drill

For small grain and forage crops the different replications of the same variety should be planted before the seed is changed. The plots may be staked out in advance to facilitate this procedure. The drill should be thoroughly cleaned out between varieties, possibly by the aid of an air bellows to dislodge seed in the corners of the drill box. Some drills are made over so that the seed box can be tipped forward on hinges to empty. In some experiments where two kinds of seed are planted in a plot, one crop may be drilled in one direction and the other at right angles to it, e.g., nurse crop studies in alfalfa.

V. Methods to Plant Small Grain Nursery Rows

Small grain nursery plots involve hand methods after the seedbed has been prepared. Rod rows 12 inches apart are generally used. At some experiment stations 18-foot rows are planted, being trimmed down at harvest time to 16 feet for wheat, 20 feet for barley, and 15 feet for oats. This enables the investigator to convert the yields in grams per plot into bushels per acre by the use of a simple factor. The rod rows may be made by the use of a sled marker with the runners spaced at the proper intervals, the ideal type being horse drawn. The rows are then opened with a wheel-hoe for hand planting. Another method is to use a sugar beet cultivator with bull tongs spaced at the proper intervals. This has proved to be very satisfactory at the Colorado station. A 12-inch furrow drill is used to mark out the rows on the Akron Station. The seed, previously weighed out, is sometimes hand-planted (scattered) in the row. A Columbia or planet Jr. planter is used in many cases to plant wheat. Modification of the Columbia drill for planting oats and barley has been suggested by Woodward and Tingey (1933) as well as by Jodon (1932). A rapid method for planting is by use of the spout drill. This is very satisfactory for genetic material where yield is not a factor. The grain is poured through the spout, all seed in the packet being planted in the row length. After a little experience the seed can be planted very uniformly. One man pushes the drill while another drops the seed. The spout-drill may be used for space-planting after a little practice. One station that uses 5-row plots for nursery studies has a horse-drawn planter. A convenient method for space-planting small grains at definite intervals, for example two inches, is to take a 6-inch board and bore holes at the proper intervals. The seeds are dibbled in these holes.

VI. Methods to Plant Row Crops

Corn will be taken as an example of a row crop. Generally a horse or hand-drawn marker is used to mark the distances between rows. When the corn is to be check-planted in hills the plots are cross-marked to give a set of squares, the intersections designating the hill locations. Hills are generally spaced 3.0 or 3.5 feet apart in all directions. Suitable alleys should be left between blocks to facilitate cultural and harvest operations. The stakes are distributed along one end of the plots. The seed sacks or envelopes, with the variety number on them, are distributed to correspond with the stakes. The numbers should be checked against the planting plan to avoid mistakes. The seed sacks may be re-distributed for each replicate. Corn is generally planted with a hand planter in yield trials. One of the most satisfactory planters is a made-over potato planter.* It is constructed to have a long, full-length tin sleeve into which the proper number of kernels is dropped into the shoe. A nail sack is convenient for carrying the seed. For planting six kernels per hill, in order to thin to three plants later, it is convenient to plant three kernels each in two jabs about one inch apart. This facilitates thinning.

B -- Field Observations and Care

VII. Value of Field Observations

Intimate knowledge of experimental plots is extremely desirable. In fact, observations during the growing period of crop may be as valuable as the yield data. Differences due to disease, irregular loss of plants, etc., may account for the variation in yield. Plot observations should be made at regular intervals. Notes should be entered in the field book at once while clear and vivid in the mind. Word descriptions should be clear and precise, being made as comparisons in terms of the check when possible. Sometimes sketches, diagrams or photographs are a better method of expression than word descriptions.

VIII. Measurement of Plant Characters

Field counts or measurements on certain plant characters given in numbers or categories make excellent comparative field records. Formal "score cards" are apt to make observations perfunctory. Hence, records should depend upon the particular crop and the needs that may arise. Some of the more important characteristics usually recorded are as follows: date emerged, stand, winter survival, date ripe, plant height, lodging, barren stalks (in corn), disease infection, etc. Some of these may be taken in quantitative measures while categories are required in other cases. When actual counts are out of the question, a scale of marks may be employed to convey the relative intensity of attack of a disease or insect pest. The numbers 1,2,3, and 4 may be used to represent, respectively, a slight, moderate, bad, or very severe attack of rust, mildew, etc. A scale of 1 to 4 is generally adequate for categorical data. Further sub-division merely leads to confusion. A very good rust scale is available in the agronomic field book used by the Division of Cereal Crops and Diseases, U.S.D.A. Yates (1934) reports a bias between different observers when a large number of counts were made on wheat culms. The bias differed from observer to observer and from sample to sample. The same individual should make all counts or at least all counts on a single replication in order to avoid this form of systematic error.

IX. Stand Counts and Estimates

In certain crops stand counts are valuable, but this depends largely upon the experiment. In forage experiments the counts are often made by the use of square yard or meter quadrats. These may be permanent quadrats in perennial crop studies. In the case of winter or spring survival counts in winter wheat, the stand percentage is usually estimated except in special tests. One person should make all the estimates due to the large personal error invariably introduced when more than one person makes them. Estimation in categories such as good, fair, and poor stands may be satisfactory. In plant-survival studies, as in winter wheat, a more precise method would be to space plant the seed in rod rows at 2-inch intervals. However spaced plants have been observed to kill worse than seeded material. Such tests are valueless for yield.

---

*Note: The type used at the Nebraska, Minnesota, and Colorado stations is the "Acme Segment" potato planter manufactured by the Potato Implement Co., Traverse City, Mich. It can be slightly modified to make an excellent planter.

X. Date Headed

There is considerable variation among workers as to the date when a crop should be considered in head. In small grains, date in head is usually a more reliable index of earliness or lateness than date ripe. This is particularly true under dryland conditions where winds may prematurely dry up a variety rather than to allow it to ripen normally. In wheat, oats, and barley, some investigators take notes on first heading, i.e., when 10 per cent of the heads are out of the boot. A plot is considered fully headed out by some workers when 75 per cent of the plants in the plot are in full head. Others use a standard as follows: (1) Oats, when the heads are half out of the boot; (2) Barley, when the beards are out of the boot; and (3) Wheat, when the heads show out of the boot. Date in silk or date in tassel are common notes in corn, date of silking being regarded as a more reliable index of relative maturity than date of tasseling. It is usual to determine the silking date and convert the data to the number of days from planting to one-half silking. The plots should be gone over at intervals of one or two days when date in head and similar notes are taken because some dates may have to be moved up and others back.

XI. Per cent Lodged

Data on the differential lodging of small grains is desirable as a measure of stiffness of straw. Sometimes after heavy rains or irrigations the soil may be loosened so that the entire plant falls over. This is not true lodging. A plant has an inherently weak straw when it bends or breaks over. It is often difficult to arrive at inherent differences because of soil heterogeneity and its influences. A variety should be considered lodged when the straw leans an angle of 45 degrees or more because, for practical purposes, grain lodged to such an extent is difficult to harvest. The per cent of grain so lodged is usually estimated regardless of the cause. However, Straw weakness can be detected before the plants lean to an angle of 45 degrees. Some investigators make notations as to whether the straw is apparently weak, medium, or strong, and denote the condition categorically by W, M, or S. Under irrigated conditions, small grains may be irrigated heavily after heading to induce lodging. In corn, the relative resistance to lodging is often reported as the percentage of plants erect at harvest. The percentages may be computed from counts of the numbers of plants erect. In the interest of uniformity a plant should be considered erect when it has not leaned more than 30 degrees from the vertical and which does not have the stalk broken below the ear. For those who wish to take more detailed records on lodged plants, it is suggested that such plants be separated into those lodged because of weak roots (leaning and down plants), and those lodged because of weak culms (plants broken below the ear).

XII. Plant Height

Two men are required to take plant height notes readily, one to make the measurements and the other to record the results. In the case of small grains such measurements are generally made just before harvest. Sometimes one measurement is taken per plot while, at other times, several plants are measured at random. One measurement per plot is enough when the heights are uniform. A convenient rule is a 1 x 1-inch stick marked at one-inch intervals to 60 inches. Height notes in corn are often taken in the fall, but can be taken almost any time after the plants have tasselled out fully. It can be accomplished with an ordinary rule about 12 feet in length, 2.5 inches wide, and marked at 3-inch intervals.

XIII. Roguing Plots

Small grain plots should be thoroughly rogued for admixtures before harvest. The plots should be gone over several times, particularly when the plants begin to head or ripen. Rogues are most conspicuous at such times. It is difficult to rogue barley out of oats because the oat plants are generally taller than barley plants. Careful work is required to rogue off-varieties and off-types within a crop. These can be detected most readily by observed differences in culm height, date of heading, color of leaves, date ripe, and whether or not awns are present. It is a safe rule to pull all plants that fail to conform to the majority of the plants in a plot.

XIV. Date Ripe

The date on which a crop ripens is important, particularly in small grains where earliness is often a desirable feature. Some of the criteria used are given below.

(a) Wheat

The grain may be considered ripe when it is hard in the morning. The straw color is not always a reliable criterion of ripeness. Those who use straw color as a criterion generally consider the grain ripe when the first nodes below the heads on the main culms have turned brown.

(b) Other Crops

In oats the plot is usually considered ripe after practically all of the heads have turned yellow. The barley crop is generally considered ripe when all green has disappeared from the heads. It is difficult to estimate date ripe on small grain that is badly rusted or lodged as it tends to ripen unevenly and often prematurely in the case of rust. In corn, date in silk is usually regarded as a more reliable index of maturity than ripening data in the fall.

C -- Methods of Harvesting Experimental Crops

XV. Difficulties in Harvesting

The time of harvesting crops often presents difficulties. Parker (1931) mentions that one might question the fairness when an early small grain variety is compared with a check variety that may ripen 10 days or more later. As a rule, plots are harvested as the varieties ripen, particularly, where there are wide differences in time of ripening. In some parts of the country, the investigator may be able to wait until the latest varieties are ripe so that the entire field may be harvested at once. Except for extreme differences in time of ripening, it is usually possible to allow the early varieties to stand without particular damage to them. It may be desirable in some instances to carry out two separate trials, grouping the early varieties in one and the late ones in the other. In the case of root and tuber crops, all varieties may be left in the ground and harvested at the same time without serious consequences. The problem in corn is rather simple because all varieties are left in the field after becoming ripe so as to dry out. In forage experiments, inclement weather may interfere with the curing process and require that the hay be turned several times. As a result, it may dry out unevenly or the leaves shatter. A possible error in weight might result.

XVI. Methods of Harvesting Field Plots

The use of farm machinery is often anticipated for large field plots.

(a) Small Grain Plots

Small grain field plots should be gone over carefully before harvest to be certain that there are no errors due to defective drilling, rodent, or other injury that might influence the yields. When small grains are badly lodged, it may be necessary to separate the varieties along the margins and push them over into their respective plots before harvest. Kiesselbach (1928) uses a binder equipped with an engine that operates the working parts. At the end of the plot, the horses are stopped but the engine continues to operate and clean out the binder. In the absence of the engine it is necessary to crank the platform and elevator canvasses by hand. The small grain shocks should be placed well within the plot to prevent chance mixtures with adjacent plots should wind scatter some of the bundles. At some stations the bundles are shocked on alternate ends of the plots. The shocks may be tied with binder twine to minimize the risk. When birds are numerous, shock covers should be provided. They can be made by sewing together ordinary burlap feed bags.

(b) Corn Yield Tests

The entire plot can be harvested without appreciable error when the plant stand is 90 per cent or better. Otherwise, it is advisable to reject at harvest all hills with less than the normal stand, and calculate yields on a perfect-stand basis. Usually the imperfect-stand hills are cut with a corn knife and removed from the plot. A record is then made of the number of perfect-stand hills that remain. Sometimes counts on barren stalks, 2-eared stalks, suckers, smutted plants, and lodged plants are made at this time. For small yield trials, actual harvesting can be done conveniently with an apple-picking bag. For large field plots, Kiesselbach (1928) uses a wagon with a flat rack with partitions built on it. A partition may be placed lengthwise through the center and each divided, for instance, into three partitions where three center rows are harvested for yield (as in 5-row plots with border rows discarded). This allows a separate compartment for each row. Three men can husk, one man being on each yield row. The compartments on the other side can be used for the next plot on the return. At the end of the field, the corn from each plot is sacked and tagged. Field weights of ear corn are sometimes taken. The corn sacks may then be either piled in small piles in a shed until air dry, or they may be tied up on wires in a drying shed (Colorado method). The latter seems to allow the corn to dry out more evenly and more quickly. Some stations now have elaborate drying equipment where the entire plot yield can be dried to a moisture-free basis in a relatively short time.

(c) Forage Experiments

Forage plots for hay are almost always cut with a mower when 1/40-acre in size or larger. The plots may be trimmed evenly on the ends before the regular cutting time. The material is then raked and removed. Borders between plots are generally disregarded for large field plots. It is an advantage to be able to start on one side of the field and mow through all the series, thus lessening the number of turns. A man should follow the mower with a fork to be sure that hay is not carried through the alley from one plot to the next. After the hay has been dried sufficiently, a side-delivery rake may be used to put it in windows, after which it may be bunched by a dump-rake or by hand. A convenient method to handle the hay from each plot is to put it on a wagon or truck on which a sling has been placed. The load is then weighed, the net weight determined, and the hay unloaded from the truck by a cable stacker. A small composite sample may be taken to dry to an air-dry basis, or it may be ground for an immediate moisture determination. For plots 1/40-acre in size or smaller, hay may be weighed conveniently by a portable platform scales on which a rack is set. For plots away from the central experiment station, a tripod and spring balance affords a good method to weigh forage plots. A large piece of canvas is equipped with snaps so that, when the hay is put on it, the sides can be gathered in and snapped to a ring. It is then readily hung to the scale.

(d) Sugar Beet Trials

In sugar beet yield trials, 4-row plots are generally used with the two center rows harvested for yield. Except in studies on stand, and certain other instances mentioned previously, the plots are commonly harvested on the basis of competitive beets, i.e., plants surrounded by plants on all sides. The tops of the other beets (non-competitives) may be chopped off with a hoe before harvest. The roots are then pulled with a standard beet puller. It is common practice to pull one replication at a time. The roots without tops are usually weighed in order to have this component for total plot yield in case this seems to be needed later. The non-competitives are then discarded. Two 20-root samples may be taken from the non-competitive beets as a sugar sample. The competitive beets are then pulled, topped, and weighed for each plot. The tare is then subtracted from the field weight of the roots. When a washer is not available, the tare may be taken in the field. The sample for tare is first weighed, the roots cleaned with steel brushes, and re-weighed. The difference in weight is the tare. It is believed desirable to calculate the tare for each plot separately.

XVII. Harvesting Small Grain Nursery Plots

Competent and continual supervision is necessary in the small grain nursery at harvest time. Some investigators clean-cultivate the alleys between series. Under such conditions the rod rows are generally trimmed down to remove border effect. In wheat, for instance, the crop is planted in 18-foot rows, one foot being trimmed from each end of the plot. A string may be stretched across the series at both ends to designate the discard area to be cut, or a 16-foot bamboo pole may be used on each center row (in 3-row plots) so that the wheat may be cut on both ends of the pole. Other investigators plant the alleys to some readily distinguishable variety, thus eliminating the border effect on the ends of the rod rows. The alleys are then removed before harvest. Hand sickles are used to cut nursery plots. The smooth-edged sickle is most widely used, but a saw-toothed sickle is satisfactory when new. Where straw yields are taken, grass shears may be used to assure an even cut. Kemp (1935) has constructed a rod-row harvester of the rotary shear type with which 2 men may cut 1500 rows per day. The harvested bundles are tied with binder twine, usually in one place. Strings should be tied with a simple, secure knot. The plot stake may be tied into the bundle or a tag attached to the string with the plot number on it. The bundles may be tied on a table. Men who tie bundles should tape their fingers. Seed plots are often sacked with large paper sacks tied over the heads to prevent mixtures. By the aid of a large funnel, 25 pound manila bags are easily placed over the heads. Sacked bundles should be put under cover as soon as possible to protect them from rain. Small grain bundles may be either shocked in the field until they are ready to thresh, or hauled to a shed and hung up to dry. A drying shed may have wires about four feet apart stretched from one end to the other at sufficient height so that the bundles can be tied to the wire with heads down. The bundles should be hung fairly wide apart when they are harvested a little green. This is particularly true for oats.

XVIII. Harvest of Corn Breeding Material

Inbred and hybrid strains of corn, which are the result of hand pollination, are usually harvested after maturity. Individual ears may be collected in the bag over the ear shoot, and all sacks from the same row tied together with binder twine, the tying being done with an ordinary sack needle. These sacks are then hung to wires in the drying shed and allowed to remain there until air-dry. This method has proved very satisfactory at the Colorado station.

D -- Threshing and Storage

XIX. Methods of Threshing Field and Increase Plots

Small grain field and increase plots are commonly threshed with the standard grain separator. Kiesselbach (1928) has found it necessary to make miner modifications to adapt them for this purpose. He lists these changes as follows: (1) Removal of the grain elevator; (2) Elimination of the self-feeder; (3) providing a hinged door at the foot of the tailings elevator for cleaning out between plots; (4) replacing the grain auger with a shaker-trough device; (5) removal of the grain-saving auger in the blower, where one exists; (6) equipment with a high-pressure air pump and tank to supply air pressure through a hose to dislodge grains when the machine is cleaned out between plots; (7) cutting several holes, with covers, into the sides of the separator at convenient places to observe the interior and to introduce air pressure to clean out the separator. Such modifications make it easier to clean out the machine between plots, thus reducing the chances for mixtures. The chances for mixture may be reduced further by threshing all plots of the same variety in succession. Seed can be saved from the last plot of the variety to be threshed. It is important to operate the machine uniformly throughout each experiment. The grain per plot is often weighed on a platform scale at the separator.

XX. Threshing Nursery Plots

Small grains in yield trials are generally threshed with small nursery threshers, while genetic material is usually threshed by hand.

(a) Nursery Threshers

Several machines that can be cleaned readily have been devised to thresh small nursery plots. According to Hayes and Garber (1927) "the chief requisites of a machine to be used for experimental purposes are as follows: It should be easily cleanable and, in so far as possible, there should be no ledges or ridges upon which seeds may lodge. The alternate threshing of different nursery crops is a desirable procedure. Each of the plots of one strain of wheat may be threshed separately in rotation and then a strain of oats may be threshed in the same way. At the Minnesota Experiment Station winter wheat is threshed alternately with barley, and spring wheat with oats. This plan helps materially to reduce the roguing of accidental mixtures from the plots." The Cornell machine designed by H. W. Teeter is very satisfactory for multiple-row plots, while the Kansas machine is widely used for rod rows. The Cornell machine has a shaker, screen, and fan. Its most serious drawback is the difficulty in cleaning it between varieties. However, it can be cleaned more readily than the Kansas machine. Recently, Vogel and Johnson (1934) have developed a new type of rod-row thresher which is a combination of an overshot cylinder and modified screenless shaker and fan of an ordinary fanning mill. The grain is further cleaned by a separate re-cleaner. It has been found satisfactory for small grains, peas, flax, and some grasses. Grain weights are taken after threshing, usually in grams for rod-row plots.

(b) Hand Threshing

In genetic material where it is desired to thresh single plants, threshing is usually done by hand. A threshing board three feet square is useful for this purpose. The frame can be made of 1 x 2-inch material over which a canvas is stretched tightly. Two blocks, about 4 x 6 inches in size, are then made and covered on both sides with corrugated rubber. These work very well for threshing wheat and other naked grains. For barley, it has been found at the Colorado station that the heads thresh out better when rolled up in a small canvas cloth (about 9 inches square) and rubbed. A piece of tin bent to form a fan can be used to blow the chaff out of the grain by striking it on the canvas. Coffman (1935) was able to thresh 100 to 150 single oat panicles per hour by the use of a light weight close-fitting leather glove on the right hand. The spikelets are stripped into a grain pan where the chaff is easily blown out.

XXI. Methods for Shelling Corn

After corn has reached an air-dry condition it is ready to shell for final determinations. Genetic material is usually shelled by hand, altho some workers use an enclosed single ear sheller. An ordinary corn sheller is very satisfactory for yield trials. It should be enclosed so that the kernels are not scattered when the ears are shelled. The air-dry weight of ear corn should first be taken for the corn from each plot. A platform scale is often used for such weights. It should be balanced frequently to keep it in adjustment. The corn is then shelled, the cobs being looked over minutely to be sure that all kernels have been recovered. The shelled corn is then weighed and recorded. A 500-gram shrinkage sample is taken at the Nebraska station and oven-dried to a constant weight. The yield of moisture-free corn is calculated from the percentage of oven-dry corn in the shrinkage sample. At the Colorado station, bushel weight is taken with the standard bushel weight tester, since bushel weight has been found to be an index of maturity. Moisture determinations are made with the Tag-Heppenstall moisture meter, one sample per plot yield.

XXII. Storage of Seed of Experimental Crops

There are probably as many methods for seed storage of experimental crops as there are experiment stations. The first requisite is a place safe from mice and insects. Cabinets with metal drawers probably afford the best storage. It is usually necessary to fumigate once or twice per year where grain weevils and other insects are troublesome. For small seed lots a crystalline compound known as "Antimot" will effectively control insects. Small grain seed is usually kept in cloth bags, especially seed saved from rod-row tests. Genetic material is commonly stored in coin envelopes. Seed corn for variety or yield tests may be stored in large bins. Genetic and breeding material may be kept either in cloth bags or in envelopes. At the Nebraska station inbred and hybrid seed corn supplies are kept in large clip envelopes (6 x 9 inches in size). These are filed in drawers in serial order. A similar plan is followed at Minnesota.

References

1. Bonnett, R. K., and Burkett, F. L. Rate of Seeding -- A factor in Variety Tests. Jour. Am. Soc. Agron., 15:161-171. 1923.
2. Coffman, F. A. A Simple Method of Threshing Single Oat Panicles. Jour. Am. Soc. Agron., 27:498. 1935.
3. Engledow, F. L., and Yule, G. U. The Principles and Practices of Yield Trials. Empire Cotton Growing Corp. 1926.
4. Hayes, H. K., and Garber, R. J. Breeding Crop Plants, McGraw-Hill, pp. 152-160. 1927.
5. Jodon, N. E. Modifications in the Columbia Drill for Seeding Oats and Barley. Jour. Am. Soc. Agron., 24:328. 1932.
6. Kemp, H. J. Mechanical Aids to Crop Experiments. Sci. Agr., 15:488-506. 1935.
7. Kiesselbach, T. A. The Mechanical Procedure of Field Experimentation. Jour. Am. Soc. Agron., 20:433-442. 1928.
8. Love, H. H., and Craig, W. T. Methods used and Results Obtained in Cereal Investigations at the Cornell Station. Jour. Am. Soc. Agron., 10:145-157. 1918.
9. Parker, W. H. The Methods Employed in Variety Tests by the National Institute of Agricultural Botany. Jour. Natl. Inst. Agr., Bot., Vol. 3, No. 1. 1931.
10. Standards for the Conduct and Interpretation of Field and Lysimeter Experiments. Jour. Am. Soc. Agron., 25:803-828. 1933.
11. Thorne, C. E. Essentials of Successful Field Experimentation. Ohio Agr. Exp. Sta. Cir. 96. 1909.

12. Vogel, O. A., and Johnson, A. J. A New Type of Nursery Thresher. Jour. Am. Soc. Agron., 26:629-630. 1934.
13. Wishart, J., and Sanders, H. G. Principles and Practice of Field Experimentation. Emp. Cotton Growing Corp. pp. 70-75, and 85-100. 1935.
14. Woodward, R. W., and Tingey, D. C. Improved Modification in the Columbia Drill. Jour. Am. Soc. Agron., 25:231. 1933.
15. Yates, F., and Watson, D. J. Observer's Bias in Sampling Observations on Wheat. Emp. Jour. Exp. Agr., 2:174-177. 1934.
16. Zavitz, C. A. Care and Management of Land used for Experiments with Farm Crops. Proc. Am. Soc. Agron., 4:122-126. 1912.

Questions for Discussion

1. What precautions are necessary in a crop rotation scheme for experimental crops?
2. How should the seedbed be prepared for experimental crops?
3. Under what conditions should experimental seeds be treated for disease?
4. What is the centgener method? Checkerboard method? Rod-row method?
5. How is corn generally planted for experimental purposes? Sugar beets?
6. How would you calibrate a drill?
7. Explain how you would lay-out, mark, and plant a wheat nursery. Give all dimensions and processes.
8. Why are field observations important? What plant measurements and notes are generally taken on small grains?
9. What different methods can be used for making stand counts?
10. At what time would you consider wheat, oats, and barley in head? Ripe?
11. How would you take lodging notes in small grains? Corn?
12. What precautions or advice should be given to your assistants when roguing plots?
13. How would you harvest small grains in a test where the varieties differed widely in date of ripening? Why?
14. How are large field plots of small grains generally harvested? Corn yield tests?
15. Give the detailed steps for harvesting sugar beet plots for yield.
16. Describe a method for harvesting forage yield tests.
17. Explain in detail how you would harvest small grain nursery plots.
18. What modifications on an ordinary grain separator are necessary to adapt it for threshing field plots to prevent mixtures?
19. What are the requisites for a small grain nursery thresher?
20. How would you hand-thresh barley heads? Wheat heads? Oat panicles?

Problems

1. It is desired to plant wheat in rod row trials at the rate of 90 lbs. per acre, the rate used by farmers in the vicinity. The nursery rows are 18 feet long and 12 inches apart. Calculate the amount of seed to weigh out in grams for each row.

2. Suppose the yield from a 16-foot rod row of wheat is 205 grams. Calculate the yield per acre.

3. The weight of shelled corn harvested is 25 lbs, on a plot 20 hills long. (a) When the hills are 36 x 36 inches, calculate the yield per acre for air dry shelled corn. (b) Calculate the yields per acre on the basis of corn with 15.5 per cent moisture when the original shelled corn contained 13.2 per cent moisture.

4. Make up a table of factors for the conversion of pounds shelled corn per plot to bushels of shelled corn per acre when 10 to 20 hills are harvested. Suppose the hills to be spaced 36 x 36 inches.

FIELD PLOT TECHNIQUE

APPENDIX

Table 1. - Area Under the Normal Curve ↓

t	A	t	A	t	A	t	A
.00	.50000	.40	.65542	.80	.78815-	1.20	.88493
.01	.50399	.41	.65910	.81	.79103	1.21	.88686
.02	.50798	.42	.66276	.82	.79389	1.22	.88877
.03	.51197	.43	.66640	.83	.79673	1.23	.89065+
.04	.51595+	.44	.67003	.84	.79955-	1.24	.89251
.05	.51994	.45	.67365-	.85	.80234	1.25	.89435-
.06	.52392	.46	.67724	.86	.80511	1.26	.89617
.07	.52790	.47	.68082	.87	.80785+	1.27	.89796
.08	.53188	.48	.68439	.88	.81057	1.28	.89973
.09	.53586	.49	.68793	.89	.81327	1.29	.90148
.10	.53983	.50	.69146	.90	.81594	1.30	.90320
.11	.54380	.51	.69497	.91	.81859	1.31	.90490
.12	.54776	.52	.69847	.92	.82121	1.32	.90658
.13	.55172	.53	.70194	.93	.82381	1.33	.90824
.14	.55567	.54	.70540	.94	.82639	1.34	.90988
.15	.55962	.55	.70884	.95	.82894	1.35	.91149
.16	.56356	.56	.71226	.96	.83147	1.36	.91309
.17	.56750	.57	.71566	.97	.83398	1.37	.91466
.18	.57142	.58	.71904	.98	.83646	1.38	.91621
.19	.57535-	.59	.72241	.99	.83891	1.39	.91774
.20	.57926	.60	.72575	1.00	.84135-	1.40	.91924
.21	.58317	.61	.72907	1.01	.84375+	1.41	.92073
.22	.58706	.62	.73237	1.02	.84614	1.42	.92220
.23	.59095+	.63	.73565+	1.03	.84850	1.43	.92364
.24	.59484	.64	.73891	1.04	.35083	1.44	.92507
.25	.59871	.65	.74215+	1.05	.85314	1.45	.92647
.26	.60257	.66	.74537	1.06	.85543	1.46	.92786
.27	.60642	.67	.74857	1.07	.85769	1.47	.92922
.28	.61026	.68	.75175-	1.08	.85993	1.48	.93056
.29	.61409	.69	.75490	1.09	.86214	1.49	.93189
.30	.61791	.70	.75804	1.10	.86433	1.50	.93319
.31	.62172	.71	.76115-	1.11	.86650	1.51	.93448
.32	.62552	.72	.76424	1.12	.86864	1.52	.93575+
.33	.62930	.73	.76731	1.13	.87076	1.53	.93699
.34	.63307	.74	.77035+	1.14	.87286	1.54	.93822
.35	.63683	.75	.77337	1.15	.87493	1.55	.93943
.36	.64058	.76	.77637	1.16	.87698	1.56	.94062
.37	.64431	.77	.77935+	1.17	.87900	1.57	.94179
.38	.64803	.78	.78231	1.18	.88100	1.58	.94295-
.39	.65173	.79	.78524	1.19	.88298	1.59	.94408

Table 1. - Area Under the Normal Curve ↓ (Cont.)

t	A	t	A	t	A	t	A
1.60	.94520	2.00	.97725+	2.40	.99180	2.80	.99745-
1.61	.94630	2.01	.97778	2.41	.99202	2.81	.99752
1.62	.94738	2.02	.97831	2.42	.99224	2.82	.99760
1.63	.94845-	2.03	.97882	2.43	.99245+	2.83	.99767
1.64	.94950	2.04	.97933	2.44	.99266	2.84	.99774
1.65	.95053	2.05	.97982	2.45	.99286	2.85	.99781
1.66	.95154	2.06	.98030	2.46	.99305+	2.86	.99788
1.67	.95254	2.07	.98077	2.47	.99324	2.87	.99795-
1.68	.95352	2.08	.98124	2.48	.99343	2.88	.99801
1.69	.95449	2.09	.98169	2.49	.99361	2.89	.99807
1.70	.95544	2.10	.98214	2.50	.99379	2.90	.99813
1.71	.95637	2.11	.98257	2.51	.99396	2.91	.99819
1.72	.95728	2.12	.98300	2.52	.99413	2.92	.99825+
1.73	.95819	2.13	.98341	2.53	.99430	2.93	.99831
1.74	.95907	2.14	.98382	2.54	.99446	2.94	.99836
1.75	.95994	2.15	.98422	2.55	.99461	2.95	.99841
1.76	.96080	2.16	.98461	2.56	.99477	2.96	.99846
1.77	.96164	2.17	.98500	2.57	.99492	2.97	.99851
1.78	.96246	2.18	.98537	2.58	.99506	2.98	.99856
1.79	.96327	2.19	.98574	2.59	.99520	2.99	.99861
1.80	.96407	2.20	.98610	2.60	.99534	3.00	.99865-
1.81	.96485+	2.21	.98645-	2.61	.99547	3.01	.99869
1.82	.96562	2.22	.98679	2.62	.99560	3.02	.99874
1.83	.96638	2.23	.98713	2.63	.99573	3.03	.99878
1.84	.96712	2.24	.98746	2.64	.99586	3.04	.99882
1.85	.96784	2.25	.98778	2.65	.99598	3.05	.99886
1.86	.96856	2.26	.98809	2.66	.99609	3.06	.99889
1.87	.96926	2.27	.98840	2.67	.99621	3.07	.99893
1.88	.96995-	2.28	.98870	2.68	.99632	3.08	.99897
1.89	.97062	2.29	.98899	2.69	.99643	3.09	.99900
1.90	.97128	2.30	.98928	2.70	.99653	3.10	.99903
1.91	.97193	2.31	.98956	2.71	.99664	3.11	.99907
1.92	.97257	2.32	.98983	2.72	.99674	3.12	.99910
1.93	.97320	2.33	.99010	2.73	.99683	3.13	.99913
1.94	.97381	2.34	.99036	2.74	.99693	3.14	.99916
1.95	.97441	2.35	.99061	2.75	.99702	3.15	.99918
1.96	.97500	2.36	.99086	2.76	.99711	3.16	.99921
1.97	.97558	2.37	.99111	2.77	.99720	3.17	.99924
1.98	.97615-	2.38	.99134	2.78	.99728	3.18	.99926
1.99	.97671	2.39	.99158	2.79	.99737	3.19	.99929

Table 1. - Area Under the Normal Curve[1] (Cont.)

t	A	t	A	t	A	t	A
3.20	.99931	3.40	.99966	3.60	.99984	3.80	.99993
3.21	.99934	3.41	.99967	3.61	.99985-	3.81	.99993
3.22	.99936	3.42	.99969	3.62	.99985+	3.82	.99993
3.23	.99938	3.43	.99970	3.63	.99986	3.83	.99994
3.24	.99940	3.44	.99971	3.64	.99986	3.84	.99994
3.25	.99942	3.45	.99972	3.65	.99987	3.85	.99994
3.26	.99944	3.46	.99973	3.66	.99987	3.86	.99994
3.27	.99946	3.47	.99974	3.67	.99988	3.87	.99995-
3.28	.99948	3.48	.99975-	3.68	.99988	3.88	.99995-
3.29	.99950	3.49	.99976	3.69	.99989	3.89	.99995+
3.30	.99952	3.50	.99977	3.70	.99989	3.90	.99995+
3.31	.99953	3.51	.99978	3.71	.99990	3.91	.99995+
3.32	.99955+	3.52	.99978	3.72	.99990	3.92	.99996
3.33	.99957	3.53	.99979	3.73	.99990	3.93	.99996
3.34	.99958	3.54	.99980	3.74	.99991	3.94	.99996
3.35	.99960	3.55	.99981	3.75	.99991	3.95	.99996
3.36	.99961	3.56	.99982	3.76	.99992	3.96	.99996
3.37	.99962	3.57	.99982	3.77	.99992	3.97	.99996
3.38	.99964	3.58	.99983	3.78	.99992	3.98	.99997
3.39	.99965+	3.59	.99984	3.79	.99993	3.99	.99997
						4.00	.99997
						4.50	.99999

[1] Table I was taken from "Tables" by L. R. Salvosa, published in "Annals of Mathematical Statistics", May 1930.

Degrees of freedom for smaller mean square	1	2	3	4	5	6	8	12	24	∞	of t
1	161.45 4052.10	199.50 4999.03	215.72 5403.49	224.57 5625.14	230.17 5764.08	233.97 5859.39	238.89 5981.34	243.91 6105.83	249.04 6234.16	254.32 6366.48	12.706 63.657
2	18.51 98.49	19.00 99.01	19.16 99.17	19.25 99.25	19.30 99.30	19.33 99.33	19.37 99.36	19.41 99.42	19.45 99.46	19.50 99.50	4.303 9.925
3	10.13 34.12	9.55 30.81	9.28 29.46	9.12 28.71	9.01 28.24	8.94 27.91	8.84 27.49	8.74 27.05	8.64 26.60	8.33 26.12	3.182 5.841
4	7.71 21.20	6.94 18.00	6.59 16.69	6.39 15.98	6.26 15.52	6.16 15.21	6.04 14.80	5.91 14.37	5.77 13.93	5.63 13.46	2.776 4.604
5	6.61 16.26	5.79 13.27	5.41 12.06	5.19 11.39	5.05 10.97	4.95 10.67	4.82 10.27	4.68 9.89	4.53 9.47	4.36 9.02	2.571 4.032
6	5.99 13.74	5.14 10.92	4.76 9.78	4.53 9.15	4.39 8.75	4.28 8.47	4.15 8.10	4.00 7.72	3.84 7.31	3.67 6.88	2.447 3.707
7	5.59 12.25	4.74 9.55	4.35 8.45	4.12 7.85	3.97 7.46	3.87 7.19	3.73 6.84	3.57 6.47	3.41 6.07	3.23 5.65	2.365 3.499
8	5.32 11.26	4.46 8.65	4.07 7.59	3.84 7.01	3.69 6.63	3.58 6.37	3.44 6.03	3.28 5.67	3.12 5.28	2.93 4.86	2.306 3.355
9	5.12 10.56	4.26 8.02	3.86 6.99	3.63 6.42	3.48 6.06	3.37 5.80	3.23 5.47	3.07 5.11	2.90 4.73	2.71 4.31	2.262 3.250
10	4.96 10.04	4.10 7.56	3.71 6.55	3.48 5.99	3.33 5.64	3.22 5.39	3.07 5.06	2.91 4.71	2.74 4.33	2.54 3.91	2.228 3.169
11	4.84 9.65	3.98 7.20	3.59 6.22	3.36 5.67	3.20 5.32	3.09 5.07	2.95 4.74	2.79 4.40	2.61 4.02	2.40 3.60	2.201 3.106
12	4.75 9.33	3.88 6.93	3.49 5.95	3.26 5.41	3.11 5.06	3.00 4.82	2.85 4.50	2.69 4.16	2.50 3.78	2.30 3.36	2.179 3.055
13	4.67 9.07	3.80 6.70	3.41 5.74	3.18 5.20	3.02 4.86	2.92 4.62	2.77 4.30	2.60 3.96	2.42 3.59	2.21 3.16	2.160 3.012

Table II. Values of F and t* (Cont.)

Degrees of freedom for smaller mean square	Degrees of freedom for greater mean square 1	2	3	4	5	6	8	12	24	∞	Values of t
14	4.60	3.74	3.34	3.11	2.96	2.85	2.70	2.53	2.35	2.13	2.145
	8.86	6.51	5.56	5.03	4.69	4.46	4.14	3.80	3.43	3.00	2.977
15	4.54	3.68	3.29	3.06	2.90	2.79	2.64	2.48	2.29	2.07	2.131
	6.68	6.36	5.42	4.89	4.56	4.32	4.00	3.67	3.29	2.87	2.947
16	4.49	3.63	3.24	3.01	2.85	2.74	2.59	2.42	2.24	2.01	2.120
	8.53	6.23	5.29	4.77	4.44	4.20	3.89	3.55	3.18	2.75	2.921
17	4.45	3.59	3.20	2.96	2.81	2.70	2.55	2.38	2.19	1.96	2.110
	8.40	6.11	5.18	4.67	4.34	4.10	3.79	3.45	3.08	2.65	2.898
18	4.41	3.55	3.16	2.93	2.77	2.66	2.51	2.34	2.15	1.92	2.101
	8.28	6.01	5.09	4.58	4.25	4.01	3.71	3.37	3.01	2.57	2.878
19	4.38	3.52	3.13	2.90	2.74	2.63	2.48	2.31	2.11	1.88	2.093
	8.18	5.93	5.01	4.50	4.17	3.94	3.63	3.30	2.92	2.49	2.861
20	4.35	3.49	3.10	2.87	2.71	2.60	2.45	2.28	2.08	1.84	2.086
	8.10	5.85	4.94	4.43	4.10	3.87	3.56	3.23	2.86	2.42	2.845
21	4.32	3.47	3.07	2.84	2.68	2.57	2.42	2.25	2.05	1.81	2.080
	8.02	5.78	4.87	4.37	4.04	3.81	3.51	3.17	2.80	2.36	2.831
22	4.30	3.44	3.05	2.82	2.66	2.55	2.40	2.23	2.03	1.78	2.074
	7.94	5.72	4.82	4.31	3.99	3.75	3.45	3.12	2.75	2.30	2.819
23	4.28	3.42	3.03	2.80	2.64	2.53	2.38	2.20	2.00	1.76	2.069
	7.88	5.66	4.76	4.26	3.94	3.71	3.41	3.07	2.70	2.26	2.807
24	4.26	3.40	3.01	2.78	2.62	2.51	2.36	2.18	1.98	1.73	2.064
	7.82	5.61	4.72	4.22	3.90	3.67	3.36	3.03	2.66	2.21	2.797
25	4.24	3.38	2.99	2.76	2.60	2.49	2.34	2.16	1.96	1.71	2.060
	7.77	5.57	4.68	4.18	3.86	3.63	3.32	2.99	2.62	2.17	2.787
26	4.22	3.37	2.98	2.74	2.59	2.47	2.32	2.15	1.95	1.69	2.056
	7.72	5.53	4.64	4.14	3.82	3.59	3.29	2.96	2.58	2.13	2.779

Table II. Values of F and t* (Cont.)

Degrees of freedom for smaller mean square	Degrees of freedom for greater mean square										Values of t
	1	2	3	4	5	6	8	12	24	∞	
27	4.21 7.68	3.35 5.49	2.96 4.60	2.73 4.11	2.57 3.78	2.46 3.56	2.30 3.26	2.13 2.93	1.93 2.55	1.67 2.10	2.052 2.771
28	4.20 7.64	3.34 5.45	2.95 4.57	2.71 4.07	2.56 3.75	2.44 3.53	2.29 3.23	2.12 2.90	1.91 2.52	1.65 2.06	2.048 2.763
29	4.18 7.60	3.33 5.42	2.93 4.54	2.70 4.04	2.54 3.73	2.43 3.50	2.28 3.20	2.10 2.87	1.90 2.49	1.64 2.03	2.045 2.756
30	4.17 7.56	3.32 5.39	2.92 4.51	2.69 4.02	2.53 3.70	2.42 3.47	2.27 3.17	2.09 2.84	1.89 2.47	1.62 2.01	2.042 2.750
35	4.12 7.42	3.26 5.27	2.87 4.40	2.64 3.91	2.48 3.59	2.37 3.57	2.22 3.07	2.04 2.74	1.83 2.37	1.57 1.90	2.030 2.724
40	4.08 7.31	3.23 5.18	2.84 4.31	2.61 3.83	2.45 3.51	2.34 3.29	2.18 2.99	2.00 2.66	1.79 2.29	1.52 1.82	2.021 2.704
45	4.06 7.23	3.21 5.11	2.81 4.25	2.58 3.77	2.42 3.45	2.31 3.23	2.15 2.94	1.97 2.61	1.76 2.23	1.48 1.75	2.014 2.690
50	4.03 7.17	3.18 5.06	2.79 4.20	2.56 3.72	2.40 3.41	2.29 3.19	2.13 2.89	1.95 2.56	1.74 2.18	1.44 1.68	2.008 2.678
60	4.00 7.08	3.15 4.98	2.76 4.13	2.52 3.65	2.37 3.34	2.25 3.12	2.10 2.82	1.92 2.50	1.70 2.12	1.39 1.60	2.000 2.660
70	3.98 7.01	3.13 4.92	2.74 4.07	2.50 3.60	2.35 3.29	2.23 3.07	2.07 2.78	1.89 2.45	1.67 2.07	1.35 1.53	1.994 2.648
80	3.96 6.96	3.11 4.88	2.72 4.04	2.49 3.56	2.33 3.26	2.21 3.04	2.06 2.74	1.88 2.42	1.65 2.03	1.32 1.49	1.990 2.638
90	3.95 6.92	3.10 4.85	2.71 4.01	2.47 3.53	2.32 3.23	2.20 3.01	2.04 2.72	1.86 2.39	1.64 2.00	1.30 1.45	1.987 2.632
100	2.94 6.90	3.09 4.82	2.70 3.98	2.46 3.51	2.30 3.21	2.19 2.99	2.03 2.69	1.85 2.37	1.63 1.98	1.28 1.43	1.984 2.626

Table II. Values of F and t* (Cont.)

Degrees of freedom for smaller mean square	Degrees of freedom for greater mean square										Values of t
	1	2	3	4	5	6	8	12	24	∞	
125	3.92 6.84	3.07 4.78	2.68 3.94	2.44 3.47	2.29 3.17	2.17 2.95	2.01 2.66	1.83 2.33	1.60 1.94	1.25 1.37	1.979 2.616
150	3.90 6.81	3.06 4.75	2.66 3.91	2.43 3.45	2.27 3.14	2.16 2.92	2.00 2.63	1.82 2.31	1.59 1.92	1.22 1.33	1.976 2.609
200	3.89 6.76	3.04 4.71	2.65 3.88	2.42 3.41	2.26 3.11	2.14 2.89	1.98 2.60	1.80 2.28	1.57 1.88	1.19 1.28	1.972 2.601
300	3.87 6.72	3.03 4.68	2.64 3.85	2.41 3.38	2.25 3.08	2.13 2.86	1.97 2.57	1.79 2.24	1.55 1.85	1.15 1.22	1.968 2.592
400	3.86 6.70	3.02 4.66	2.63 3.82	2.40 3.37	2.24 3.06	2.12 2.85	1.96 2.56	1.78 2.23	1.54 1.84	1.13 1.19	1.966 2.588
500	3.86 6.69	3.01 4.65	2.62 3.82	2.39 3.36	2.23 3.05	2.11 2.84	1.96 2.55	1.77 2.22	1.54 1.83	1.11 1.16	1.965 2.586
1000	3.85 6.66	3.00 4.63	2.61 3.80	2.38 3.34	2.22 3.04	2.10 2.82	1.95 2.53	1.76 2.20	1.53 1.81	1.08 1.11	1.962 2.581
∞	3.84 6.64	2.99 4.60	2.60 3.78	2.37 3.32	2.21 3.02	2.09 2.80	1.94 2.51	1.75 2.18	1.52 1.78	1.00 1.00	1.960 2.576

*This table was compiled by G. W. Snedecor from the values of t and z in tables IV and VI of R. A. Fisher's "Statistical Methods for Research Workers". First published by G. W. Snedecor in "Analysis of Variance and Covariance" and reproduced here with his permission.

(4474-37)

Table III.[1] Table of χ^2

n.	P = .99	.98	.95	.90	.80	.70	.50	.30	.20	.10	.05	.02	.01
1	.000157	.000628	.00393	.0158	.0642	.148	.455	1.074	1.642	2.706	3.841	5.412	6.635
2	.0201	.0404	.103	.211	.446	.713	1.386	2.408	3.219	4.605	5.991	7.824	9.210
3	.115	.185	.352	.584	1.005	1.424	2.366	3.665	4.642	6.251	7.815	9.837	11.341
4	.297	.429	.711	1.064	1.649	2.195	3.357	4.878	5.989	7.779	9.488	11.668	13.277
5	.554	.752	1.145	1.610	2.343	3.000	4.351	6.064	7.289	9.236	11.070	13.388	15.086
6	.872	1.134	1.635	2.204	3.070	3.828	5.348	7.231	8.558	10.645	12.592	15.033	16.812
7	1.239	1.564	2.167	2.833	3.822	4.671	6.346	8.383	9.803	12.017	14.067	16.622	18.475
8	1.646	2.032	2.733	3.490	4.594	5.527	7.344	9.524	11.030	13.362	15.507	18.168	20.090
9	2.088	2.532	3.325	4.168	5.380	6.393	8.343	10.656	12.242	14.684	16.919	19.679	21.666
10	2.558	3.059	3.940	4.865	6.179	7.267	9.342	11.781	13.442	15.987	18.307	21.161	23.209
11	3.053	3.609	4.575	5.578	6.989	8.148	10.341	12.899	14.631	17.275	19.675	22.618	24.725
12	3.571	4.178	5.226	6.304	7.807	9.034	11.341	14.011	15.812	18.549	21.026	24.054	26.217
13	4.107	4.765	5.892	7.042	8.634	9.926	12.340	15.119	16.985	19.812	22.362	25.472	27.688
14	4.660	5.368	6.571	7.790	9.467	10.821	13.339	16.222	18.151	21.064	23.685	26.873	29.141
15	5.229	5.985	7.261	8.547	10.307	11.721	14.339	17.322	19.311	22.307	24.996	28.259	30.578
16	5.812	6.614	7.962	9.312	11.152	12.624	15.338	18.418	20.465	23.542	26.296	29.633	32.000
17	6.408	7.255	8.672	10.085	12.002	13.531	16.338	19.511	21.615	24.769	27.587	30.995	33.409
18	7.015	7.906	9.390	10.865	12.857	14.440	17.338	20.601	22.760	25.989	28.869	32.346	34.805
19	7.633	8.567	10.117	11.651	13.716	15.352	18.338	21.689	23.900	27.204	30.144	33.687	36.191
20	8.260	9.237	10.851	12.443	14.578	16.266	19.337	22.775	25.038	28.412	31.410	35.020	37.566
21	8.897	9.915	11.591	13.240	15.445	17.182	20.337	23.858	26.171	29.615	32.671	36.343	38.932
22	9.542	10.600	12.338	14.041	16.314	18.101	21.337	24.939	27.301	30.813	33.924	37.659	40.289
23	10.196	11.293	13.091	14.848	17.187	19.021	22.337	26.018	28.429	32.007	35.172	38.968	41.638
24	10.856	11.992	13.848	15.659	18.062	19.943	23.337	27.096	29.553	33.196	36.415	40.270	42.980
25	11.524	12.697	14.611	16.473	18.940	20.867	24.337	28.172	30.675	34.382	37.652	41.566	44.314
26	12.198	13.409	15.379	17.292	19.820	21.792	25.336	29.246	31.795	35.563	38.885	42.856	45.642
27	12.879	14.125	16.151	18.114	20.703	22.719	26.336	30.319	32.912	36.741	40.113	44.140	46.963
28	13.565	14.837	16.928	18.939	21.588	23.647	27.336	31.391	34.027	37.916	41.337	45.419	48.278
29	14.256	15.574	17.708	19.768	22.475	24.577	28.336	32.461	35.139	39.087	42.557	46.693	49.588
30	14.953	16.306	18.493	20.599	23.364	25.508	29.336	33.530	36.250	40.256	43.773	47.962	50.892

For larger values of n, the expression $\sqrt{2\chi^2} - \sqrt{2n - 1}$ may be used as a normal deviate with unit variance.

Table IV. Neparian or Hyperbolic Logarithms[1]

	0	1	2	3	4	5	6	7	8	9	1	2	3	4	5	6	7	8	9
1.0	0.0000	0100	0198	0296	0392	0488	0583	0677	0770	0862	10	19	29	38	48	57	67	76	86
1.1	0.0953	1044	1133	1222	1310	1398	1484	1570	1655	1740	9	17	26	35	44	52	61	70	78
1.2	0.1823	1906	1989	2070	2151	2231	2311	2390	2469	2546	8	16	24	32	40	48	56	64	72
1.3	0.2624	2700	2776	2852	2927	3001	3075	3148	3221	3293	7	15	22	30	37	44	52	59	67
1.4	0.3365	3436	3507	3577	3646	3716	3784	3853	3920	3988	7	14	21	28	35	41	48	55	62
1.5	0.4055	4121	4187	4253	4318	4383	4447	4511	4574	4637	6	13	19	26	32	39	45	52	58
1.6	0.4700	4762	4824	4886	4947	5008	5068	5128	5188	5247	6	12	18	24	30	36	42	48	55
1.7	0.5306	5365	5423	5481	5539	5596	5653	5710	5766	5822	6	11	17	23	29	34	40	46	51
1.8	0.5878	5933	5988	6043	6098	6152	6206	6259	6313	6366	5	11	16	22	27	32	38	43	49
1.9	0.6419	6471	6523	6575	6627	6678	6729	6780	6831	6881	5	10	15	20	26	31	36	41	46
2.0	0.6931	6981	7031	7080	7129	7178	7227	7275	7324	7372	5	10	15	20	24	29	34	39	44
2.1	0.7419	7467	7514	7561	7608	7655	7701	7747	7793	7839	5	9	14	19	23	28	33	37	42
2.2	0.7885	7930	7975	8020	8065	8109	8154	8198	8242	8286	4	9	13	18	22	27	31	36	40
2.3	0.8329	8372	8416	8459	8502	8544	8587	8629	8671	8713	4	9	13	17	21	26	30	34	38
2.4	0.8755	8796	8838	8879	8920	8961	9002	9042	9083	9123	4	8	12	16	20	24	29	33	37
2.5	0.9163	9203	9243	9282	9322	9361	9400	9439	9478	9517	4	8	12	16	20	24	27	31	35
2.6	0.9555	9594	9632	9670	9708	9746	9783	9821	9858	9895	4	8	11	15	19	23	26	30	34
2.7	0.9933	9969	0006	0043	0080	0116	0152	0188	0225	0260	4	7	11	15	18	22	25	29	33
2.8	1.0296	0332	0367	0403	0438	0473	0508	0543	0578	0613	4	7	11	14	18	21	25	28	32
2.9	1.0647	0682	0716	0750	0784	0818	0852	0886	0919	0953	3	7	10	14	17	20	24	27	31
3.0	1.0986	1019	1053	1086	1119	1151	1184	1217	1249	1282	3	7	10	13	16	20	23	26	30
3.1	1.1314	1346	1378	1410	1442	1474	1506	1537	1569	1600	3	6	10	13	16	19	22	25	29
3.2	1.1632	1663	1694	1725	1756	1787	1817	1848	1878	1909	3	6	9	12	15	18	21	25	28
3.3	1.1939	1969	2000	2030	2060	2090	2119	2149	2179	2208	3	6	9	12	15	18	21	24	27
3.4	1.2238	2267	2296	2326	2355	2384	2413	2442	2470	2499	3	6		12	15	17	20	23	26
3.5	1.2528	2556	2585	2613	2641	2669	2698	2726	2654	2782	3	6		11	14	17	20	22	25
3.6	1.2809	2837	2865	2892	2920	2947	2975	3002	3029	3056	3	5	9	11	14	16	19	22	25
3.7	1.3083	3110	3137	3164	3191	3218	3244	3271	3297	3324	3	5	8	11	13	16	19	21	24
3.8	1.3350	3376	3403	3429	3455	3481	3507	3533	3558	3584	3	5	8	10	13	16	18	21	23
3.9	1.3610	3635	3661	3686	3712	3737	3762	3788	3813	3838	3	5	8	10	13	15	18	20	23
4.0	1.3863	3888	3913	3938	3962	3987	4012	4036	4061	4085	2	5	7	10	12	15	17	20	22
4.1	1.4110	4134	4159	4183	4207	4231	4255	4279	4303	4327	2	5	7	10	12	14	17	19	22
4.2	1.4351	4375	4398	4422	4446	4469	4493	4516	4540	4563	2	5	7	9	12	14	16	19	21
4.3	1.4586	4609	4633	4656	4679	4702	4725	4748	4770	4793	2	5	7	9	12	14	16	18	21
4.4	1.4816	4839	4861	4884	4907	4929	4951	4974	4996	5019	2	5	7	9	11	14	16	18	20
4.5	1.5041	5063	5085	5107	5129	5151	5173	5195	5217	5239	2	4	7	9	11	13	15	18	20
4.6	1.5261	5282	5304	5326	5347	5369	5390	5412	5433	5454	2	4	6	9	11	13	15	17	19

Table IV. Neparian or Hyperbolic Logarithms[1] (Con

	0	1	2	3	4	5	6	7	8	9	1	2	3
4.7	1.5476	5497	5518	5539	5560	5581	5602	5623	5644	5665	2	4	6
4.8	1.5686	5707	5728	5748	5769	5790	5810	5831	5851	5872	2	4	6
4.9	1.5892	5913	5933	5953	5974	5994	6014	6034	6054	6074	2	4	6
5.0	1.6094	6114	6134	6154	6174	6194	6214	6233	6253	6273	2	4	6
5.1	1.6292	6312	6332	6351	6371	6390	6409	6429	6448	6467	2	4	6
5.2	1.6487	6506	6525	6544	6563	6582	6601	6620	6639	6658	2	4	6
5.3	1.6677	6696	6715	6734	6752	6771	6790	6808	6827	6845	2	4	6
5.4	1.6864	6882	6901	6919	6938	6956	6974	6993	7011	7029	2	4	5
5.5	1.7047	7066	7084	7102	7120	7138	7156	7174	7192	7210	2	4	5
5.6	1.7228	7246	7263	7281	7299	7317	7334	7352	7370	7387	2	4	5
5.7	1.7405	7422	7440	7457	7475	7492	7509	7527	7544	7561	2	3	5
5.8	1.7579	7596	7613	7630	7647	7664	7681	7699	7716	7733	2	3	5
5.9	1.7750	7766	7783	7800	7817	7834	7851	7867	7884	7901	2	3	5
6.0	1.7918	7934	7951	7967	7984	8001	8017	8034	8050	8066	2	3	5
6.1	1.8083	8099	8116	8132	8148	8165	8181	8197	8213	8229	2	3	5
6.2	1.8245	8262	8278	8294	8310	8326	8342	8358	8374	8390	2	3	5
6.3	1.8405	8421	8437	8453	8469	8485	8500	8516	8532	8547	2	3	5
6.4	1.8563	8579	8594	8610	8625	8641	8656	8672	8687	8703	2	3	5
6.5	1.8718	8733	8749	8764	8779	8795	8810	8825	8840	8856	2	3	5
6.6	1.8871	8886	8901	8916	8931	8946	8961	8976	8991	9006	2	3	5
6.7	1.9021	9036	9051	9066	9081	9095	9110	9125	9140	9155	1	3	4
6.8	1.9169	9184	9199	9213	9228	9242	9257	9272	9286	9301	1	3	4
6.9	1.9315	9330	9344	9359	9373	9387	9402	9416	9430	9445	1	3	4
7.0	1.9459	9473	9488	9502	9516	9530	9544	9559	9573	9587	1	3	4
7.1	1.9601	9615	9629	9643	9657	9671	9685	9699	9713	9727	1	3	4
7.2	1.9741	9755	9769	9782	9796	9810	9824	9838	9851	9865	1	3	4
7.3	1.9879	9892	9906	9920	9933	9947	9961	9974	9988	0001	1	3	4
7.4	2.0015	0028	0042	0055	0069	0082	0096	0109	0122	0136	1	3	
7.5	2.0149	0162	0176	0189	0202	0215	0229	0242	0255	0268	1	3	
7.6	2.0281	0295	0308	0321	0334	0347	0360	0373	0386	0399	1	3	4
7.7	2.0412	0425	0438	0451	0464	0477	0490	0503	0516	0528	1	3	
7.8	2.0541	0554	0567	0580	0592	0605	0618	0631	0643	0656	1	3	4
7.9	2.0669	0681	0694	0707	0719	0732	0744	0757	0769	0782	1	3	4
8.0	2.0794	0807	0819	0832	0844	0857	0869	0882	0894	0906	1	3	4
8.1	2.0919	0931	0943	0956	0968	0980	0992	1005	1017	1029	1	2	4
8.2	2.1041	1054	1066	1078	1090	1102	1114	1126	1138	1150	1	2	4
8.3	2.1163	1175	1187	1199	1211	1223	1235	1247	1258	1270	1	2	4

Table IV. Neparian or Hyperbolic Logarithms[1] (Cont.)

	0	1	2	3	4	5	6	7	8	9	1	2	3	4	5	6	7	8	9
8.4	2.1282	1294	1306	1318	1330	1342	1353	1365	1377	1389	1	2	4	5	6	7	8	10	11
8.5	2.1401	1412	1424	1436	1448	1459	1471	1483	1494	1506	1	2	4	5	6	7	8	9	11
8.6	2.1518	1529	1541	1552	1564	1576	1587	1599	1610	1622	1	2	3	5	6	7	8	9	10
8.7	2.1633	1645	1656	1668	1679	1691	1702	1713	1725	1736	1	2	3	5	6	7	8	9	10
8.8	2.1748	1759	1770	1782	1793	1804	1815	1827	1838	1849	1	2	3	5	6	7	8	9	10
8.9	2.1861	1872	1883	1894	1905	1917	1928	1939	1950	1961	1	2	3	4	6	7	8	9	10
9.0	2.1972	1983	1994	2006	2017	2028	2039	2050	2061	2072	1	2	3	4	6	7	8	9	10
9.1	2.2083	2094	2105	2116	2127	2138	2148	2159	2170	2181	1	2	3	4	5	7	8	9	10
9.2	2.2192	2203	2214	2225	2235	2246	2257	2268	2279	2289	1	2	3	4	5	6	8	9	10
9.3	2.2300	2311	2322	2332	2343	2354	2364	2375	2386	2396	1	2	3	4	5	6	7	9	10
9.4	2.2407	2418	2428	2439	2450	2460	2471	2431	2492	2502	1	2	3	4	5	6	7	8	10
9.5	2.2513	2523	2534	2544	2555	2565	2576	2536	2597	2607	1	2	3	4	5	6	7	8	9
9.6	2.2618	2628	2638	2649	2659	2670	2680	2690	2701	2711	1	2	3	4	5	6	7	8	9
9.7	2.2721	2732	2742	2752	2762	2773	2783	2793	2803	2814	1	2	3	4	5	6	7	8	9
9.8	2.2824	2834	2844	2854	2865	2875	2885	2895	2905	2915	1	2	3	4	5	6	7	8	9
9.9	2.2925	2935	2946	2956	2966	2976	2986	2996	3006	3016	1	2	3	4	5	6	7	8	9

Table of Neperian Logarithms of 10^{+n}

n	1	2	3	4	5	6	7	8	9
$\log_e 10^n$	2.3026	4.6052	6.9078	9.2103	11.5129	13.8155	16.1181	18.4207	20.7233

Table of Neperian Logarithms of 10^{-n}

n	1	2	3	4	5	6	7	8	9
$\log_e 10^{-n}$	3.6974	5.3948	7.0922	10.7897	12.4871	14.1845	17.8819	19.5793	21.276

[1]This table is reproduced from "Four Figure Mathematical Tables" by the late J. T. Bottomley and published by Macmillan and Co., Ltd. (London). The consent of the publishers and representatives of the author have been obtained.

transformed into degrees of an angle. Angles of
the body of the table corresponding to observed per-
and top. (Each angle ending in 5 is followed by a +
he last decimal is dropped).

Chester I. Bliss of the Institute for Plant Protec-
ed by permission of the author.

0.03	0.04	0.05	0.06	0.07	0.08	0.09
0.99	1.15-	1.28	1.40	1.52	1.62	1.72
2.07	2.14	2.22	2.29	2.36	2.43	2.50
2.75-	2.81	2.87	2.92	2.98	3.03	3.09
3.29	3.34	3.39	3.44	3.49	3.53	3.58
3.76	3.80	3.85-	3.89	3.93	3.97	4.01
4.17	4.21	4.25+	4.29	4.33	4.37	4.40
4.55+	4.59	4.62	4.66	4.69	4.73	4.76
4.90	4.93	4.97	5.00	5.03	5.07	5.10
5.23	5.26	5.29	5.32	5.35+	5.38	5.41
5.53	5.56	5.59	5.62	5.65+	5.68	5.71

0.3	0.4	0.5	0.6	0.7	0.8	0.9
6.55-	6.80	7.04	7.27	7.49	7.71	7.92
8.72	8.91	9.10	9.28	9.46	9.63	9.81
10.47	10.63	10.78	10.94	11.09	11.24	11.39
11.97	12.11	12.25-	12.39	12.52	12.66	12.79
13.31	13.44	13.56	13.69	13.81	13.94	14.06
14.54	14.65+	14.77	14.89	15.00	15.12	15.23
15.68	15.79	15.89	16.00	16.11	16.22	16.32
16.74	16.85-	16.95+	17.05+	17.16	17.26	17.36
17.76	17.85+	17.95+	18.05-	18.15-	18.24	18.34
18.72	18.81	18.91	19.00	19.09	19.19	19.28
19.64	19.73	19.82	19.91	20.00	20.09	20.13
20.53	20.62	20.70	20.79	20.88	20.96	21.05-
21.39	21.47	21.56	21.64	21.72	21.81	21.89
22.22	22.30	22.38	22.46	22.55-	22.63	22.7[illegible]
23.03	23.11	23.19	23.26	23.34	23.42	23.50
23.81	23.89	23.97	24.04	24.12	24.20	24.27
24.58	24.65+	24.73	24.80	24.88	24.95+	25.03
25.33	25.40	25.48	25.55-	25.62	25.70	25.77
26.06	26.13	26.21	26.28	26.35-	26.42	26.49
26.78	26.85+	26.92	26.99	27.06	27.13	27.20
27.49	27.56	27.63	27.69	27.76	27.83	27.90
28.18	28.25-	28.32	28.38	28.45+	28.52	28.59
28.86	28.93	29.00	29.06	29.13	29.20	29.27
29.53	29.60	29.67	29.73	29.80	29.87	29.93

Table V. Values of percentages transformed into degrees of an angle. Angles of equal information are given in the body of the table corresponding to observed percentages along the left margin and top. (Continued)

	0.0	0.1	0.2	0.3	0.4	0.5	0.6	0.7	0.8	0.9
25	30.00	30.07	30.13	30.20	30.26	30.33	30.40	30.46	30.53	30.59
26	30.66	30.72	30.79	30.85+	30.92	30.98	31.05-	31.11	31.18	31.24
27	31.31	31.37	31.44	31.50	31.56	31.63	31.69	31.76	31.82	31.88
28	31.95-	32.01	32.08	32.14	32.20	32.27	32.33	32.39	32.46	32.52
29	32.58	32.65-	32.71	32.77	32.83	32.90	32.96	33.02	33.09	33.15-
30	33.21	33.27	33.34	33.40	33.46	33.52	33.58	33.65-	33.71	33.77
31	33.83	33.89	33.96	34.02	34.08	34.14	34.20	34.27	34.33	34.39
32	34.45-	34.51	34.57	34.63	34.70	34.76	34.82	34.88	34.94	35.00
33	35.06	35.12	35.18	35.24	35.30	35.37	35.43	35.49	35.55-	35.61
34	35.67	35.73	35.79	35.85-	35.91	35.97	36.03	36.09	36.15+	36.21
35	36.27	36.33	36.39	36.45+	36.51	36.57	36.63	36.69	36.75+	36.81
36	36.87	36.93	36.99	37.05-	37.11	37.17	37.23	37.29	37.35	37.41
37	37.47	37.52	37.58	37.64	37.70	37.76	37.82	37.88	37.94	38.00
38	38.06	38.12	38.17	38.23	38.29	38.35+	38.41	38.47	38.53	38.59
39	38.65-	38.70	38.76	38.82	38.88	38.94	39.00	39.06	39.11	39.17
40	39.23	39.29	39.35-	39.41	39.47	39.52	39.58	39.64	39.70	39.76
41	39.82	39.87	39.93	39.99	40.05-	40.11	40.16	40.22	40.28	40.34
42	40.40	40.46	40.51	40.57	40.63	40.69	40.74	40.80	40.86	40.92
43	40.98	41.03	41.09	41.15-	41.21	41.27	41.32	41.38	41.44	41.50
44	41.55+	41.61	41.67	41.73	41.78	41.84	41.90	41.96	42.02	42.07
45	42.13	42.19	42.25-	42.30	42.36	42.42	42.48	42.53	42.59	42.65-
46	42.71	42.76	42.82	42.88	42.94	42.99	43.05-	43.11	43.17	43.22
47	43.28	43.34	43.39	43.45+	43.51	43.57	43.62	43.68	43.74	43.80
48	43.85+	43.91	43.97	44.03	44.08	44.14	44.20	44.25+	44.31	44.37
49	44.43	44.48	44.54	44.60	44.66	44.71	44.77	44.83	44.89	44.94
50	45.00	45.06	45.11	45.17	45.23	45.29	45.34	45.40	45.46	45.52
51	45.57	45.63	45.69	45.75-	45.80	45.86	45.92	45.97	46.03	46.09
52	46.15-	46.20	46.26	46.32	46.38	46.43	46.49	46.55-	46.61	46.66
53	46.72	46.78	46.83	46.89	46.95+	47.01	47.06	47.12	47.18	47.24
54	47.29	47.35+	47.41	47.47	47.52	47.58	47.64	47.70	47.75+	47.81
55	47.87	47.93	47.98	48.04	48.10	48.16	48.22	48.27	48.33	48.39
56	48.45	48.50	48.56	48.62	48.68	48.73	48.79	48.85+	48.91	48.97
57	49.02	49.08	49.14	49.20	49.26	49.31	49.37	49.43	49.49	49.54
58	49.60	49.66	49.72	49.78	49.84	49.89	49.95+	50.01	50.07	50.13
59	50.18	50.24	50.30	50.36	50.42	50.48	50.53	50.59	50.65+	50.71
60	50.77	50.83	50.89	50.94	51.00	51.06	51.12	51.18	51.24	51.30
61	51.35+	51.41	51.47	51.53	51.59	51.65-	51.71	51.77	51.83	51.88
62	51.94	52.00	52.06	52.12	52.18	52.24	52.30	52.36	52.42	52.48
63	52.53	52.59	52.65+	52.71	52.77	52.83	52.89	52.95+	53.01	53.07
64	53.13	53.19	53.25-	53.31	53.37	53.43	53.49	53.55-	53.61	53.67

Table V. Values of percentages transformed into degrees of an angle. Angles of equal information are given in the body of the table corresponding to observed percentages along the left margin and top. (Continued)

	0.0	0.1	0.2	0.3	0.4	0.5	0.6	0.7	0.8	0.9
65	53.73	53.79	53.85-	53.91	53.97	54.03	54.09	54.15+	54.21	54.27
66	54.33	54.39	54.45+	54.51	54.57	54.63	54.70	54.76	54.82	54.88
67	54.94	55.00	55.06	55.12	55.18	55.24	55.30	55.37	55.43	55.49
68	55.55+	55.61	55.67	55.73	55.80	55.86	55.92	55.98	56.04	56.11
69	56.17	56.23	56.29	56.35+	56.42	56.48	56.54	56.60	56.66	56.73
70	56.79	56.85+	56.91	56.98	57.04	57.10	57.17	57.23	57.29	57.35+
71	57.42	57.48	57.54	57.61	57.67	57.73	57.80	57.86	57.92	57.99
72	58.05+	58.12	58.18	58.24	58.31	58.37	58.44	58.50	58.56	58.63
73	58.69	58.76	58.82	58.89	58.95+	59.02	59.08	59.15-	59.21	59.28
74	59.34	59.41	59.47	59.54	59.60	59.67	59.74	59.80	59.87	59.93
75	60.00	60.07	60.13	60.20	60.27	60.33	60.40	60.47	60.53	60.60
76	60.67	60.73	60.80	60.87	60.94	61.00	61.07	61.14	61.21	61.27
77	61.34	61.41	61.48	61.55-	61.62	61.68	61.75+	61.82	61.89	61.96
78	62.03	62.10	62.17	62.24	62.31	62.37	62.44	62.51	62.58	62.65+
79	62.72	62.80	62.87	62.94	63.01	63.08	63.15-	63.22	63.29	63.36
80	63.44	63.51	63.58	63.65+	63.72	63.79	63.87	63.94	64.01	64.08
81	64.16	64.23	64.30	64.38	64.45+	64.52	64.60	64.67	64.75-	64.82
82	64.90	64.97	65.05-	65.12	65.20	65.27	65.35-	65.42	65.50	65.57
83	65.65-	65.73	65.80	65.88	65.96	66.03	66.11	66.19	66.27	66.34
84	66.42	66.50	66.58	66.66	66.74	66.81	66.89	66.97	67.05+	67.13
85	67.21	67.29	67.37	67.45+	67.54	67.62	67.70	67.78	67.86	67.94
86	68.03	68.11	68.19	68.28	68.36	68.44	68.53	68.61	68.70	68.78
87	68.87	68.95+	69.04	69.12	69.21	69.30	69.38	69.47	69.56	69.64
88	69.73	69.82	69.91	70.00	70.09	70.18	70.27	70.36	70.45	70.54
89	70.63	70.72	70.81	70.91	71.00	71.09	71.19	71.28	71.37	71.47
90	71.56	71.66	71.76	71.85+	71.95+	72.05-	72.15-	72.24-	72.34	72.44
91	72.54	72.64	72.74	72.84	72.95-	73.05-	73.15+	73.26	73.36	73.46
92	73.57	73.68	73.78	73.89	74.00	74.11	74.21	74.32	74.44	74.55-
93	74.66	74.77	74.88	75.00	75.11	75.23	75.35-	75.46	75.58	75.70
94	75.82	75.94	76.06	76.19	76.31	76.44	76.56	76.69	76.82	76.95-
95	77.08	77.21	77.34	77.48	77.61	77.75+	77.89	78.03	78.17	78.32
96	78.46	78.61	78.76	78.91	79.06	79.22	79.37	79.53	79.69	79.86
97	80.02	80.19	80.37	80.54	80.72	80.90	81.09	81.28	81.47	81.67
98	81.87	82.08	82.29	82.51	82.73	82.96	83.20	83.45+	83.71	83.96

Table V. Values of percentages transformed into degrees of an angle. Angles of equal information are given in the body of the table corresponding to observed percentages along the left margin and top. (Continued)

	0.00	0.01	0.02	0.03	0.04	0.05	0.06	0.07	C.08	0.09
99.0	84.26	84.29	84.32	84.35-	84.38	84.41	84.44	84.47	84.50	84.53
99.1	84.56	84.59	84.62	84.65-	84.68	84.71	84.74	84.77	84.80	84.84
99.2	84.87	84.90	84.93	84.97	85.00	85.03	85.07	85.10	85.13	85.17
99.3	85.20	85.24	85.27	85.31	85.34	85.38	85.41	85.45-	85.48	85.52
99.4	85.56	85.60	85.63	85.67	85.71	85.75-	85.79	85.83	85.87	85.91
99.5	85.95-	85.99	86.03	86.07	86.11	86.15	86.20	86.24	86.28	86.33
99.6	86.37	86.42	86.47	86.51	86.56	86.61	86.66	86.71	86.76	86.81
99.7	86.86	86.91	86.97	87.02	87.08	87.13	87.19	87.25+	87.31	87.37
99.8	87.44	87.50	87.57	87.64	87.71	87.78	87.86	87.93	88.01	88.10
99.9	88.19	88.28	88.38	88.48	88.60	88.72	88.85+	89.01	89.19	89.43
100.0	90.00	-----	-----	-----	-----	-----	-----	-----	-----	-----

(507-38)

Numbers 1-30 inc., 20 series

I	II	III	IV	V	VI	VII	VIII	IX	X	XI	XII	XIII	XIV	XV	XVI	XVII	XVIII	XIX	XX
29	12	15	20	6	13	16	23	1	10	2	20	15	29	4	18	8	7	6	14
27	1	26	7	22	26	17	9	20	15	29	6	25	26	2	15	13	30	3	22
5	23	23	15	26	22	19	28	4	7	12	5	12	28	5	16	4	14	7	19
20	3	24	30	1	21	22	4	3	13	6	7	16	3	16	21	29	6	14	3
10	25	9	27	25	3	3	14	7	5	30	1	23	23	13	25	20	4	20	2
21	16	12	4	2	23	26	5	24	21	26	3	5	11	17	12	1	15	24	27
11	13	1	13	9	28	20	2	29	29	13	24	26	21	9	22	11	22	11	29
7	19	28	2	8	16	30	13	22	11	19	30	21	6	7	30	21	23	25	1
14	20	19	12	5	1	17	24	16	4	20	13	4	7	1	10	23	27	28	12
25	5	8	23	27	4	12	25	6	12	8	25	13	16	30	5	14	1	18	23
2	21	17	26	20	5	7	8	23	6	28	2	6	15	26	19	22	19	1	26
18	26	11	9	21	18	24	1	12	25	4	17	1	18	6	14	12	24	21	4
4	9	2	21	29	8	5	22	5	18	27	27	10	4	11	4	26	17	10	18
23	17	30	18	13	10	25	26	17	1	24	18	18	12	27	13	18	18	13	13
30	10	21	16	10	29	4	19	8	26	5	15	22	10	20	7	2	26	16	9
8	27	25	1	3	30	21	29	11	23	7	14	2	27	12	8	16	8	9	5
17	6	16	8	17	17	27	12	30	20	11	11	27	1	14	29	3	2	27	16
13	30	3	14	30	11	9	30	13	2	21	26	3	14	22	9	19	12	29	8
26	14	10	29	11	15	28	16	27	27	3	23	17	19	10	2	28	13	22	20
22	11	5	19	7	9	13	3	21	30	16	9	28	5	25	23	7	25	17	21
24	28	7	11	4	25	6	10	14	8	15	10	11	13	15	17	17	5	26	11
9	2	6	25	12	7	2	11	10	14	23	4	8	22	18	11	25	21	8	25
28	24	29	3	28	6	3	15	13	3	9	19	7	8	8	3	6	11	5	6
15	3	13	5	24	2	24	17	2	19	22	12	30	2	21	6	10	10	19	17
16	22	20	17	14	14	11	27	26	24	14	16	14	20	23	24	30	3	2	28
12	7	22	6	15	12	10	6	25	16	10	8	19	9	29	20	27	9	4	30
6	29	4	22	23	24	29	20	23	28	1	21	9	30	23	28	15	16	15	24
3	4	18	10	16	19	25	13	19	9	17	22	29	17	19	26	24	20	23	7
1	15	27	28	13	27	18	7	9	22	25	28	20	24	24	1	9	29	12	15
19	18	14	24	19	20	1	21	18	17	18	29	24	25	3	27	5	28	30	10

SUBJECT INDEX

CPSIA information can be obtained
at www.ICGtesting.com
Printed in the USA
BVHW08s1031210918
528173BV00022B/1219/P